AF564873

Next Generation Libraries in the Networked Digital Environment

NIPA® GENX ELECTRONIC RESOURCES & SOLUTIONS P. LTD.
New Delhi-110 034

About the Editors

Dr. Bhanu Partap is presently working as Assistant Librarian in Nehru Library, CCS Haryana Agricultural University, Hisar (HR), India since July 2013. Before joining the current position, he has served many institutions situated in Punjab, Haryana, Uttarakhand and Chandigarh between September 2005 to July 2013. He has received MLISc. and M.A. (Mass Comm.) from Kurukshetra University, Kurukshetra and received M.Phil. (LIS) from Alagappa University, Karaikudi (Tamilnadu), India. He has passed the exam of UGC-NET in June 2010 and also received the degree of Ph.D. in Library & Information Science from Kurukshetra University, Kurukshetra. Apart from this, he has also received PGDCA, PGDJMC, and CC Yoga from Kurukshetra University, Kurukshetra. He has about 19 years of professional experience in the field of librarianship. He has contributed around 120 articles in the National and International journals, conference proceedings and edited books, and has also authored/edited 13 books. Apart from this, he has coordinated many national workshops and training programs for students, faculty and library professionals. He is a life member of Haryana Library Association (HLA), Indian Library Association (ILA) and Association of Agricultural Librarians and Documentalists of India (AALDI). Dr. Bhanu Partap has also awarded by Indian Library Association with P.V. Verghese Award- 2022 for the best paper entitled "Awareness and Use of CeRA Consortium among the Students of Uttar Banga Krishi Viswavidyalaya, Pundibari, Cooch Behar (W.B.), India: A Case Study" published in the Journal of Indian Library Association, Vol. 58(4), October-December, 2022.

Dr. Renu Bala is presently working as a Librarian at PIET Sanskriti Senior Secondary School, NFL Township, Panipat, Haryana. She holds M. Lib. & Inf. Sc., PGDCA and Ph.D (LIS). She has an extensive professional exposure of more than 22 years of library services and serving at various esteemed organizations. She has published a good number of research papers in journals, conference proceedings and edited books. She has attended around 70 workshops in different capacities and has 21 presentations in various professional activities.

Next Generation Libraries in the Networked Digital Environment

Bhanu Partap
Assistant Librarian
CCS Haryana Agricultural University
Hisar- 125 004 (India)

Renu Bala
Librarian
PIET Sanskriti Senior Secondary School
NFL Township, Panipat, Haryana- 132 103 (India)

NIPA® GENX ELECTRONIC RESOURCES & SOLUTIONS P. LTD.
New Delhi-110 034

**NIPA® GENX ELECTRONIC
RESOURCES & SOLUTIONS P. LTD.**

101,103, Vikas Surya Plaza, CU Block
L.S.C. Market, Pitam Pura, New Delhi-110 034
Ph : +91-11-43860225, Mob.: +91 9717133558, 9540816132
E-mail: newindiapublishingagency@gmail.com
Website: www.nipaersources.com

Print ISBN: 978-93-58873-71-9
ebook ISBN: 978-93-58876-62-8

Composed and Designed by NIPA®.

Preface

In an age where information flows ceaselessly and technology evolves at an unprecedented pace, libraries stand at the crossroads of tradition and innovation. The present volume "Next Generation Libraries in the Networked Digital Environment" emerges as a timely and essential exploration of this dynamic landscape. In other words, it can be said that in this rapidly evolving landscape of information and knowledge management "Next Generation Libraries in the Networked Digital Environment" emerges as a pivotal contribution to the discourse surrounding modern libraries. The prelude to this volume sets the stage for an exploration of how libraries are not merely repositories of books but dynamic entities that enhance their services through Information and Communication Technology (ICT). Furthermore, this volume brings together the insights of respected authors who share a common vision; to redefine the role of libraries as vital hubs of knowledge, community engagement, and sustainable practices in our increasingly digital world.

The transformation of libraries into next-generation knowledge centres is not merely a response to technological advancements; it is a commitment to embracing holistic and sustainable strategies that cater to diverse user needs. Each chapter in this book serves as a stepping stone toward understanding the myriad ways in which libraries can adapt to, and thrive in, a networked digital environment.

From integrating Information and Communication Technologies (ICT) to enhancing library services, to exploring the principles of a circular economy in library management, our contributors delve into crucial themes that resonate with current and future library practices. The implications of social media, Open Educational Resources (OER), and the innovative use of robotics highlight a shared goal: to foster inclusivity and accessibility in library services while promoting sustainable development.

Moreover, the thoughtful analyses presented within this volume address critical issues such as digital literacy, disaster management, and the essential role of libraries in community upliftment. The juxtaposition of theory and practice, exemplified through case studies like that of the Nehru Library, CCSHAU, Hisar and the Bir Chandra State Central Library, MBBU, Agartala provides practical insights for librarians seeking to implement forward-thinking strategies in their own contexts.

As we navigate the complexities of Web 4.0, this book emphasizes not only the importance of adapting to technological change but also the necessity of integrating sustainable practices that respect our environment. The chapters on green libraries and eco-friendly initiatives reflect a growing recognition of our responsibility to contribute positively to society and the planet.

The authors delve into the relevance and impact of ICT in enriching library services, showcasing innovative practices that bridge traditional methods with cutting-edge solutions. This integration is not just about technology; it represents a fundamental shift in how libraries engage with their users, ensuring that they remain relevant in an age characterized by rapid digital transformation.

Ultimately, "Next Generation Libraries in the Networked Digital Environment" is more than a scholarly examination; it is a call to action. It challenges library professionals, educators, and policymakers to re-imagine the library's role within the community and to seize the opportunity to lead in the digital era. As you journey through this collection of intellectuals output, you may find inspiration and actionable insights that empower you to contribute to the evolution of libraries as catalysts for knowledge, innovation, and sustainability.

In conclusion, "Next Generation Libraries in the Networked Digital Environment" serves as both a reflective examination and a forward-looking guide for professionals in the field. It challenges us to envision libraries as vibrant ecosystems that not only adapt to change but also lead it towards cultivating knowledge, supporting sustainable development goals, and ultimately enriching the communities they serve. As we embark on this journey through the pages of this volume, let us be inspired by the potential of libraries to shape a more informed and equitable future.

The authors who have contributed to this volume are seasoned library professionals, knowledgeable educators, and dedicated researchers. The editors wish to clarify that the opinions presented in the chapters belong solely to the respective authors, who bear full responsibility for their content. In striving to uphold the quality and standards of this book, we have made every effort to preserve the original ideas and perspectives of the authors. We encourage readers to share their thoughts and suggestions, as they will be invaluable for future enhancements. This book aims to serve as a reference for the Library and Information Science (LIS) community, providing insights into the challenges and opportunities presented by technological transformation and digital environment in libraries and information centres, along with other related topics.

Dr. Bhanu Partap
Dr. Renu Bala

Acknowledgement

The completion of "Next Generation Libraries in the Networked Digital Environment" would not have been possible without the invaluable contributions and unwavering support of numerous individuals and organizations. First and foremost, we extend our heartfelt gratitude to the esteemed authors who have generously shared their expertise, insights, and innovative ideas. Your expertise and insights have enriched this volume for advancing the field of librarianship and inspired countless readers.

We also wish to thank the editorial team for their dedication and meticulous efforts in curating and refining the content, ensuring that the ideas presented remain true to the authors' visions while maintaining the highest standards of quality.

Our appreciation goes to the institutions and organizations that supported the contributors, providing the necessary resources and encouragement for their research and writing endeavours. A special acknowledgment goes to the libraries and institutions that served as case studies. Your willingness to share best practices and lessons learned exemplifies the spirit of collaboration that is essential for the growth and evolution of modern libraries.

Additionally, we acknowledge the vital role of our readers. Your feedback and engagement are essential as we strive to enhance the discourse around the challenges and opportunities faced by libraries in the digital age. We hope this book serves as a valuable resource in your endeavours to navigate the ever-evolving landscape of libraries in the digital age.

We are deeply grateful to our families and friends for their unwavering support and understanding throughout this project. Your encouragement has been a constant source of motivation during challenging times.

Finally, the editors would like to express their sincere appreciation for the tremendous efforts of NIPA® Genx Electronic Resources & Solutions P. Ltd., New Delhi, in ensuring the timely and high-quality publication of this book.

Together, let us continue to embrace innovation and champion the vital role that libraries play in shaping a more informed and equitable world. Thank you all for being part of this journey towards reimagining the future of libraries.

Dr. Bhanu Partap
Dr. Renu Bala

Contents

List of Contributors

Abhi Sharma
Research Scholar, Department of Library & Information Science, University of Jammu, Jammu - 180 006, J&K, India, *E-mail: a8hisharma@gmail.com*

Abdul Rajak Ahmed
Department of Mass Communication, Mizoram University
Aizawl -796 004, Mizoram, *E-mail: abdulrajakahmed22@gmail.com*

Anju Saini
Librarian, SNRL Jairam Girls College, Lohar Majra, Kurukshetra-136 119
Haryana, India, *E-mail: anjusaini6080@gmail.com*

Dr. Anuj Kumar Singh
Assistant Professor-Library, Government Girls Degree College,
BLW, Varanasi - 221 001, (U.P.), India, *E-mail: aps.kumbh@gmail.com*

Amit Kumar
Assistant Librarian, M.S. Randhawa Library, Punjab Agricultural University
Ludhiana - 141 004, Punjab, India, *E-mail: 14ak25@gmail.com*

Amit Sharma
Research Scholar, Department of Library & Information Science, University of Jammu, Jammu - 180 006, J&K, India, *E-mail: amit.smvdu@gmail.com*

Archanagita Saikia
M. Lib. I. Sc., Gauhati University, Guwahati, Assam - 781 014, India
Email: fftmc.archana@gmail.com

Basudeb Jana
Assistant Librarian, Kalinga University, Raipur - 492 101, Chhattisgarh
India, *E-mail: basujana5@gmail.com*

Dr. Bhanu Partap
Assistant Libraria, CCS Haryana Agricultural University
Hisar - 125 004, Haryana, India, *E-mail: bpartaps2005@gmail.com*

Diksha
Junior Research Fellow, Department of Library & Information Science University of Jammu, Jammu-180 006, J&K, India
E-mail: dikshaphd23@gmail.com

Hatneihoi Changsan
Teaching Associate, DLIS, Cotton University, Guwahati - 781 001, Assam India, *E-mail: hatneihoichangsan@gmail.com*

Ar. Jagriti Pathak
Assistant Professor, Royal School of Architecture, Guwahati - 781 035 Assam, India, *E-mail: ar.jagriti108@gmail.com*

Ar. Jayantajit Buragohain
Junior Planner, Town and Country Planning, Guwahati - 781 035, Assam, India, *E-mail: jntburagohain@gmail.com*

Jitamoni Bhattacharyya
Librarian, Biswanath College, Assam - 784176, Former guest faculty member (Biswanath college of Agriculture), *E-mail: jitamoni096@gmail.com*

Dr. Madhu Midha
Deputy Librarian and Head (KRC), IKG Punjab Technical University Jalandhar- Kapurthala Highway, Near Science City, Jalandhar - 144 603 Punjab, India, *E-mail: ptu_madhu@yahoo.com*

Mahalakshmi K R
Research Scholar, Department of Studies and Research in Library & Information Science, Tumkur University, Tumakuru - 572 103, Karnataka India
and Librarian, Atomic Energy Central School, Kaiga Karwar - 581 400 Karnataka, India, *E-mail: mahalakshmikr1993@gmail.com*

Mathias Swargiary
PhD Research Scholar, DLIS, Cotton University, Guwahati - 781 001, Assam India, *E-mail: mathias95912@gmail.com*

Dr. Meghna Dhar
Assistant Professor, Department of Library & Information Science University of Jammu, Jammu - 180 006, J&K, India
E-mail: dhar_meghna@rediffmail.com

Mrunalini Dhondiram Gadade
Librarian, Christ College, Pune - 411 014, Maharashtra, India
E-mail: mrunalini.gadade0706@gmail.com

Muthuraj Anbalagan
Assistant Librarian, Amrita School of Engineering, Amrita Vishwa Vidyapeetham, Amaravati Campus, Guntur - 522 503, Andhra Pradesh, India
E-mail: a_muthuraj@av.amrita.edu

Dr. Navkiran Kaur
Assistant Professor, Department of Library & Information Science
Punjabi University, Patiala - 147 002, Punjab, India
E-mail: navkiran1907@gmail.com

Dr. P. Krishnamoorthy
Assistant Librarian, R.M.K. College of Engineering and Technology
Puduvoyal, Thiruvallur - 601 206, Chennai, Tamil Nadu, India
E-mail: krishlib@rmkcet.ac.in

Padmavati Tubachi
Librarian, DPM's Shree Mallikarjun & Shri Chetan Manju Desai College
Delem, Canacona - 403 702, Goa, India
E-mail: padmavati.tubachi@shreemallikarjuncollege.ac.in

Dr. Piyali Sarkar
Assistant Headmistress (PGT), Barabisha High School (H.S.),
Alipurduar - 736 207 West Bengal, India, *E-mail: piu12sarkar@gmail.com*

Prince Buragohain
Department of Library and Information Science, Mizoram University
Aizwal-796 004, Mizoram, India, *E-mail: buragohainprince@gmail.com*

Priya Ranjan Mahalik
Associate Librarian, International Institute of Information Technology
Bhubaneswar - 751 003, Odisha, India, *E-mail: priyaranjan@iiit-bh.ac.in*

Dr. Rabindra K. Mahapatra
Chief Librarian (Retd), Odisha University of Agriculture & Technology
Bhubaneswar - 751 003, Odisha, India, *E-mail: rkmahapatra60@gmail.com*

Dr. Rajendra Babu. H
Assistant Professor, Department of Studies and Research in Library & Information Science, Tumkur University, Tumakuru - 572 103, Karnataka India, *E-mail: hrajendra.babu@gmail.com*

Dr. Rajinder Kaur Gill
Librarian, Arya Kanya Mahavidyalya, Shahabad Markanda, Kurukshetra Haryana, India- 136 135, *E-mail: rajinder18gill@gmail.com*

Rajshree Dutta
MLIS, PGDCA, LIS Intern, Central Reference Library, Kolkata - 700027 West Bengal, India, *E-mail: rajsheedutta35@gmail.com*

Rashmi Shewale
Assistant Librarian, National Institute of Technology, Goa - 403 703, India *E-mail: shewalerashmi@gmail.com*

Dr. Renu Bala
Librarian, PIET Sanskriti Senior Secondary School, NFL Township Panipat - 132 103, Haryana, India, *E-mail: renu.bala149@gmail.com*

Rohan Makwana
Library Associate, INFLIBNET Centre (An IUC of UGC), InfoCity Gandhinagar - 382 007, Gujarat, India, *E-mail: raviraj.rohan@yahoo.com*

Sagar Shivaji Kumbhar
Librarian, Gokhale Education Society's, Arts, Commerce & Science College Shreewardhan, Raigad - 402 110, Maharashtra, India
E-mail: kumbhar020@gmail.com

Sagarika Medhi
PhD Research Scholar, DLIS, Cotton University, Guwahati - 781 001, Assam India, *E-mail: samudrika1230@gmail.com*

Sanjeev Kumar
Associate Librarian, Mohinder Singh Randhawa Library
Punjab Agricultural University, Ludhiana - 141 004, Punjab, India
E-mail: srishti88@yahoo.com

Dr. Sonia Sharma
Assistant Professor, Department of Education, Lovely Professional University, Phagwara - 144 411, Punjab, India
E-mail: soniasharma7oct@gmail.com

Srikanth H. G.
Librarian, Soundarya Institute of Management and Science
Soundarya Nagar, Sidedahalli, Bengaluru - 560 073, Karnataka, India
E-mail: srikanth.hg@soundaryainstitutions.in

Subhankar Ray Chowdhury
Senior Library Assistant, NSHM Knowledge Campus, Kolkata, West Bengal India- 700 041, *E-mail: src2509@gmail.com*

Subrata Chakraborty
Library Assistant, NSHM Knowledge Campus, Kolkata - 700 041 West Bengal, India, *E-mail: subrata311289@gmail.com*

Dr. Sucheta Bhattacharjee
Assistant Librarian, Techno India University, Agartala - 799 004, Tripura India, *E-mail: sucheta.aus@gmail.com*

Sudhir Baburav Halvegar
Librarian, MES Vasant Joshi College, Zuari Nagar, Vasco - 403 726, Goa India, *E-mail: halvegarsb@rediffmail.com*

Dr. Sudip Bhattacharjee
Assistant Professor, Dept. of Library and Information Science Maharaja Bir Bikram University, Agartala - 799 004, Tripura, India *E-mail: sudipbht12@gmail.com*

Sukhdeep Kaur
Librarian, Guru Gobind Singh College for Women, Chandigarh - 160 019 India
and, Researcher Department of Library & Information Science Punjabi University, Patiala - 147 002, Punjab, India
E-mail: sukhdeepkaur05@gmail.com

Dr. Suman Singh
Assistant Professor (SOH), Library Information Science, Uttar Pradesh Rajashri Tandon Open University, Prayagraj - 211 013, Uttar Pradesh, India *E-mail: drsumansingh33333@gmail.com*

Sunil Madhukar Gauns
Librarian Grade-I, GVM's Dr. DadaVaidya College of Education Ponda - 403 401, Goa, India, *E-mail: 1479smg@gmail.com*

Sutapa Chakraborty
MLISc., Rabindra Bharati University, Kolkata - 700 050, West Bengal, India *E-mail: sutapapaul333@gmail.com*

Dr. Swapan Khan
Librarian, Narasinha Dutt College, Howrah - 711 101, West Bengal, India *E-mail: khanswapan@gmail.com*

Tarali Barman
Guest Faculty, DLIS, Cotton University, Guwahati - 781 001, Assam, India
E-mail: barman.tarali@rediffmail.com

Yamini P. Galapure
Librarian, Gokhale Education Society's, SMRK-BK-AK Mahila Mahavidyalaya, Nashik - 422 005, Maharashtra, India
E-mail: yaminigalapure.smrk@gmail.com

1

Next Generation Libraries in Digital Environment: Prelude to Practice

Dr. Rabindra K. Mahapatra[1] and Priya Ranjan Mahalik[2]

[1]Chief Librarian (Retd), Odisha University of Agriculture & Technology Bhubaneswar - 751 003, Odisha, India, E-mail: rkmahapatra60@gmail.com

[2]Associate Librarian, International Institute of Information Technology, Bhubaneswar - 751 003, Odisha, India, E-mail: priyaranjan@iiit-bh.ac.in

Abstract

The paper discusses the state-of-the-art digital libraries in the present day environment. The definition and concept of digital libraries has been elaborated in detail. Brief history of digital library has been discussed. Since the title of the book is more concerned with next generation of digital libraries, three types of digital library generations such as first generation starting from initial period of computer and digital activities, the second generation having the existence and use of Internet and the third generation digital libraries with the dynamism of open access system, open sources softwares and remote access to libraries have been discussed in detail. Again, the next generation libraries with web based technologies are reflected. The usage and existence of digital libraries with the emerging technology of artificial intelligence has also been depicted. Finally, the future of digital libraries as per the opinion of authors has been discussed in brief.

Keywords: *Digital libraries, Net-Gen libraries, Emerging technologies, Digital library generations, Artificial intelligence.*

1. Introduction

The present library system has been designated as a treasure of knowledge in virtual environment. The concept of modern library has been considered as the source of high level-information using web technology, digitization, networking,e-resources and multimedia. The rapid development of modern information and communication technologies and smart usage of computers in library and information system and services has given rise to new generation of libraries. The last two decades have radically transformed the library system with new avenues of information transfer in virtual world. An incredible amount of growth in increasing new interfaces have been developed utilising latest technologies to optimise the delivery of contents and services in the context of web environment. The emerging new vision of library interfaces has involved web technologies, digital collection of texts and images, documents, genealogy of records and other materials of importance for the user community. The major shift of new technology has developed the subscription of e-journals, e-books, and e-databases collection in libraries. Such library interface is compelled to provide equal access to the contents or these collection to all users irrespective of area or locality. Another remarkable trend in library technology field is the emergence of open-source software movement that enabled to transform the automation and digital library interface. Open Source Software efforts such as OAI, Grenstone, DSpace, Koha etc, have drastically transformed the library technology scenario in terms of sophistication and functionality. This article/ chapter is dealt with the state of digital library system with stages of generation of digital libraries and future directions.

2. Digital Library: Concept and Definitions

The concept of digital library has been evolved since the Vannevar Bush has designed '*As We May Think*' in 1945. But actually, the concept of a digital library came into existence during 1990s. Before that, many institutions acquired collections of purchased electronic resources or even digitized materials, but did not refer to them as digital collections (Schwartz, 2000). Prominent libraries developed electronic contents and term it as electronic library. Libraries whose information resources are made available to others termed as virtual library. The library resources available through a network called as "network-accessible libraries," or "libraries without walls" These were the terms used to describe the new phenomenon in the early phase of digital library development. The term "digital library" came from the National Information Infrastructure Initiative (Bearman, 2007; Lagoze, 2010). In May 1995, the *Communications of the ACM* devoted an entire issue to digital libraries.

The American Digital Library Federation has defined the digital library as "Digital libraries are organizations that provide the resources, including the specialized staff, to select, structure, offer intellectual access to, interpret, distribute, preserve the integrity of, and ensure the persistence over time of collection of digital works so that they are readily and economically available for use by a defined community or set of communities" (Waters, D.J).

A research community focused the digital contents of the library using technology. "The evolution of the concepts of digital libraries demonstrates a shift from the early focus on the traditional library model and the system-centered approach to emphasizing the complexity of these systems, their multiple dimensions, and the social context of use.Many researchers emphasize that the term 'digital library' evokes different meanings to different people" (Borgman, 1999; Fox, 1995; Schwartz, 2000). "The diversity of opinions stems from the fact that the early concepts of digital libraries combined the missions, techniques, and cultures of traditional libraries with the capabilities and cultures of computing and telecommunications" (Marchionini, 2000).

The term Digital Libraries is not new now. The D-Lib Working Group on Digital Library Metrics chaired by Barry M. Leiner in 1998 defined the digital library as follows:

"The Digital Library is the collection of services and the collection of information objects and their organization, structure, and presentation that supports users in dealing with information objects available directly or indirectly via electronic/digital means" (Leiner, 1998a). Digital library is a focused collection of digital objects, including text, video, and audio, along with methods for access and retrieval, and for selection, organization, and maintenance of the collection.

The Digital Library Federation (DLF) presents a definition that "de-emphasizes the digital nature of collections but stresses the functions and services offered by libraries as organizations". The DLF describes digital libraries as "organizations that provide resources, including the specialized staff, to select, structure, offer intellectual access to, interpret, distribute, preserve the integrity of, and ensure the persistence over time of collections of digital works so that they are readily and economically available for use by a defined community or set of communities" (Waters, 1998, para. 3). This definition reflects the conviction that, with time, there will be less emphasis on the digital nature of material and more on traditional roles of libraries. The DLF definition includes distinct traditional library rolessuch as selection, intellectual control, providing access, and preservationapplied to digital works.

3. Generation of Digital Libraries

The digital libraries have been developed in course of time coping with the emerging technologies.There are basically three generation of digital libraries which are discussed below.

3.1. First Generation of Digital Libraries

The first generation of digital libraries started with online catalogues. Though this is not complete in its sphere, but substantial adoption has been made in good libraries who have converted their hard catalogues into online. The most important facilitator is MARC and numerous international cataloguing cooperative organisations. Libraries have adopted online cataloguing and showed that their libraries are digitized. The evolution of digital libraries was not spontaneous but was taken place due to contribution of many disciplines. A variety of information systems that have been grown leading to formation of digital libraries. Most of the systems are dealt with their original conception, even though not in their original technological solutions. It can be observed that most of the information systems are still practising in one form or other using newer terminological solutions. The first generation of digital libraries is traced back to the use of computer and microprocessor technologies. Again, the digital storage devices have also been used by reduction of its size and enhancement of the capacity of storage. Then came magnetic storge media for large volumes of data and optical storage media called Compact Disc (CD). The Digital Video Disk was introduced replacing CDs. The smallest portable semiconductor-based storage devices or USB devices came to the market which is still very popular.

3.2. Second Generation of Digital Libraries

The second generation of digital libraries is best characterised by the use of Internet as the primary means of delivering traditional library services. The most prominent task before libraries was to develop access to the online cataloguing through Telnet, Gopher, Mosaic and later on by the present web-browsing technologies. Subsequently, efforts were taken up vigorously for growing electronic reference materials in digital formats. The digital collections developed during the period was to change the image of the library rather helping the teachers, researchers and students. The earliest digital collections were basically done for selected rare and special material collections to illuminate the cultural and historical significance of the regional or local collections. At this period, it was more of experiments than exhibitions of library collections.

Creation of online databases through information retrieval was "an important landmark to the digital libraries. The first databases were primarily bibliographic in nature and were online version of existing indexing and abstracting services. The databases used data files generated in the process of electronic phototypesetting of printed abstracting and indexing services and other primary journals. Subsequently, full textual information, news, statistics, etc called full-text databases emerged. Till 1989, there were almost 1700 full text sources of sixteen online systems" (Arora, 1996).

A number of software packages were introduced in the market for computer storage, indexing and retrieval. Micro CDS/ISIS was one of such non-numerical information retrieval system developed by UNESCO in 1985 was extensively used in libraries. There was an emerging need for DBMS such as Ingres, Microsoft Access, MS-SQL Server, MS-FoxPro, MySQL, ORACLE contributed to the digital libraries. Digital library imaging system was developed leading to digital document imaging system that helps scanning and storage of images of documents in digital formats. Several documents imaging software packages were made available in the market to make libraries digital.

3.3. Third Generation of Digital Libraries

The open access movement was a great thrust on digital library system. The concept of information retrieval began as a movement to facilitate access to digitized pre-prints to users through remote access electronically. Repositories or archives are developed by prominent universities and research institutes. ArXiv is a digital depository in physics, mathematics, biology and computer science. Subsequently, many universities and research institutes and technical institutes of higher learning developed their own information retrieval for dissemination of e-resources for research and academic interest. Domain specific information retrieval were generated to support searching of multiple repositories. Launching of OAI and its Protocol of Metadata Harvesting (PMH) is the most significant initiative for development of information retrieval systems. A number of OAI-complaint open source software for IRs led to wide spread establishment of IRs in leading institutions and universities. The publishing of open access journals led to the need for author's web sites and self-archiving. Major publishers like Elsevier, Sage, Springer, etc, allowed their authors to self-archive their research articles. The invention of World Wide Web by Tim Berners-Lee dramatically transformed the whole spectrum of information use, sharing and dissemination in library environment. The beginning of full text digital library emerged. The growth of electronic resources of all libraries in leading institutions in full text formats became the usual practice which is supported by the internet. Leading publishers all over the world started converting their publications of books and journals into e-books and e-journals and deliver their contents in different formats. The new technologies help significantly for delivery/ dissemination and usage of e-resources making success of digital libraries.

Jewell mentions "the extent to which leading research libraries have adopted highly complex processes involved in identifying, evaluating, negotiating access to and supporting the use of electronic information that is commercially supplied by third parties" (Jewell, 2001). Pitschmann finds "an evolving and highly sophisticated understanding of pitfalls, opportunities and real costs involved for libraries that choose to organise access to free internet resources" (Pitschmann, 2001). Smith recommends that digital libraries that are digital reformatting materials from their collections in the 1990s are doing so strategically rather than experimentally to profile selected rare and special collections to support teaching and research needs (Smith, 2001).

4. New Generation of Libraries and Web Technologies

The web technology has transformed the libraries into new generation of interface. With the increasing digital literacy of the user community their expectations have witnessed a great change to obtain resources through web technologies. Taking advantages of some latest search techniques and methods, the libraries started to offer better and dynamic interaction with library users. The library environment is replete with diverse resources that demand a new interface with aggregation of several contents and resources in a single mode. There was a greater need for a single repository with digital contents. Such a new generation of library interfaces demand better web technology. Using such technology the users feel satisfied to access their desired information. Once web gives more contents, the expectation of users also increases. Today's library interfaces must go beyond helping discover the available resources, they must make the contents available in the most convenient way. The expectations of users fulfilled when they find the full text of contents are available online. Libraries became able to create seamless path for the users as they navigate from the tools that have been provided for resource discovery to the actual information.

5. Digital Libraries and Artificial Intelligence

In order to build concerted next generation digital libraries, Artificial Intelligence (AI) plays a very important role. Digital libraries are such information system in which the integration of AI has made significance changes in access and use of information. The basic function of AI is to provide the tools and techniques to effectively manage and sort vast resources in digital libraries. The integration of AI in digital libraries helps enhance user experience by providing personalised services efficiently. AI uses technologies like Computer Vision, Machine Learning and Natural Language Processing to enhance the organisation, discovery, and retrieval of information in digital libraries. The natural language processing tool in AI categories and classifies the desired information on its contents, making it easier for the users to find what they are looking for. The most spectacular advantages of AI in digital libraries is the ability to personalise contents to suit individual user's needs and preferences. Machine learning algorithms can assess the user behaviour and needs and provide the exact contents to suit their information requirements. Computer vision algorithms example pictures and videos, recognise persons and objects and retrieve on the basis of visual characteristics. Since preservation is one of the important objectives of digital libraries, AI provides various solutions on preservation of library materials on digitization and restoration. Since AI appears to be the latest technology it is imperative on the part of the Librarian to incorporate it in digital library operations. It is acknowledged that AI has reshaped the pattern of access, organise and retrieve information in effective and efficient manner for which its application in digital libraries is highly recommended. AI is invaluable tool for digital libraries that can enhance the user experience and contribute to the preservation of information for future generation. With the advancement of technology, the digital libraries may adopt AI for achieving its objectives.

6. Future of Digital Libraries

The digital revolution has assuaged the burden on physical storage of the library. The adoption of digital repositories and cloud-based solutions, libraries can grow its digital collections and preserve huge information resources in digital formats. Moreover, libraries will continue to broaden its scope and sphere for collaboration, research, innovation and academic pursuits. Even though e-resources in digital formats continue to reign the information world, the physical space will also continue to attract the users for personal and quiet reading and usage. AI tools with access to vast digital archives and databases have made the information retrieval system more efficient and convenient enabling the scholars to use the resources across the globe. The digital platform can make more vibrant the information sharing allowing the scholars to engage in real time collaboration irrespective of geographical boundaries.

Another major aspect is the improvement of digital literacy of the scholars that make them more competent to use vast sea of information. The personal cognition of users will continue to evolve to adopt to new digital technology. The cognitive changes of the users force to gain interactive learning tools and adaptive learning platforms. Librarian shall be responsible to remain vigilant in mobilising the influence of digital technology on cognitive changes of the users that improve their learning and teaching strategies.

The digital repositories and online archives have already harnessed the perpetual access to information in a democratic way. Libraries have made it possible in changing the collection, preservation of rare informative materials and publication of scholarly usage. But, there are also challenges to maintain the digital contents. The present technological obsolesces and fragility of digital records in different formats requires robust preservation strategies to ensure the longevity of digital collections. In order to ensure perennial and qualitative preservation safeguarding the digital resources for future generations, the libraries need to develop digital preservation initiatives in collaboration with archival technology experts and archival institutions for designing and maintaining long term digital preservation of assets.

Till the end of 2023, there has been a significant growth in the number of digital libraries. The future is very bright as the new technology of AI integration will certainly reshape the digital libraries. Again, the integration of AR and VR methods toenable intense learning and provide interactive contents much better to the scholars and user community. In future, the digital libraries will play a very crucial role in global collaboration and interoperability, permitting the research communities to share open access resources in larger scale.

7. Conclusions

If we look at the present library system and services, the place and role of digital libraries are very much prevalent in all types of libraries. The academic libraries especially in universities and institutes of higher learning in India have almost developed their own digital libraries. Considering the frequency and huge usage of information by the user community the digital libraries have become the inevitable part. The digital contents are the key of modern libraries. The users demand emphasise ease of access, friendly usage, effective search and retrieval and instant and multiple availability of information from anywhere, high volume of storage of multiform information, effective search, and retrieval, accessible through multiple points at any time, instantaneous downloading, faster addition of information etc prompted the development of digital libraries. Hence, the development and growth of digital libraries was spontaneous and it became a worldwide phenomenon. The latest technologies have invigorated the trend of digital libraries. While the web technology has made the services of digital libraries easy and widespread, at the same time the emergence of artificial intelligence has improved the functioning and information retrieval system of digital libraries. The digital libraries will continue to become an indispensable organisation for community resources and will remain as the hub of information, knowledge, and learning. The journey of digital libraries will continue further in fulfilling the ever-increasing information needs of the user community that will further inspire,innovate and illuminate the information and services in future.

References

Alguliyev, R., & Ismayilova, N. (2017). Prospects of next Generation Digital Libraries. Problems of Information Society, 8(1), 3-9.

Arora, Jagdish. (1996). Online information retrieval system: Databases, data networks and online search services in India. In: Librarianship Today and Tomorrow (Dr. S.C. Verma Festschrift) India (eds. U.C.Sharma and M.R. Rawtani). New Delhi, Ess Ess Publications, p.173-192,

Audunson, R. A., & Shuva, N. Z. (2016). Digital library education in Europe: a survey. SAGE Open, 6(1), 2158244015622538. Retrieved 12.8.2024

Bamgbade, B. J., Akintola, B. A., Agbenu, D. O., Ayeni, C. O., Fagbami, & Abubakar, (2015). 5. Comparative analysis and benefits of digital library over traditional library. World Scientific News, 24.

Breeding, Marshall (2007). The birth of a new generation of library interfaces. Computers in Libraries, 27 (9), 34-37.

Breeding, Marshall (2002). Preserving digital information. Information Today, 19, 5.

Bush, Vannevar (1945). "As We May Think" (PDF). The Atlantic Monthly: 101–108. Archived from the original (PDF) on 21 April 2018. Retrieved 30 April 2018.

Cain, Mark (2003). Managing technology: Being a library of record in a digital age. Journal of Academic Librarianship, 29, 6.

Fox, Edward A. (1998). The digital libraries initiative: Update and discussion. Bulletin of the American Society for Information Science, 26 (1). Archived from the original on 3 April 2018. Retrieved 20.8.2024.

Gert, Janet (2000). Selection for preservation in the digital age. Library Resources & Technical Services, 44(2), 97–104.

Hinze, Annika, Dana McKay, Nicholas Vanderschantz, Claire Timpany, & Sally Jo Cunningha (2012). Book selection behavior in the physical library: implications for ebook collections. In Proceedings of the 12th ACM/IEEE-CS joint conference on Digital Libraries, pp. 305-314.

Hirtle, Peter B. (2011). Digital preservation and copyright. Archived 2008-04-19 at the Wayback Machine. Stanford University Libraries. Retrieved on 12.08.2024

Jewell, Timothy, D. (2001). Selection and preservation of commercially available electronic resources: Issues and Practices. http://www.clir.org/pubs/abstract/Pub99abst.html. Retrieved on 12.08.2024

Koehler, Amy E. C. (2013). Some thoughts on the meaning of open access for university library technical services. Serials Review, 32 (1), 17–21. doi:10.1080/00987913.2006.10765020. S2CID 220292279. Retrieved on 13.08.24

Lancaster, F. W., ed. (1980). The role of the library in an electronic society. Proceedings of the 1979 clinic on library applications of data processing, April 1979, University of Illinois. Urbana-Champaign, IL: University of Illinois, Graduate School of Library Science.

Ltaper, Thomas H. (2001). Where Next? Long-Term Considerations for Digital Initiatives. Kentucky Libraries, 65(2), 12–18.

Lynch, Clifford (2005). where do we go from here? the next decade for digital libraries. D-Lib Magazine, 11 (7/8). doi:10.1045/july2005-lynch. Retrieved on 13.08.24.

Pitsehmann, Louis, A. (2001). Building a sustainable collection of free third party web resources. http://www.clir.org/pubs/abstract/Pub98abst.html. Retrieved on 12.08.24

Pomerantz, Jeffrey, & Marchionini, Gary (2007). The digital library as place. Journal of Documentation, 63 (4), 505-533.

Pymm, Bob. (2006). Building collections for all time: The issue of significance. Australian Academic & Research Libraries, 37(1), 61–73.

Ross, Seamus (2006). Approaching digital preservation holistically. In Moss, Michael; Currall, James (eds.), Information Management and Preservation, Oxford: Chandos Press, pp. 115–153.

Schatz, Bruce (1996). Chen, Hsinchun (ed.). Building large-scale digital libraries. IEEE Computer, 29 (5), 22–25. doi:10.1109/2.493453. hdl:10150/106127 Retrieved on 12.08.24

Schatz, Bruce (1997). Information retrieval in digital libraries: Bringing search to the net. Science, 275 (5298), 327–334. doi: 10.1126/science.275.5298.327. PMID 8994022.

Smith, Abby (2001). Strategies for building digitized collection. http://www.clir.org/pubs/abstract/Pub101abst.html. Retrieved on 12.08.24

Stocker, Gerfried (1 January 2014). Beyond archives (or the Internet 100 years before the Internet). In Magalhães, Ana Gonçalves; Beiguelman, Giselle (eds.). Possible Futures: Art, Museums and Digital Archives. EditoraPeirópolis LTDA.

Van Le, Christopher (Spring 2010). Opening the doors to digital libraries: A proposal to exempt digital libraries from the copyright act, Case Western Reserve Journal of Law, Technology & The Internet, 1.2, 135.

2

Enhancing Library Services Through ICT: Relevance and Impact

Amit Sharma[1] and Dr. Meghna Dhar[2]

[1]Research Scholar, Department of Library & Information Science, University of Jammu -180 006, Jammu & Kashmir, India, E-mail: amit.smvdu@gmail.com

[2]Assistant Professor, Department of Library & Information Science, University of Jammu -180 006, Jammu & Kashmir, India
E-mail: dhar_meghna@rediffmail.com

Abstract

The Information and Communication Technology (ICT) revolution in the last one and a half decades has a radical and extensive impact on all aspects of professional endeavours particularly in the knowledge and information sectors. The internet has added up a new dimension to information technology, giving birth to rich concepts such as knowledge management, digital library, archiving and e-learning of indigenous culture and heritage. It further helps in improving accountability, competence and effectiveness in library activities. ICT has revolutionized information access and sharing. The paper explores the impact of ICT on library services, collection, library staff and users. It further emphasizes its necessity and applications in Library and Information Centres.

Keywords: *ICT, Benefits of ICT, WEB-OPAC, OPAC, CAS, SDI, e-References, Library software, Library services, Library users.*

1. Introduction

Information and Communication Technologies (ICT) has transformed the functioning of modern libraries in the 21st century. The applications of ICT to data, information and knowledge resources have given rise to data centers, digital libraries, and knowledge repositories. New software's that are based on international standards have emerged to undertake a wide range of jobs that suit the requirements of individual libraries. Rapid advances in the application of ICT have given rise to 'Open Standards and Open Source Software's. Libraries can benefit a great deal now by using these standards and software, as during the last twenty five years only rich libraries could afford to use commercial software. As a result the transformation of libraries was not universal in India. Now with the advent of open source software, we are expecting many libraries are going to be modernized. This transformation is not going to be limited to libraries only but to users as well who belong to different disciplines and walks of life due to social networking which is emerging as a major activity of libraries around the world. Libraries

now are able to use resources of other libraries through national and international networks in order to serve their users well.

The term Information Communication Technology (ICT) was used as synonymous with Information Technology, a few years ago. Nowadays, due to vast advancement of Communication Technologies and particularly the Internet, the term is more popular as ICT. In fact, ICT is a tool developed with the help of a set of technologies. ICTs also refer to the different infrastructures that are used in these processes, their applications and numerous services these infrastructures render. ICTs comprise of three elements such as media of communication (e.g. radio, television, internet), information machine (e.g. computer, mobile, tab, laptop), telecommunication technologies and equipments like satellites, OFC, phone, facsimile machines etc.

2. Review of Existing Literature

Literature search plays a very important role in research activities, as it forms the very first step of research pursuit. Some of collected literature related to the topic of research has been reviewed as under:

(Khan & Khayal, 2024) revealed the use, benefits, satisfaction, and dissatisfaction level of the users with regard to ICT based services. They also investigated further areas to clarify the reason for dissatisfaction and taking suggestions from respondents about Information Communication Technology (ICT) based services of the libraries. The collected data were analyzed and inferences were made based on standard statistical methods. A total of 205 questionnaires were distributed out of which 156 responses were received with a response rate of 76.10%. Most of the library users are male. Unavailability of ICT based library resources and services, lack of financial support, shortage of ICT trained staff and poor Library infrastructure are the main hurdles before the academic libraries of Arunachal Pradesh. Based on the study and suggestions given by the respondents, it has been stated that the use of ICT based services is almost satisfactory. However, the libraries need to take some more steps to maximize the use of ICT. Adequate training should be given to the staff of the library. The library authority should be taken users feedback to understand the problems. Maximum number of users suggested that the library should organise workshop, Seminar, training for user from time to time to solve the challenges faced by users. 24X7 power backup is the major issue which needs to take care by the library authority during library hours. In the present age, the ICT based services cannot be ignored or delayed. The overall ICT based services must be taken into importance and, this will lead to overall development of library system as well as the educational system.

(Melbha and Melvin, 2023) studied the teacher's perception and challenges of using ICT in teaching and learning. The study was focused on college teachers of Kanyakumari District, Tamilnadu. It is stated that as ICT plays an important role in day to day life , therefore it is important to provide ICT facilities in classroom and usage of ICT by teachers for the future generation. The study has revealed that the teachers should spend some time to equip themselves through both professional and selflearning. The employer should be provided ICT tools in their institutions and support for attending workshops and training to cope up with the change in Information and Communication Technology (ICT) scenario in educational system.

(Jones & Umoh, 2022) described that in this era of globalization, in which the world is connected; information gains its power through permanent storage and wide distribution, which could be achieved through ICT. ICT has continued to be a driving force in the issue of national advancement and development. Information is power, as power is effectively an infinite resource and a vital tool needed for the sustainable development of all sectors in any nation. The study concluded that ICT is a generic term that refers to the technologies that are used to collect, store, edit, and communicate information in various formats. Libraries are mostly used to manage, disseminate, and preserve knowledge. The study also recommended that in order to achieve sustainable development the government should adopt the role of ICT in the provision of library services, allocate resources for the creation of digital libraries, provide training and conduct awareness programs to emphasize the value of digital libraries.

(Rahman *et al.*, 2020) in this paper aimed to show the status of information communication technology (ICT) application in some selected University libraries and Special Libraries of Assam. The paper focuses on the application of ICT and the ICT based services to the libraries. This paper discusses about different ICT tools and their use in the library operations, how the library services are developed with the use ICT have discoursed in the study. The present study discusses various library operations using ICT. The purpose of this study is to express the usefulness of the different ICT tools and techniques for making the access of information quicker and accurate to the user community as well as efficient management of the systems and services of the library.

(Mlay & Sabi, 2019) in their paper regarding Information and communication technologies have opened up new opportunities for cultivating a reading culture, where people go beyond instrumental to leisure reading, in developing countries. But the research on information and communication technology (ICT) and reading has been conducted mainly in western countries. This paper explores the potential of ICT-based interventions for fostering a reading habit among university students in Uganda. A cross-sectional survey design was employed and data were collected using a self-designed instrument from eight universities in Uganda. Structural equation analysis using Partial Least Squares was used for statistical analysis of the data. Major findings from 337 valid responses indicate that ICT usage, home culture, school culture, and disposable income significantly impact the reading habits of university students in Uganda.

(Japhet *et al.*, 2018) aimed at examining the information needs of users of Lagos State University (LASU) library and services at their disposal and to determine the level of satisfaction derived from library use. The study shows that the attitudinal pattern of the users to the use of the library was pleasant but there is need for attitude changes in some areas, such as (attitude) of library staff towards users, provision of information resources, and creating a conducive environment for students reading, and online resources among others. The student's reaction to the library resources was generally fair and their satisfaction level was also fair. The findings revealed that LASU Library users were more interested in information related to teaching, learning and research activities. The investigation also revealed that the attitudes of respondents were favourable in the use of LASU library resources but there is need for improvement in such areas as library services and provision of adequate space for readers. It was found that majority of the respondents sought information from journals, textbooks, reference books rather than other library materials. The major obstacle faced by the users while seeking information included lack of ICT infrastructure, resource sharing facilities and inadequate ICT skills.

(Nworie *et al.*, 2017) in this paper studied professional statues of librarians in relation to the challenges of using ICTs in provision of information services in the library. On this, competencies of librarians, challenges of librarians to use ICTs, and challenges of professional practice of librarians in academic libraries were studied. 82% of the librarians responded to the questionnaire. Analytically, a significant mean (X) of 3.06 were derived showing that, the statements made are accepted. This paper therefore recommended a way forward to ameliorate the challenges, and concluded that for librarians to live up to their professional statues, they need to work in ICT induced environment, as such they need to be proficiently skilled on new technologies, and also, be competent in information handlings.

3. ICT in Library and Information Science

The term ICT is extensively used in Library and Information Centres and Library and Information Science Education (LISE). ICT acts as a tool for Information management and use in Library and Information Centres. ICT is used for Information process and management and it is not restricted to a particular discipline. The application of ICT has spread its roots in different knowledge areas. The use of ICT has increased in the field of engineering, education, medical science, economics, business, management, Library and Information Science, etc.).

4. Need of Information and Communication Technology

The relevance for establishing the prerequisites for the development of the information society are adequate infrastructure, and appropriate regulatory framework and standards. Facilitating access of the developing countries to information and communication technology through human resources, technology transfer, production, trade and research and development have to be given high priority. The emerging trends of Information and Communication Technology (ICT) and its application explore the opportunity to make more efficient the functions and services of library and information centers. This paradigm shift have not only specified new exploratory and innovative thoughts of managing the functions and services of Library and Information Centers but also create an unprecedented challenging environment for professionals and information officers to fulfill the astounding information desire of users and also to manage the unimaginable proliferation of electronic information. Electronic information is becoming a challenging issue for information professionals and practitioner because most of the libraries are now moving towards automation or the partial/complete digitization without giving due consideration to the post technology deployment, services and other management issues. ICT in Library and Information Centers have made easier production; storage, access and easy dissemination of electronic information. Consequently, the user is becoming more demanding in terms of availability of specific, measurable, accurate and timely information. The libraries have found it very difficult to acquire, arrange, process and disseminate information in traditional ways. So, Librarians are compelled to plan, organize and communicate the huge Information Communication Technologies. For this, libraries need to be equipped with ICTs tools and products to store, retrieve and disseminate information effectively to the end users.

5. Impact of ICT on Library and Information Services

Libraries are experiencing a significant impact of IT on information processing, sources and services. Rapid technological developments enabled libraries not only to improve the quality of existing services but also to offer a wide range of new services to users. Today, libraries have wide range of opportunities and variety of challenges offered by the rapid development and wide application of Information Communication Technology (ICT). ICT has become integral part of all library operations and information services. This movement of using ICT in Libraries has dramatically changed the way by which data acquisition, processing and access has been carried out and the libraries have provided information services. As this technology provides librarians with new choices, new opportunities and new challenges, there have been a phenomenal progress in the uses of ICT applications in library operations and information services in the developed countries of world. ICT usage in Indian Libraries has also gained considerable momentum in recent years, which is going to increase at much faster rate in future.

The benefits of using ICT in Library and Information Centers may be consolidated as:

- Improved efficiency of library operations,
- Improved quality of existing services,
- Introduced new services,
- Improved collection management,
- Improved accuracy and control,
- Increased ability to share resource,
- Improved images,
- Improved management information,
- Eliminate duplication of efforts,
- Rapid communication,
- Perform tasks not possible by the manual system,
- Improved uniformity and standardization,
- Improved user's satisfaction.

The Fig 1.1 indicates the benefits of using ICT in Library and Information Centers

Fig. 1.1 Showing Benefits of Using ICTs in Library & Information Centres

6. Impact of Information and Communication Technology on Library Environment

Rapid advances in ICT have brought revolutionary changes in the concepts, organization, functioning and management of libraries. The impact of these changes is affecting all the aspects of library operations, information resources and services, staff skills requirement and users expectations. Some of the notable impacts on libraries, staff, and users are:

- **Impact on Library Collection:** Library collection goes beyond the print materials and includes the CDs/DVDs, audio & video cassettes, e-books e-journals and e-databases. The traditional paper as a medium of storage is getting replaced with electronic media. In India, many national and international origin library management software (i.e., Ventura, LibSys, E-Granthalaya, Library Manager, SOUL etc. commercial software and Koha, NewGenLib etc. Open Source Software) are available for library automation. In India, UGC-INFONET, DELNET, and INDEST etc. consortiums are in action. Internet technology provides a wide scope for communication and information search across the globe.

- **Impact on Library Staff:** Information and communication technology has changed the duties, responsibility, and functions of the library professionals. The changing environment forces the librarian to become ICT skilled, dynamic, ready to accept the changes and challenges and outfit the requirements of library users, because, without change, management library could not be survived.
- **Impact on Library Users:** Library users can access the worldwide information through their desktops without any time and distance limitations (24 x 7 x 365 days). Current users need to possess basic technical skills to access the information in electronic media. In the age of technology, users have multiple sources of information such as the Internet, commercial and non-commercial information service providers. As a result, the library is not the only source of information provided for users.

7. ICT Application in Library and Information Centres

Nowadays in the information era, in order to avoid obsolescence of information, library professionals are applying advanced technologies to enable its user community to get the right information to right user at the right time. In order to achieve this objective, libraries are automating their services. Many Libraries mainly concentrated on the house keeping functions consisting of acquisition, serial control, cataloguing, circulation, reference etc. In some libraries it has expanded on the library management system to incorporate OPAC's, Web OPAC's, CD- ROM Networks, and DTP Office Automation etc. Library Management Software like Koha, E- Libsys, Dspace, Granthalaya, Greenstone, SOUL etc. allows the library and information centers to give better management of resources and services to the users of the library. The following are the various ICT applications in Library and Information Centers:-

Acquisition System

Acquisition system is concerned with the finding book selection sources, searching and selection of books required, placing of orders with a wide range of suppliers, keeping track of orders, receiving and recording of materials, settlement of invoices, accessing and maintaining detailed accounts of funds. Requisition of payment; order follow up; receiving and accessioning of free gift items are also dealt with. Here, the computer is of immense help to the library staff as in computerized information systems the bibliographical details of the material are to be entered only once and at different stages for different jobs. The computer, through the integrated software takes relevant bibliographical details for each book at each stage. Similarly in a computerized library it is easier and automatic to maintain bill register, expenditure register, vendor directory etc. in computer's memory. Online queries may be made on titles, orders, invoices, vendors and budget heads. Various reports such as approval request form; purchase orders; overdue notices; budget and expenditure analysis; payment requisition report, accession register, and bill register can be generated and printed on requirement. It helps in

- Creating purchase orders,
- Provision of information on outstanding orders,
- Pre-order searching,
- Preparation of vouchers for payment to the vendors,
- Maintenance of fund accounts and printing of fund reports.

It can also help to find list of orders by order number, by title, language, by country or by any other date element which has been preserved in the system, claims or cancellation notices to vendor in case the ordered item is not received in the library within a stipulated period, Information notices to the requesters of documents intimating them that the items they requested for the purchase have been received in the library, provision for converting the currency from one to another.

Cataloguing System

Cataloguing is one of the most important functions, which link's user requirements to the documents in the library. In cataloguing we take into account the various approaches of user in searching for a book in the library and make catalog entries in order to ensure that each major search approach of a user is satisfied. Computerized catalogue is the most efficient tool in retrieving information about the documents in a library easily and quickly. The cataloguing system maintains titles-in-process file of all items that are accessioned. They are than catalogued. Catalogues are also produced by data import, or by network down loading or by direct data entry. Details about a book or any other library material can be searched by either item number or by a combination of fields such as keywords, title, author, classification number, class number or by subject heading. In addition, this system covers catalogue maintenance; thesaurus construction; authority files; and holding updates. Cataloguing system provides the facility to provide Current Awareness Service (CAS), Selective Dissemination of Information (SDI) and special bibliographies, and lists of fresh arrivals. It makes possible import/export of bibliographic data in CCF, MARC and UNIMARC formats; and automates the stock verification process.

Circulation System

Circulation procedure in a conventional system is very lengthy process and much of the time of staff is consumed in repetitive works. The use of technological devices such as computers, barcode scanners, RFID Tags, RFID staff station, self check in/out kiosks and Integrated Library Management Software in circulation helps in performing these routine operations easily and quickly. An automated circulation system provides information about the location of the item on loan, at the bindery, on reserve, being re-catalogued, details of items on loan to a borrower; record of reserves alerting the library staff on return of a reserved item by a borrower and prints a book available notice to the one reserved; prints recall notices for items on long term loan; renewal of loan; calculation of fines, printing of fine notices; analysis of summary statistics; printing due date slips; automatically generating orders for lost books, and provisions for interlibrary loan transactions. An automated circulation system provides:

- Information about a given book to whom it is issued and when it is due.
- Information about a given borrower what books are issued to him and when they are due.
- Information regarding overdue books.
- Information regarding books placed on reserve by the borrowers.
- Information about circulation statistics.

Serial Control System

Serials include periodicals, newspapers, manuals, journals, proceedings, transactions etc., of societies and monographic series. Serials are distinguished from monographs by their ongoing nature. The continuing nature of serials subscriptions creates problems and makes it a complex process requiring a separate control system. An automatic serial control system provides the following benefits:

- Ordering new journals; Renewal/discontinuation; sending reminders; receiving the journals;
- Preparation of a list of periodicals received; preparation of list of periodicals cancelled; preparation of a list of holdings; list of holdings with their status (i.e. on shelf, in binding, in circulation etc.);
- Bindery management; recording and accessioning bound volumes, and;
- Keeping track of amount spend on subscriptions, binding etc.; estimation of budget for next year; announcement of the missing serials for recording the same.

The nature of serials (multi-component publications arriving in parts over a long period of time). The complexities associated with the acquisition processes, continuous monitoring of the serial issues for which payment has been made in advance, and claiming the issues that fail to arrive to the library. The need for maintaining records to monitor receipt of current issues and for continuous budgetary control.

Article Indexing

The article indexing system facilitates indexing and abstracting of articles from various journals, technical reports, conference proceedings, monographs etc. It includes scanning of articles, entry of citation, and online searches on author, keywords/descriptors and even word based free text searches. This system also provides periodic documentation lists, personalized SDI, bibliographies on specific subjects etc.

ICT Based Library Services

It is established fact that information and communication technology (ICT) has brought the revolution in the field of library and information science, as we have already mentioned that all library house-keeping activities like acquisition, procurement, processing, organization and dissemination of information services are based on computer, communication, reprographic and digital technologies, which has facilitated the following ICT based library services being provided to the users more effectively.

- Reference Service
- Bibliographic Service
- Referral Service
- New Arrivals alert service
- Online reservation service
- Online Public Access Catalogue- OPAC
- Current Awareness Service (CAS)
- Document Delivery Service (DDS)
- Selective Dissemination of information (SDI)
- Barcode / RFID Based Service
- E-Employment Alert Service
- E-Book Service
- E-Journals & Magazine Service
- E-Databases Service

- E-Newspapers Service
- E-Project/Report/Thesis Service
- User Orientation Program
- In-House Personal Training
- Resources Sharing and other services.

Such facilities, being extended to the academicians, researchers and library users to find their required information without any barriers, are possible only because of information and communication technology (ICT) use in library and information centers or libraries.

Some library users are adopting electronic habits, making increasing use of the new ICT including computers, the Internet, the Web, Intranet, Extranet, and other technologies. As a result, library users are placing new demands on their libraries. They require access to the latest information, updated information resources and access to ICT facilities that they could use in their work.

Use of ICT in libraries enhances user's satisfaction. It provides numerous benefits to library users. Some of the benefits are given as under:

- Speedy and easy access to information
- Remote access to users
- Round the clock 24×7 access to users
- Access to unlimited information from different sources
- Increased flexibility
- Facilitates the reformatting and repackaging of data from different sources.

Libraries are also provide various ICT-based services to their users, including the following.

Web-based Online Public Access Catalogues (Web-OPAC)

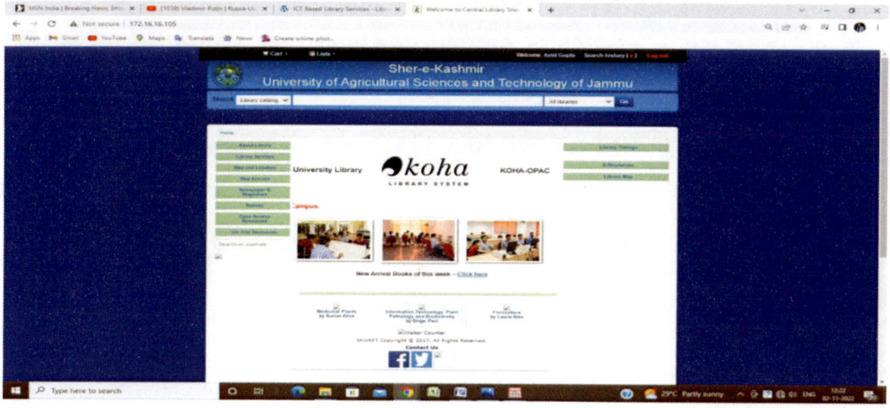

Plate 1: Screenshot showing Web OPAC (SKUAST-Jammu Central Library, Jammu & Kashmir) (*Source:* http//www.skuast.org/)

The Internet and Web-based technologies have made it possible for libraries to provide access to their catalogues on the local intranets, extranets and sometimes via the Internet. This arrangement, especially when the OPAC is available on the Internet, makes it possible for library users to access the facilities from anywhere in the world and for 24X7. This is possible because most library software systems now include Web-based interfaces to OPACs, as opposed to telnet-based access systems. Library users also find it easier to learn and use the OPACs from different library systems since they only have to know how to use one universal access client, the Web browser. Web-based OPACs also allow for linking to other information resources such as tables of content, full-text documents, and works/titles by the same author.

Digital Library Service

Digital library provides a variety of digital information sources. It reduces the physical space, the user can access to information remotely and it also provides access to distributed information resources. Its advantage is that it has the ability to handle multilingual content. Using ICTs librarians are creating digital libraries, that is libraries where some or all of the holdings are available in electronic form, and the services of the library are also made available electronically - frequently over the Internet so that users can access them remotely (Rosenberg, 2005). Digital libraries are made up of digital collections including document surrogates like bibliographic records and indexes in addition to full-text documents, videos, images some of which cannot be represented or distributed in printed formats. These digital works include both internal and external resources. In an academic environment, a digital library can provide students with access to educational materials, i.e. solved and unsolved problem sets, courseware modules (drills, simulations, models, virtual lab benches, and class presentation materials; while in a National library environment, digital libraries opens up the information resources for access by library users located across the country. In most countries, national libraries are located in the capital cities and therefore, access to their resources is restricted mainly to those that can afford to travel to the capital city. Digital library resources of national libraries can be accessed even from remote places.

Electronic Document Delivery Service

The libraries are implementing ICT-based Inter-Library Lending (ILL) using networks to deliver copies of journal articles and other documents in digital format like PDF (Portable Document Format) to the users' desktops. It allows users to access information that may not be available in their local libraries. This service is especially valuable for researchers and scholars in remote areas who may not have easy access to certain resources.

Institutional Repository Service

An Institutional Repository (IR) is a system that universities use to handle and share digital content created by their community members. It's essentially a collection of services designed to organize, manage, and ensure long-term preservation of these digital materials. This includes making sure that the content remains accessible and properly cited. Typically, an IR will house a variety of materials, including theses, dissertations, research reports, exam papers, conference papers, newsletters, and journal articles authored by faculty and staff. With the help of information and communication technologies (ICTs), universities can make these resources available in full text via their internal networks, external networks, or the broader Internet.

The main goals of an IR are:

1. To provide open access to the university's digital materials like research reports and seminar papers.
2. To ensure long-term storage and preservation of these digital assets.
3. To manage and provide access to "grey" literature, which includes documents like research reports and technical papers that may not be published elsewhere.

Librarians play a crucial role in institutional repositories by applying their expertise to manage these digital resources, ensuring they are preserved and accessible over time.

Current Awareness Service (CAS)

Current Awareness Services (CAS) is crucial for keeping users informed about the latest developments in their fields of interest. These services can be quite straight forward, such as providing copies of journal table of contents or bulletins that feature bibliographic details of recent articles organized by subject. Libraries now create these bulletins using specific search strategies on databases, either from CD-ROMs or online sources, and periodically generate the desired results. With proper copyright permissions, these results can be stored locally and shared through various means, including online (internet, intranet), offline (print, CD-ROM), or email formats.

Many journals offer their tables of contents for free on their websites, and some even provide free email updates. Additionally, numerous electronic publishing platforms and portals deliver current information via email to subscribers. For instance, the New York Times allows users to register for daily news summaries.

The Internet has spurred significant advancements in the creation and distribution of current awareness services. Tools like Listservs, weblogs (blogs), webzines, and e-newsletters are now commonly used. Listservs offer updates on the latest topics, discussions, and advice. Weblogs, or blogs, review and compile the latest relevant information in specific subject areas. Examples of web-based current awareness services include the NSDL Scout Report for Math, Engineering, and Technology (available at http://scout.wisc.edu/Reports/NSDL/MET/Current/) and Free Pint (www.freepint.com). For example: OKMS Institutional Repository of SKUAST Jammu.

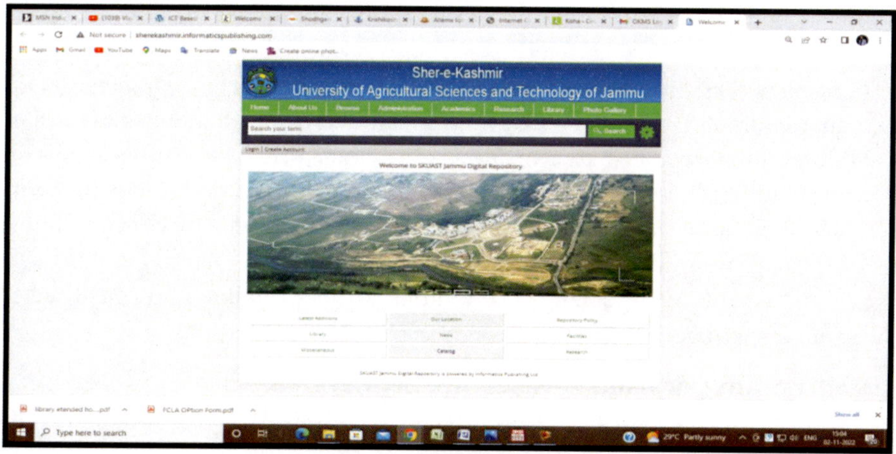

Plate 2: Screenshot showing OKMS - Institutional Repository of SKUAST Jammu (*Source:* https://www.skuast.org)

Audio-Visual Services

Audio-Visual Services encompass a range of materials used for information, education, and entertainment. Libraries, especially media libraries and larger academic or public libraries, often house a variety of materials, including music, films, pictures, and photographs. Historically, older media formats like Long Play (LP) records and slide tapes have been largely replaced by more modern technologies. Today, audio and video content is commonly stored on CDs, Video CDs (VCDs), and Digital Video Discs (DVDs). These new multimedia formats offer advantages such as higher storage capacity, random access to different parts of the content, and greater durability compared to older audio and video tapes.

Many libraries offer these multimedia resources for borrowing. Advances in storage media, compression, and encryption technologies have further revolutionized how multimedia content is stored and shared. Large volumes of multimedia documents can now be kept on hard drives and accessed via the internet. Software like QuickTime Player and Microsoft Media Player allows users to play these files directly from their browsers. Users can explore various types of hardware, software, and file formats involved in the creation, storage, distribution, and use of digital multimedia documents, enhancing their ability to access and utilize these resources effectively.

Online User Education

Online User Education involves libraries leveraging Information and Communication Technologies (ICTs), especially the internet, to provide educational resources and training for their users. These educational programs often include online tutorials or CD-ROM-based guides on how to search for information online, as well as virtual tours of the library's collections. These resources are typically accessible through intranets, extranets, or the broader internet. By using ICTs, libraries can overcome challenges associated with traditional lecture-based or in-person library orientation programs. This includes difficulties like managing large groups of students, dealing with limited staff, or the constraints of time when trying to cover a lot of information. ICTs also allow students to engage with the educational material at their own pace and on their own schedule, providing a more flexible and self-directed learning experience.

Readers' Advisory and E-Reference Services

Readers' Advisory and E-Reference Services utilize Information and Communication Technologies (ICTs) to enhance library services online. Through these technologies, libraries can offer web-based versions of traditional advisory and reference services. This includes informing users about new library acquisitions, providing updates and announcements, and offering Selective Dissemination of Information (SDI) to match users' interests. Additionally, ICTs facilitate virtual interactions between users and library staff through tools like Virtual Reference Desks, allowing users to get real-time assistance without needing to visit the library in person. For academic institutions with distance learning programs, ICT-based advisory services are especially valuable, as they enable libraries to support remote students with guidance and resources tailored to their needs.

Electronic Theses and Dissertations (ETDs)

Electronic Theses and Dissertations (ETDs) are an important component of institutional repositories, particularly within university libraries, as they provide access to full-text versions of theses and dissertations. Before the advent of Information and Communication Technologies (ICTs), accessing these documents remotely was nearly impossible. In developed countries, such resources were often available on microfiche and microfilm. However, in sub-Saharan Africa, the only way to access theses and dissertations was by physically visiting the libraries where they were kept. This meant that these resources were largely restricted to students and researchers within the host country, limiting their accessibility.

The introduction of ICTs has transformed this situation. Now, university libraries in various regions, including Africa, can offer digital access to ETDs. For instance, universities in India have developed projects to make full-text copies of ETDs available online through databases like Shodhganga and Krishikosh. This shift has made it much easier for researchers and students worldwide to access these valuable resources from remote locations.

Plate 3: Screenshot showing Webpage of Shodhganga Online Repository for ETDs (*Source:* https://www.shodhganga.inflibnet.ac.in)

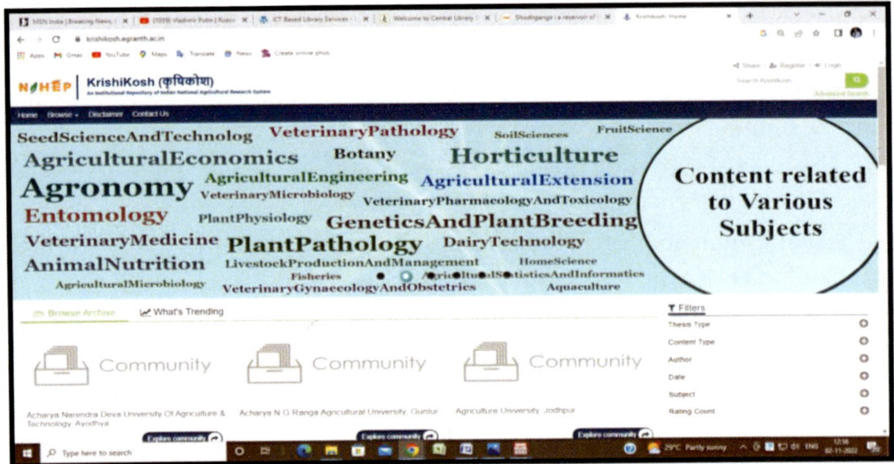

Plate 4: Screenshot showing Webpage of Krishikosh Online Repository for ETDs in Agricultural and allied subjects (*Source:* https://www.krishikosh.egranth.ac.in)

Online Chat Services

Online chat refers to real-time, text-based communication over the internet. In libraries, it is used for instant reference help and consulting services. This can involve one-on-one communication or broadcasting messages to multiple users at once.

Electronic Books Service

Electronic books, or e-books, are digital versions of books that offer round-the-clock global access to reliable information .Key aspects for e-books in academic libraries include content quality, software and hardware compatibility, digital rights management, access protocols, and privacy concerns. They make it easy for users to find and access specific research materials promptly and efficiently.

Electronic Journals

An electronic journal is any type of digital publication—such as a journal, magazine, or newsletter—that is accessible online. These publications can be reached through various technologies like the World Wide Web, Gopher, FTP, telnet, email, or listservs. Often, publishers who sell print journal subscriptions will also provide access to the digital version at no extra cost. Examples of publishers offering electronic journals include Emerald, Elsevier, Sage, Springer, EBSCO, J-Gate, and John Wiley.

Electronic Mail (E-mail) Service

Email is a widely used tool for sending and receiving messages over the internet. In libraries, it helps with information dissemination and communication, including services like Electronic Document Delivery Service (EDDS). Email is a versatile and popular web-based medium that library professionals can leverage for various functions, including interacting with publishers and vendors. Its efficiency and ease of use make it a valuable asset in the realm of information dissemination and service delivery.

Internet Service

Internet constitutes a vast, interconnected global infrastructure, facilitating seamless communication and data exchange between computing systems and individuals across the world. Functioning as a critical repository of knowledge, it enables advanced, high-bandwidth processing and dissemination of information. By linking academic institutions and transforming traditional library systems, it has redefined the paradigm of information access and global connectivity.

Reprographic & Micrographic Service

These technologies remain prevalent in libraries worldwide. A significant number of research libraries are equipped with reprographic machines, offering on-demand photocopying services for any requested document. Microform, an umbrella term for various high-density information carriers, encompasses media such as microfilm and other optical formats used for the compact recording and storage of optically encoded data. This includes micro images of printed documents, bit patterns, and holographic representations, enabling efficient archival and retrieval of vast quantities of information.

Document Scanning Services

The scanner plays a pivotal role in the modernization of library systems. It is instrumental in digitizing textual content, images, and book pages, significantly contributing to the development of digital and virtual libraries. By enabling the efficient conversion of physical materials into digital formats, scanners facilitate enhanced access, preservation, and dissemination of library resources in a modern, technology-driven environment.

Library Network Service

The main role of a network is to link computers and other communication devices so that data can be transferred quickly from one place to another. Networks enable multiple users to share a common connection and communicate with one another. There are different types of networks: a Local Area Network (LAN) is used for connecting devices within a limited area, such as a library or office, to facilitate resource sharing and communication. On the other hand, a Wide Area Network (WAN) covers a larger geographic region, such as a country or state, or can even span across several locations. In India, major WANs include INFLIBNET, DELNET, NICNET, INDONET, ADINET, and among others.

Open Source Software Service

Open Source Software (OSS) is computer software that is freely available and allows users to modify and customize its source code. Recent developments in OSS include Integrated Library Systems (ILS) like Koha, Digital Library Software like Greenstone, and DSpace; Content Management Software, like Moodle.

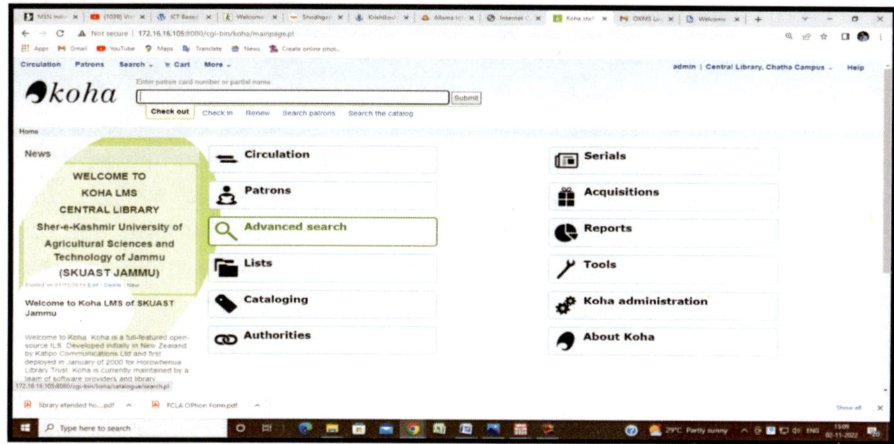

Plate 5: Screenshot showing Homepage of KOHA - Open Source Library Management Software (*Source:* https://172.16.16.105)

Library Portal Service

A library portal is a single access point combining the library catalogues subscribed database, electronic journals etc. Library portal meets the individual needs of the users and the portal is now the standard interface to generate library resources and services through single access and management point for users. Librarians are becoming increasingly aware that the multiplication of electronic resources is a problem for end users. Users find it difficult to have the most appropriate database or resource to search for their information need. Library portal reduces the barrier of users to remember various log-on. Example: Allama Iqbal Library, University of Kashmir.

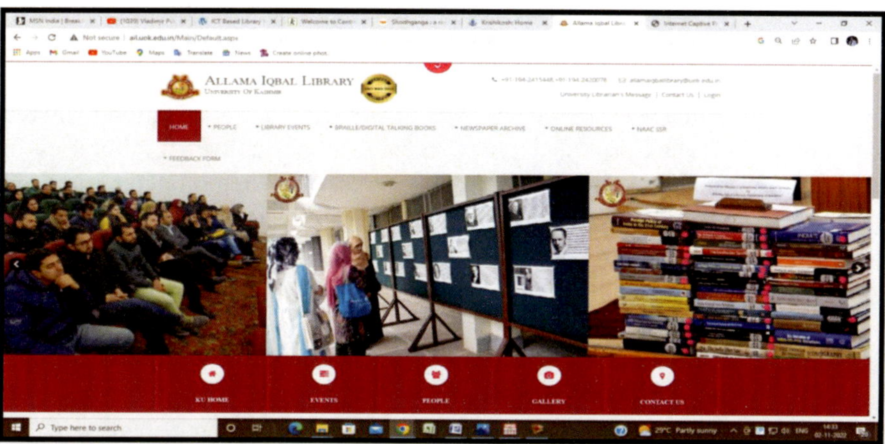

Plate 6: Screenshot showing Library Portal of Allama Iqbal Library, University of Kashmir (*Source:* https://www.kashmiruniversity.net)

Ask-A-Librarian Service

Ask-A-Librarian services provide an advanced, internet-based question-and-answer platform that connects users with experts skilled in specialized subject areas and precision searches. Typically, these services offer a web-based form or an email submission option, allowing users to easily submit queries. Once received, the question is assigned to a knowledgeable expert who provides a response. The expert's reply includes either direct factual information or a list of relevant resources. The response is delivered either via email or posted online for the user to access after a set period.

Bulletin Board Service

A bulletin board is an electronic communications forum that hosts posted messages and articles connected to a common subject or theme or interest. It allows users to call in and either leaves or retrieves messages. The messages may be directed to all users of the bulletin board or only to particular users. But all messages can be read by all users. Several libraries are using bulletin boards for their web-based library services. The bulletin board system is also used as an interactive interface to invite suggestions on activities and services of a library. It can also be used as an interface to distribute library services. A Bulletin Board System, or BBS, is a computer system running software that allows users to connect and log in to the system using a terminal. Once logged in, a user can perform functions such as uploading and downloading software and data, reading news and bulletins, and exchanging messages with other users, either through electronic mail or in public message boards.

8. Conclusion

Information and Communication Technology (ICT) has revolutionized library operations, enabling extensive networking, creation, and access to remote electronic databases. This transformation is gradually shifting libraries from being primarily book-centered to becoming information-centered institutions. ICT has provided unprecedented opportunities for libraries to implement electronic networks, streamlining access to remote, computerized information services and resources with far greater ease than was previously possible. Libraries are increasingly leveraging internet technologies, seamlessly integrating them into their daily functions to enhance service delivery. In the modern ICT era, libraries, once seen merely as repositories of knowledge, have evolved into dynamic information hubs. Tasks that once required significant manual effort are now efficiently executed through ICT, significantly improving effectiveness and productivity. Library management, administration, and technical processes have been greatly simplified, allowing for a greater volume of work to be accomplished with less effort. ICT, as the foundation for results-oriented management, enables libraries to achieve more in a shorter time, transforming how knowledge and resources are managed and delivered.

References

Babu, B., Vinayagamoorthy, P., & Gopalakrishnan, S. (2007). ICT Skills among Librarians in Engineering Educational Institutions in Tamil Nadu. DESIDOC Journal of Library & Information Technology, 27(6), 55–64. https://doi.org/10.14429/djlit.27.6.145

Baby, M. (1998). Use of information technology in information handling in libraries. Kelpro Bulletin, 2(1), 9–12.

Balachandar, P., & Ramesh, R. (2019). Information and Communication Technology (ICT) in library operations resources and services in university libraries. Journal of Recent Academic Research, 1(7), 472–474.

Balasubramanian, P. (2011). Library Information and Communication Technology. Deep and Deep Publications.

Japhet, A. A., Oduwole, A., Olumide, O. H., & Oluwaseyi, M. J. (2018). Information need, resources availability, and ICT proficiency: Experience at the Lagos State University Library. World Digital Libraries– An International Journal, 11(1). https://doi.org/10.18329/09757597/2018/11102

Jones, F. P., & Umoh, M. (2022). The Role of ICT in the provision of library services: A panacea for sustainable development. Intercontinental Academic Journal of Library and Information Science, 4(1), 45–56.

Khan, T.H. & Khayal, R. (2024). A case study on the use of ICT-based Services in the academic libraries of Arunachal Pradesh, India. Juni Khayat, 14(2), 30-36.

Mlay, S. V., & Sabi, H. M. (2019). A structural analysis of ICT-based intervention on reading habits in Uganda. The Information Society, 35(1), 26–35. https://doi.org/10.1080/01972243.2018.1542645

Melbha, D. Ashlin, & Melvin, C.L. Jeba (2023). Opportunities and challenges of using ICT in teaching and learning–teachers' perception. Library Herald, 61(2), 104-115

Nworie, J. C., Irunegbo, C. G., Omorodion, E. E., & Opara, I. O. (2017). The concept of professionalism and librarians challenges of ICT utilization in academic libraries: A case study. Information Technologist, 14(2).

Rahman, M., Choudhury, T. A., & Barooah, P. K. (2020). Status of Information Communication Technology (ICT) in some selected university libraries and special libraries of Assam: A comparative study. Library Philosophy and Practice, 1–26.

3

Integrating Circular Economy Principles in Library Management: A Conceptual Framework

Diksha[1] and Dr. Meghna Dhar[2]

[1]*Junior Research Fellow, Department of Library & Information Science, University of Jammu - 180 006, Jammu, J&K, India, E-mail: dikshaphd23@gmail.com, Orcid Id: 0009-0000-5265-0708*

[2]*Assistant Professor, Department of Library & Information Science, University of Jammu, Jammu - 180 006, J&K, India, E-mail: dhar_meghna@rediffmail.com*

Abstract

The integration of Circular Economy (CE) principles into library management offers a transformative approach to sustainability, aligning with the growing need for resource efficiency and environmental stewardship. This paper presents a conceptual framework for incorporating CE principles into library operations, focusing on core components such as resource optimization, collection management, sustainable facility management, community engagement, and policy development. By analyzing case studies of Gujarat Vidyapith (GV) and Anna Centenary Library, the paper illustrates practical applications of CE principles. GV, guided by Mahatma Gandhi's sustainability-oriented principles, has implemented comprehensive environmental initiatives including water harvesting, waste management, and energy-efficient technologies. In contrast, Anna Centenary Library, Asia's first LEED Gold-rated library, exemplifies advanced sustainable design through its use of eco-friendly materials and energy-efficient systems. These case studies demonstrate how libraries can not only reduce their environmental footprint but also serve as educational hubs for promoting sustainability. The paper underscores the potential of libraries to lead by example in the adoption of CE principles, contributing to broader sustainability goals and inspiring similar initiatives globally.

Keywords: *Circular economy, Library management, Sustainability, Resource optimization, Green libraries, LEED certification*

1. Introduction

There is an emergence of the concept of a Circular Economy as one of the transformational approaches toward sustainability, where greater emphasis is laid on rethinking and redesigning products and services for maximum resource efficiency. In contrast to the traditional linear model ("take, make, dispose"), CE emphasizes product reuse, recycling, and cascading with the clear ambition of closing the loop of product lifecycles through continuous recovery of resources. The key to this solution involves changing the paradigm, highly relevant to resolving pressing environmental issues facing the planet such as resource depletion, waste management, and climate change.

Conventionally regarded as information stores, libraries are increasing awareness of their potential to create an impact toward sustainable practices. Centres of information and meeting points for communities, libraries are at an advantageous position for leading by example in integrating the principles of CE into their system. This paper examines the current status of integrating the circular economy principles within library management and discusses a proposed conceptual framework that aims to guide libraries in adopting circular practices. The paper addresses the contribution that libraries might make to a more sustainable future by exploring the intersection between CE and library management through innovative practices and strategic plans.

2. Review of Existing Literature

Literature search plays a very important role in research activities, as it forms the very first step of research pursuit. Some of the earlier studies identified are reviewed as under.

2.1. Overview of Circular Economy Principles

Circular Economy (CE) represents a fundamental shift from traditional linear economic models, which often lead to resource depletion and environmental degradation. CE emphasizes principles such as designing for longevity, maintaining product value, and minimizing waste through strategies like reuse, recycling, and cascading. The goal is to create a closed-loop system where materials and resources are continually cycled back into the economy, thus reducing the reliance on finite resources and mitigating environmental impacts (Geissdoerfer *et al.*, 2017).

Key Principles of Circular Economy

1. **Design for Longevity**: Products are designed to last longer, reducing the need for frequent replacements and minimizing waste.
2. **Maintain Value**: Ensuring that materials and products retain their value through repair, refurbishment, and remanufacturing.

3. **Resource Efficiency**: Utilizing resources in a way that minimizes environmental impact, including efficient use of materials and energy.
4. **Waste Minimization**: Reducing, reusing, and recycling materials to prevent waste from ending up in landfills (Murray *et al.*, 2017).

2.2. Application of CE in Different Sectors

CE principles have been applied across various sectors, including manufacturing, fashion, and construction, demonstrating their versatility and potential for impact. For instance, in the manufacturing sector, companies are redesigning products for easier disassembly and material recovery. In the fashion industry, brands are focusing on using recycled materials and promoting clothing recycling programs (Kirchherr *et al.*, 2017).

Examples of CE Application

1. **Manufacturing**: Companies like Philips are implementing take-back schemes for medical equipment, ensuring that components are recycled or reused.
2. **Fashion**: Brands such as Patagonia and H&M are incorporating recycled materials into their products and promoting clothing recycling initiatives.
3. **Construction**: Building projects are increasingly using recycled materials and designing structures for easy disassembly and material recovery (Bocken *et al.*, 2016).

2.3. Role of Libraries in Sustainable Development

Libraries have long been pivotal in supporting community education and access to information. In recent years, their role has expanded to include promoting sustainability and environmental stewardship. Libraries serve as educational hubs where communities can learn about sustainability practices and the principles of CE.

Key Contributions

1. **Educational Role**: Libraries offer programs and resources that educate the public on sustainability and environmental issues.
2. **Community Engagement**: Libraries engage with local communities to promote green practices and sustainable living.
3. **Resource Sharing**: Libraries support the sharing economy by providing access to materials and resources that can be reused or recycled (Aytac, 2017).

2.4. Previous Studies on CE in Library Contexts

While the application of CE principles in libraries is an emerging field, several studies have explored related aspects, such as green libraries and sustainability initiatives. (Gupta, 2020) highlighted the role of libraries in reducing their environmental footprint through energy-efficient practices and the adoption of sustainable materials. (Hauke, 2019) investigated how libraries are implementing green practices, such as energy-efficient lighting and sustainable building designs. (Mashroofa, 2022) explored how libraries contribute to environmental education by providing resources and programs aligned with sustainability goals. (David-West & Wali, 2023) documented successful examples of libraries integrating sustainability principles, demonstrating the potential for broader application.

The literature review analyzed library management focusing on the application of the principles of the Circular Economy, looking into key strategies and applications in diverse industries. It investigated the contribution of libraries to sustainability and identifies those earlier studies that investigate CE practice applications within the context of a library setting.

3. Conceptual Framework

The Conceptual Framework for Integrating Circular Economy Principles into Library Management systematically transforms conventional library practices to more sustainable and resource-efficient operations; it is grounded in the guiding principles of the circular economy, adapted into the unique setting a library offers through its services: collections management, facility operations, and community engagement.

Core Components of the Framework

The framework consists of five core components:

1. **Resource Optimization**: Libraries can also maximize the use of resources by adopting practices that lengthen the life of materials and reduce their wastage. This includes adapting more and more to going digital, reusing the physical source whenever possible, and utilizing energy-efficient technologies.
2. **Collection Management**: A CE-inspired collection management would focus better on the priorities of digital collection that may reduce the demand for physical materials and thus decrease most of the environmental impacts related to the production and distribution of physical materials. Regarding the physical collections, libraries could also consider take-back schemes, donating material that is out-dated and establishing that materials withdrawn be properly recycled.

3. **Sustainable Facility Management**: Libraries would be able to minimize their environmental footprint by adopting energy-efficient lighting and HVAC systems; other than that, they will incorporate renewable energy sources, cool roofs, and minimum water use, in their construction. Green certifications like LEED can guide these efforts.
4. **Community Engagement and Education**: The library stands at the forefront within the community for issues on sustainability by offering programs to sensitize the public on CE principles, workshops on practices that are sustainable, as well as resources related to environmental concerns. Libraries may partner with local organizations to further promote greater community-wide sustainability efforts.
5. **Policy and Strategic Planning**: Integrating CE into library management requires strategic planning and policy development. Libraries should establish sustainability goals, develop policies that support resource-efficient operations, and create action plans for continuous improvement.

3.1. Resource Optimization

Resource optimization in library management involves maximizing the utility and lifespan of resources while minimizing waste.

- **Digital Collections and E-Resources**: The transition to digital collections is a critical element in reducing the environmental impact of libraries. Digital resources, such as e-books and online journals, not only decrease the need for physical storage but also eliminate the waste associated with the production and disposal of printed materials. Libraries can further optimize resource use by promoting open access resources, which are freely available and reduce the duplication of materials across institutions (Noh, 2021).
- **Material Reuse and Recycling**: Libraries can implement programs to encourage the reuse of materials, such as book exchanges, where patrons can donate and take books for free. Additionally, libraries can establish recycling programs for outdated materials, ensuring that paper, plastics, and electronics are disposed of in an environmentally responsible manner (Solemanpharcy & Gaffar, 2023).
- **Energy Efficiency**: Reducing energy consumption is a crucial aspect of resource optimization. Libraries can invest in energy-efficient lighting, such as LED bulbs, and install motion sensors to reduce electricity use in low-traffic areas. Upgrading HVAC systems to more efficient models and incorporating renewable energy sources, such as solar panels, can significantly reduce a library's carbon footprint (Ismail *et al.*, 2022).

3.2. Collection Management

Collection management under circular economy principles focuses on curating and maintaining library resources in a way that supports sustainability and minimizes redundancy.

- **Prioritization of Digital Collections**: As libraries increasingly shift towards digital resources, the need for extensive physical collections diminishes. Prioritizing digital collections not only conserves physical space but also reduces the environmental impact of printing, shipping, and storing materials. Libraries can also explore digital lending options, which allow multiple users to access the same resource without the need for additional copies (Thorpe & Gunton, 2022).
- **Sustainable Physical Collection Practices**: For physical collections, libraries can implement strategies such as rotating collections to ensure that materials are used efficiently and do not become obsolete. Libraries can also engage in collaborative collection development, where institutions share resources, thus reducing duplication and conserving resources. When deaccessioning materials, libraries should prioritize donation to other institutions or community groups before considering recycling or disposal (Amanina & Ilham, 2024).
- **Take-Back and Donation Programs**: Libraries can establish take-back programs where patrons can return outdated or damaged books, which the library can then recycle or refurbish. Additionally, libraries can donate withdrawn materials to schools, prisons, or international organizations, ensuring that books continue to serve a purpose even after they are no longer needed in the library's collection (Bangani & Dube, 2023).

3.3. Sustainable Facility Management

Sustainable facility management in the framework of circular economy principles aims to reduce environmental impact through efficient building practices and resource use.

- **Green Building Practices**: Incorporating green building practices into library construction and renovation projects is essential for reducing environmental impact. Libraries can achieve this by using sustainable materials, incorporating natural lighting, and ensuring that buildings are designed for energy efficiency. Green roofs, rainwater harvesting systems, and energy-efficient insulation are examples of how libraries can reduce their ecological footprint(Gupta, 2020).

- **Energy and Water Conservation**: Libraries can adopt technologies that minimize energy and water use. For example, energy management systems can monitor and control energy use throughout the building, while low-flow faucets and toilets can reduce water consumption. Regular audits of energy and water use can help libraries identify areas for improvement and implement targeted conservation strategies (Tank, 2023).
- **LEED Certification**: Libraries can pursue Leadership in Energy and Environmental Design (LEED) certification as a benchmark for their sustainability efforts. LEED certification involves meeting stringent criteria across areas such as energy use, water efficiency, and indoor environmental quality. Achieving LEED certification not only demonstrates a library's commitment to sustainability but also provides a framework for continuous improvement (Hauke, 2019).

3.4. Community Engagement and Education

Libraries are not only repositories of knowledge but also central hubs for community interaction and education. By leveraging their role within the community, libraries can significantly contribute to promoting Circular Economy (CE) principles and sustainability practices among the public.

3.4.1. Public Programs and Workshops

- **Sustainability Workshops**: Libraries can organize workshops and educational sessions focused on sustainability topics such as recycling, composting, energy conservation, and sustainable living. These programs can be tailored to different age groups and interests, providing practical knowledge that community members can apply in their daily lives. For example, a library might host a workshop on creating zero-waste homes or an event on the benefits of sustainable gardening practices (Ajani *et al.*, 2024).
- **Community Events and Challenges**: Libraries can engage the community through events and challenges that promote sustainable practices. For instance, libraries could organize a "Green Week" with daily activities related to reducing waste, conserving resources, and learning about CE. Libraries can also host challenges such as a "Plastic-Free Month," encouraging patrons to reduce their plastic use and sharing tips and resources to help them succeed.
- **Collaboration with Local Organizations**: Libraries can partner with local environmental groups, businesses, and government agencies to expand the reach and impact of their sustainability programs.

By collaborating with organizations that are already active in the sustainability space, libraries can offer joint events, share resources, and amplify messages related to CE and environmental stewardship. These partnerships can also help libraries stay informed about local sustainability initiatives and provide their patrons with relevant and timely information.

3.4.2. Educational Resources and Collections

- **Curated Collections on Sustainability**: Libraries can create special collections focused on sustainability and CE topics, making it easier for patrons to access information on these subjects. These collections could include books, e-resources, magazines, and multimedia content covering topics like renewable energy, sustainable agriculture, green technology, and environmental policy. By highlighting these resources, libraries can encourage patrons to learn more about the principles and practices of CE.
- **Resource Guides and Toolkits**: In addition to curated collections, libraries can develop resource guides and toolkits that provide practical advice on implementing CE principles. These guides could include tips on reducing household waste, recommendations for sustainable products, and step-by-step instructions for projects like building a home composting system. Libraries can also offer digital toolkits that patrons can access online, ensuring that these resources are available to a wider audience.
- **Educational Exhibits and Displays**: Libraries can use their physical space to create educational exhibits and displays that promote sustainability. For example, a library might create a display on the lifecycle of everyday products, illustrating how materials can be reused, repurposed, or recycled instead of being discarded. These exhibits can serve as visual reminders of the importance of CE and inspire patrons to think more critically about their consumption habits.

3.4.3. Digital Literacy and Online Engagement

- **Online Courses and Webinars**: To reach a broader audience, libraries can offer online courses and webinars on CE topics. These digital offerings can be especially valuable for patrons who are unable to attend in-person events or who prefer to learn at their own pace. Topics could range from introductory courses on sustainability to more advanced sessions on specific aspects of CE, such as circular product design or sustainable supply chains.

- **Social Media Campaigns**: Libraries can use social media platforms to promote sustainability and CE principles, sharing tips, resources, and success stories with their followers. Regular posts can highlight library events, feature sustainable products, or share facts about environmental issues. Social media campaigns can also encourage patrons to participate in library challenges, share their experiences, and engage with the community online(Hamid *et al.*, 2017).
- **Virtual Collaboration Platforms**: Libraries can create or participate in virtual platforms that facilitate collaboration on sustainability projects. These platforms can connect patrons with local environmental initiatives, provide forums for discussing CE topics, and enable the sharing of resources and ideas. By fostering an online community focused on sustainability, libraries can help drive collective action and support local CE efforts.

3.5. Policy and Strategic Planning

For libraries to effectively integrate Circular Economy (CE) principles into their operations, it is essential to develop robust policies and strategic plans that guide sustainability efforts and ensure continuous improvement.

3.5.1. Establishing Sustainability Goals

- **Long-Term Vision and Mission**: Libraries should begin by defining a long-term vision and mission that includes a commitment to sustainability and CE principles. This vision should articulate the library's role in promoting environmental stewardship and outline how CE will be integrated into all aspects of library operations. By aligning sustainability goals with the library's core mission, libraries can ensure that these efforts are prioritized and embedded in the institution's culture.
- **Specific, Measurable Objectives**: Sustainability goals should be specific, measurable, achievable, relevant, and time-bound (SMART). For example, a library might set a goal to reduce energy consumption by 20% within five years or to achieve zero waste by a certain date. These objectives should be supported by detailed action plans that outline the steps needed to achieve them, as well as metrics for tracking progress.
- **Stakeholder Engagement**: Engaging stakeholders, including library staff, patrons, and community members, is crucial for successful policy development. Libraries should involve these groups in the goal-setting process, soliciting feedback and ideas on sustainability initiatives. This inclusive approach not only helps to build support for CE efforts but also ensures that the policies reflect the needs and values of the community.

3.5.2. Developing Comprehensive Policies

- **Sustainability Policy**: A comprehensive sustainability policy serves as the foundation for integrating CE principles into library management. This policy should cover key areas such as resource management, energy use, waste reduction, and procurement practices. It should also outline the library's commitment to continuous improvement, specifying how the institution will regularly assess and update its sustainability efforts.
- **Procurement and Supplier Policies**: Libraries can implement procurement policies that prioritize sustainable products and services. This might include purchasing recycled paper, energy-efficient equipment, and environmentally friendly cleaning products. Libraries can also develop supplier policies that require vendors to adhere to sustainability standards, such as using minimal packaging or offering take-back programs for used products.
- **Waste Management Policies**: Effective waste management is a critical component of CE in libraries. Libraries should develop policies that promote waste reduction, reuse, and recycling across all operations. These policies might include guidelines for reducing paper use, procedures for recycling electronic waste, and protocols for managing hazardous materials. Libraries can also establish zero-waste goals and work towards achieving them through targeted initiatives (Amusan & Issa, 2022).

4. Implementation Guidelines

Integrating Circular Economy (CE) principles into library management requires a thoughtful approach and a structured implementation plan. The following guidelines provide a step-by-step approach to help libraries transition towards more sustainable practices.

4.1. Assessing Current Practices

Before implementing CE principles, libraries should conduct a thorough assessment of their current operations, resources, and practices. This includes evaluating:

- **Resource Utilization:** Analyze how materials and resources are used within the library, including energy consumption, material usage, and waste production.
- **Collection Management:** Review current collection practices, including the balance between physical and digital resources, and strategies for handling outdated or damaged materials.

- **Facility Operations:** Assess the environmental impact of building operations, including energy use, water consumption, and waste management.

4.2. Developing a Sustainability Action Plan

Based on the assessment, libraries should develop a comprehensive sustainability action plan that outlines:

- **Short-Term and Long-Term Goals:** Establish clear, actionable goals for implementing CE principles. Short-term goals might include initiating a recycling program, while long-term goals could involve achieving LEED certification or transitioning to a fully digital collection.
- **Action Steps:** Define specific actions needed to achieve each goal. For example, if a goal is to reduce energy consumption, action steps might include upgrading to energy-efficient lighting and installing motion sensors.
- **Timeline and Milestones:** Create a timeline for implementing the action plan, including key milestones and deadlines for achieving specific targets.

4.3. Engaging Stakeholders

Successful implementation of CE principles requires the support and involvement of various stakeholders:

- **Library Staff:** Engage library staff in the sustainability efforts by providing training on CE principles and involving them in the development and execution of the action plan.
- **Patrons:** Educate library patrons about the library's sustainability initiatives and encourage their participation in programs such as recycling or donation drives.
- **Community Partners:** Collaborate with local organizations, businesses, and government agencies to support sustainability efforts and leverage additional resources and expertise.

4.4. Monitoring and Evaluation

To ensure the effectiveness of the sustainability initiatives, libraries should establish a system for monitoring and evaluating progress:

- **Performance Metrics:** Develop metrics to track progress towards sustainability goals. Metrics might include energy consumption rates, waste diversion rates, or the number of patrons participating in sustainability programs.

- **Regular Reviews:** Conduct regular reviews to assess the effectiveness of implemented practices and identify areas for improvement. Adjust the action plan as needed based on these evaluations.
- **Reporting:** Provide transparent reporting on sustainability achievements and challenges. Share progress with stakeholders through reports, newsletters, or public announcements.

4.5. Continuous Improvement

CE is an evolving field, and libraries should adopt a mindset of continuous improvement:

- **Stay Informed:** Keep abreast of new developments in sustainability practices. Attend conferences, participate in workshops, and engage with professional networks to stay updated.
- **Innovate:** Experiment with new ideas and technologies that can further enhance sustainability efforts. Encourage innovation within the library community to find creative solutions to sustainability challenges.
- **Feedback Loop:** Establish a feedback loop with stakeholders to gather input on sustainability initiatives and make necessary adjustments to improve outcomes.

5. Case Studies

To illustrate the practical application of Circular Economy (CE) principles in libraries, the following case studies highlight successful examples of libraries integrating sustainability practices:

5.1. Gujarat Vidyapith Library, Ahmedabad

Gujarat Vidyapith (GV), founded by Mahatma Gandhi in 1920, operates across several campuses in Ahmedabad and rural Gujarat. Although Gandhi was not explicitly an environmentalist, his principles, such as non-possession (Apaigrah), self-reliance, and limiting wants, resonates with modern sustainability ideals. Over the past 7–8 years, GV has embraced these principles to enhance its sustainability practices, aligning with contemporary concepts of sustainable development (Shukla & Khimani, 2015).

GV's library and campus initiatives include:

- **Environmental Education:** GV integrates sustainability into its graduate and postgraduate programs, fostering awareness and understanding of environmental issues among students.
- **Water Harvesting and Waste Management:** The campus features water harvesting structures and waste management systems, including bio-gas and composting research.

- **Energy Efficiency:** The use of LED and solar lighting, along with energy-efficient appliances and green architecture in new constructions, such as an auditorium, reflects GV's commitment to reducing its environmental impact.
- **Self-Sufficiency:** The campus promotes food self-sufficiency through the cultivation of food grains and milk production, demonstrating a comprehensive approach to sustainability.

GV's library and campus practices exemplify how integrating CE principles can align with traditional values to achieve modern sustainability goals.

5.2. Anna Centenary Library, Chennai

Anna Centenary Library, inaugurated in 2010, is notable for being Asia's first LEED Gold-rated library. This green library serves as a leading example of integrating sustainability into library design and operations (Elavazhagan & Kohila, 2016).

Key features include:

- **Building Design:** Located on an 8-acre site, the library incorporates modern technologies for energy efficiency, including optimal use of natural light and ventilation. The construction cost of ₹172 crores underscores the significant investment in sustainable architecture.
- **Green Features:** The library's design utilizes eco-friendly materials and energy-efficient systems, contributing to its LEED Gold certification. The emphasis on natural lighting and ventilation minimizes the need for artificial lighting and climate control.

Anna Centenary Library demonstrates how a contemporary library can successfully incorporate CE principles through thoughtful design and operational practices, setting a benchmark for green libraries in India.

6. Conclusion

The study highlights the significant impact of integrating Circular Economy (CE) principles into library management. Embracing CE practices allows libraries to move beyond traditional linear models, focusing on resource efficiency, waste reduction, and sustainability. The proposed framework emphasizes optimizing resources, managing collections sustainably, and adopting green practices within facilities. Libraries, as central community hubs, have a unique opportunity to lead in sustainability through digital resource expansion, energy-efficient technologies, and environmentally-friendly building practices.

Successful case studies, such as the Gujarat Vidyapith Library and Anna Centenary Library, showcase practical applications of CE principles, providing valuable insights and strategies for other institutions. These examples illustrate the tangible benefits of CE integration, including enhanced resource management and community engagement.In conclusion, by adopting CE principles, libraries can significantly contribute to environmental stewardship while enhancing their role in the community. This approach not only supports sustainability but also enriches the library's value to its patrons, setting a powerful example for others to follow in the pursuit of a more sustainable future.

References

Ajani, Y. A., Muhammed, R. B., Oladokun, B. D., Folami, S. O., & Onyebinama, C. O. (2024). The green library advocacy: Need to engage the international communities in a climate change action. Business Information Review, 02663821241265913. https://doi.org/10.1177/02663821241265913

Amanina, F., & Ilham, Z. (2024). Placemaking strategies in greening Universiti Malaya Main Library. Urban Resilience and Sustainability, 2(1), 76–92. https://doi.org/10.3934/urs.2024005

Amusan, B. B., & Issa, A. O. (2022). ICT waste management in federal university libraries in Nigeria: A survey report. In Handbook of Research on Emerging Trends and Technologies in Librarianship (pp. 22–33). IGI Global. https://doi.org/10.4018/978-1-7998-9094-2.ch002

Aytac, S. (2017). Library Environment Sustainability Progress Index (LESPI): Benchmarking Libraries' Progress Towards Sustainable Development. https://library.ifla.org/id/eprint/2443/

Bangani, S., & Dube, L. (2023). Academic libraries and the actualisation of Sustainable Development Goals two, three and thirteen. Journal of Librarianship and Information Science, 09610006231174650. https://doi.org/10.1177/09610006231174650

Bocken, N. M. P., de Pauw, I., Bakker, C., & van der Grinten, B. (2016). Product design and business model strategies for a circular economy. Journal of Industrial and Production Engineering, 33(5), 308–320. https://doi.org/10.1080/21681015.2016.1172124

David-West, B., & Wali, N. B. (2023). Green initiative and environmental sustainability in South Nigerian academic libraries. Library Progress (International). Bpas Publication. https://acspublisher.com/journals/index.php/lp/article/view/12642

Elavazhagan, K., & Kohila, G. T. (2016). Green library initiatives in India: Anna Centenary Library - a case study. http://repository.kln.ac.lk/handle/123456789/14926

Geissdoerfer, M., Savaget, P., Bocken, N. M. P., & Hultink, E. J. (2017). The circular economy – A new sustainability paradigm? Journal of Cleaner Production, 143, 757–768. https://doi.org/10.1016/j.jclepro.2016.12.048

Gupta, S. (2020). Green Library: A strategic approach to environmental sustainability (SSRN Scholarly Paper 3851100). https://papers.ssrn.com/abstract=3851100

Hamid, S., Ijab, M. T., Sulaiman, H., Md. Anwar, R., & Norman, A. A. (2017). Social media for environmental sustainability awareness in higher education. International Journal of Sustainability in Higher Education, 18(4), 474–491. https://doi.org/10.1108/IJSHE-01-2015-0010

Hauke, P. (2019). Green Libraries Towards Green Sustainable Development.

Ismail, F. Z., Yaman, R., & Razali, K. N. (2022). Formulating an assessment tool for the implementation of green initiatives in library. IOP Conference Series: Earth and Environmental Science, 1067(1), 012021. https://doi.org/10.1088/1755-1315/1067/1/012021

Kirchherr, J., Reike, D., & Hekkert, M. (2017). Conceptualizing the circular economy: An analysis of 114 definitions. Resources, Conservation and Recycling, 127, 221–232. https://doi.org/10.1016/j.resconrec.2017.09.005

Mashroofa, M. M. (2022). Contribution of academic libraries towards sustainable development goals. http://ir.lib.seu.ac.lk/handle/123456789/6033

Murray, A., Skene, K., & Haynes, K. (2017). The circular economy: An interdisciplinary exploration of the concept and application in a global context. Journal of Business Ethics. https://link.springer.com/article/10.1007/S10551-015-2693-2

Noh, Y. (2021). Study on the perception of South Korean librarians of the UN Sustainable Development Goals (SDGs) and the strategy to support libraries. Profesional de La Información, 30(4), Article 4. https://doi.org/10.3145/epi.2021.jul.04

Shukla, N., & Khimani, R. (2015). Gujarat Vidyapith: On the Way to Greening of the University (pp. 229–249). https://doi.org/10.1007/978-3-319-11961-8_18

Solemanpharcy, M., & Gaffar, S. K. A. (2023, March 1). The role of Green Libraries for Sustainable Development Goals and Environmental Education: A study on Global perspective. https://openurl.ebsco.com/contentitem/gcd:170910862?sid=ebsco:plink:crawler&id=ebsco:gcd:170910862

Tank, S. (2023). Libraries and Sustainable Development Goals: Case study of Atmiya University Library and Learning Centre. https://ir.inflibnet.ac.in:8443/ir/handle/1944/2434

Thorpe, C., & Gunton, L. (2022). Assessing the United Nation's sustainable development goals in academic libraries. Journal of Librarianship and Information Science, 54(2), 208–215. https://doi.org/10.1177/09610006211005528

4

Expanding Educational Influence: Utilizing Social Media in the Libraries

Abhi Sharma[1] and Diksha[2]

[1]Research Scholar, Department of Library & Information Science, University of Jammu - 180 006, Jammu, J&K, India E-mail: a8hisharma@gmail.com Orcid Id: 0009-0009-8475-7257

[2]Junior Research Fellow, Department of Library & Information Science, University of Jammu, Jammu - 180 006, J&K, India E-mail id- dikshaphd23@gmail.com, Orcid Id: 0009-0000-5265-0708

Abstract

The integration of social media and libraries has led to the exploration of various potential improvements in user engagement, resource promotion, and service enhancement. Libraries, being at the forefront of people's digital literacy improvement, play a crucial role in keeping individuals informed about the latest technological advancements. The present paper aims to examine the ramifications of social media utilization in libraries and provide valuable insights into how libraries can extend their educational influence and outreach through the employment of social media platforms. It additionally discusses the challenges and anticipates destiny patterns of social media software within the library context, emphasizing the want for libraries to prioritize ethical issues and data privacy. Multiple studies have conducted an analysis of social technology implementations in libraries in order to gain a deeper understanding of the effects of social media learning. The paper attempts to furnish libraries with guidance on how to effectively employ social media for educational purposes, adjust to the digital era, and augment their academic impact. As libraries adapt to the digital age the significance of social media in shaping education and disseminating knowledge will remain crucial. Thus, Social media can be a powerful tool for libraries to increase their educational impact.

***Keywords**: Social media, Digital platforms, Virtual learning, Online users Digital literacy.*

1. Introduction

In this time of technological advancement, social media has crept into the daily existence of individuals. It is Facebook, Twitter and Instagram which are standard ways of communication to bring people together. The aspect of social media as an auxiliary means of user engagement is also appreciated and embraced by libraries making it part of marketing communication. Using social media, libraries are able to efficiently disseminate information and operate online call centers. There are social media companies which have helped libraries extend their reach especially to young people who are known to be technology-oriented. Thanks to social media, users are able to find value in the services offered by the libraries and this turns out to be beneficial.

1.1. The Role of Libraries in Promoting Digital Literacy

In modern conditions, libraries additionally undertake the responsibility for the promotion of information literacy and information skills. This is applicable to those who want to develop their digital skills irrespective of whether it involves looking for information on the web, research or creative work using digital technology, or social video engagement. Moreover, social engagement which is part of being digitally literate, can improve the learning experience of library users by allowing them to interact with content and resources, hence creating an online community with the content. Also, social media can help in the cooperation and exchange of information among the users of the library: creating a community and motivating users to continue learning new things. Libraries are therefore slowly turning a trend for several users where they can sharpen their digital literacy skills and keep pace with recent advancement in technology. Encouraging digital literacy in libraries ensures that every citizen will be better equipped to take an active, creative and helpful role within the community.

1.2. Social Media as a Platform for User Engagement

Social media is a vibrant platform that offers endless possibilities for human beings, companies, and corporations in reaching out to their wanted audience. Its attribute of two-way verbal exchange encourages immediate interaction and discussion between the content material creators and their audiences. With all sorts of content codes at hand and audience segmentation equipment, the use of social media empowers innovation with better relevance and engagement toward specific target audience options. These sites can allow for real-time interaction: users can comment on content while activities are unfolding, have real-time chats, and discuss to create a sense of community and immediacy. User generated content, such as reviews, testimonials, and user submissions can be a potent means for audience engagement and be made to feel valued. The viral effect can give more reach and engagement to the content in social media. In those systems, analytics tools will allow content creators to music performance metrics, enabling them to adjust their strategies. Contests, demanding situations, and giveaways on social media may incentivize consumer participation and build excitement for a brand or product. Trending topics and hash tags will help content material gain visibility, reaching a wider audience of the targeted marketplace. It is long-term to build up an online community that can be trusted for a brand or cause to have consumer participation for an extended period. Emotional connections wherein content material evokes feelings will boost up engagement and interest at a very significant level. When using social media, the main concerns should be the privacy of facts and ethical issues with the help of respecting consumers' consent and protection of their private information.

1.3. Social Media Platforms Used for Library Learning

Libraries and educational institutions have increasingly grown to be utilizing social media systems to decorate studying, engage with their communities, and sell their offerings and assets. From this kind of perspective, using social media these days has grown to turn out to be a very significant device within the arms of libraries with the purpose of keeping their users engaged and knowledgeable.

Facebook is especially good for creating book clubs and study groups, while the brevity of Twitter's character limit is best for sharing quick updates and for staying in the know through hashtags. Instagram is a great way to display new arrivals and offer behind-the-scenes peeks, and Pinterest allows libraries to interact visually with customers who like to do this. YouTube would be more useful where interactive visual learning takes place with its channels for video tutorials and virtual tours. Research and instructional libraries can also share their research resources and can become professional relationship members with LinkedIn involvement in discussions that are of importance to their set areas of disciplines.

Libraries are able to leverage the multimedia platform Tumblr for collections and services to a younger demographic market through engaging gifs, snap shots, and articles. In contrast, Snapchat is an effective tool in which to create interest and excitement about library events and services through the use of sharing time-limited content that creates a sense of urgency. They have utilized most of the modern technologies, from podcasts to digital and augmented reality, in providing interactive learning stories for his or her clients and have leveraged open structures, such as OER Commons, to bridge the training gap by sharing advanced nice educational resources with everybody. Through libraries, they are able to rent a variety of systems to reach their target audience for the purposes of promoting the services provided. Each one of them offers unlimited prospects to share any content and connect with people. If used appropriately, the infrastructure will ultimately allow the libraries to create an interactive and interesting virtual presence that will go a long way in helping them attain new customers while maintaining closer ties with existing ones.

2. Previous Studies on Social Media Learning in Libraries

In latest years, there was a developing interest in the use of social media in libraries as a tool for gaining knowledge of and engagement. This emerging trend has triggered researchers and practitioners to explore the potential blessings and challenges of integrating social media into library services. To shed mild at the impact of social media mastering in libraries, numerous case studies were conducted to examine successful implementations of social media in libraries. These case researches provide treasured insights into the ways wherein libraries can efficaciously leverage social media systems to decorate gaining knowledge of opportunities for his or her patrons.

(Minocha, 2012) discussed in the paper that social media platforms provide a means for gathering and sharing information, enabling collaboration, and changing the behaviors and expectations of students, educators, researchers, employers, and funding bodies. The challenges and opportunities of using social media tools in education are discussed through case studies and examples from empirically-grounded research on their role in student learning and engagement, teaching, and research discourse, including practical outcomes such as frameworks and design guidance.

(Neal, 2012) edited the book that offers practical implementation strategies for academics to use social media in their professional and research endeavors, based on applied research findings and successful experiences of contributors in academic settings.

(du Toit & Mulatiningsih, 2013) Social media provides libraries with immediate and direct communication with their members regardless of location. This paper examines the implementation of social media in libraries, with a focus on the SABC Media Libraries as a practical example. It provides tips and resources for librarians to update their skills in order to meet the changing needs of users in the social media era.

(Taylor, 2014) in their article explored the functionality of social media to reinforce the bond among libraries and their customers, and shows strategies wherein libraries can make use of social media for outreach and issuer. The authors intend to create a centralized on-line platform for libraries to percentage high-quality practices, case research, and updates on social media utilization.

(Shulman *et al.*, 2015) The Twitter Engagement of the San Francisco Public Library was scrutinized and evaluated, which delved into the library's approach to community engagement on the social media platform. The library's tweet content material, frequency, and person interactions have been all put to the take a look at. The effects confirmed that Twitter served as an powerful device in improving the library's visibility, fostering customer engagement, and promoting events with excellent success. Through this study, the San Francisco Public Library has demonstrated that social media may be an invaluable resource for institutions seeking to establish a strong connection with their target market.

(Green *et al.*, 2018) in their study delves into a case that scrutinizes a library's Instagram marketing campaign with the objective of selling its digital property, especially e-books and databases. Researchers analytically evaluated the sort of content material that changed into published, the engagement metrics, as well as the marketing campaign's effect on the usage of sources. The consequences of this take a look at indicate that a nicely-designed Instagram campaign can appreciably heighten awareness and usage of virtual sources. Therefore, it could be inferred that Instagram is a amazing tool that libraries can make use of to reach out to their shoppers and promote their virtual assets efficiently. It is imperative for libraries to embrace social media structures consisting of Instagram as it could assist them reach a wider target audience whilst offering better get entry to their assets.

(Danbaki *et al.*, 2020) in their study investigates the potential of social media as a tool for promoting social integration in libraries. Social integration involves incorporating newcomers or minorities into a host society's social structure, and is one of three key dimensions of the immigrant experience. It highlights several challenges that librarians face in using social media, such as privacy concerns and low technology penetration. However, the authors argue that social media has the potential to transform library services and enhance social integration, and recommend that libraries design services that fully employ social media's sharing capacity, and formulate a viable Social Media Policy to guide its integration.

(Rose, 2020) In their study, the researchers delved into the non-public Facebook institution for college kids and faculty of a community college library. The group sports, discussions, and the library's role in supplying instructional guide have been tested. Through this investigation, the researchers determined that the Facebook group became a powerful tool in cultivating a feel of network and providing instructional resources. This study underscored the significance of social media in enhancing the educational experience, and it confirmed that the Facebook organization may be utilized as an effective platform for communique and collaboration among students and college. The findings of this case examine have huge implications for the use of social media in education, and that they offer treasured insights into the capacity blessings of on line groups in facilitating instructional fulfilment. Therefore, it's far recommended that instructional institutions include social media as a tool for boosting pupil engagement and getting to know outcomes.

(Maziriri *et al.*, 2020) The problem of investigation in this situation examine the YouTube channel of an educational library, which has been closely analysed to decide the efficacy of video tutorials in fostering facts literacy competencies. They delved into numerous components including viewer engagement, comments from the audience, and the affect of such tutorials at the learning outcomes of the users. The findings display that the usage of well-crafted video tutorials on YouTube can indeed make contributions drastically to improving the records literacy abilities of library-goers. This paves the manner for greater progressive and engaging approaches to impart information and getting to know to the hundreds through the use of digital media. It underscores the need to explore and harness the capacity of these new technologies in creating a more knowledgeable and conscious society.

(Ćirić & Ćirić, 2021) The present case examine delves into the Public Library's reaction to the COVID-19 pandemic, specifically its edition to the crisis by using providing digital programs thru social media structures. The library's usage of systems which includes Facebook Live and Zoom for programming became examined by way of researchers. They determined that the library's virtual programming played a crucial position in maintaining engagement with the community and persevering with to serve them at some point of these attempting instances.

(Oyetola *et al.*, 2023) examined the emergence and impact of social media technologies (SMTs) on academic library services in developing countries, emphasizing the need for their adoption to enhance service delivery. The research highlights the low awareness and implementation of SMTs in these libraries and recommends training and expertise development to better integrate these technologies.

These case studies offer actual-global examples of ways libraries have harnessed the energy of social media to achieve several goals, from network engagement to resource advertising and marketing and schooling. They could function treasured references for researchers and practitioners searching for to understand and replicate successful social media techniques in library settings.

3. Effective Use of Social Media for Library Services

To make the most of social media, it's crucial to set clear goals and apprehend your audience. Tailoring content material to their hobbies and wishes and choosing the right platforms will increase engagement and fulfilment. In addition to these essential practices, there are several other techniques you may employ to maximise the blessings of social media for your library offerings.

1. Create a content material calendar: Plan out your social media posts beforehand of time to ensure a constant and organized presence on-line.
2. Monitor analytics: Keep music of your social media metrics to measure the success of your content and alter your approach consequently.
3. Encourage consumer-generated content: Engage with your network by way of encouraging them to percentage their reports and critiques on social media.
4. Collaborate with different corporations: Partner with other libraries, community groups, or local companies to make bigger your reach and create valuable content.
5. Use visuals: Incorporate pleasing photographs, videos, and photos to make your social media content material greater engaging.
6. Be responsive: Monitor your social media bills and reply promptly to remarks and messages from buyers.
7. Use hashtags: Tag your social media posts with applicable hashtags to growth visibility and connect to large audiences.
8. Share behind-the-scenes content material: Give your fans a glimpse into the inner workings of your library by way of sharing photographs and tales from behind the curtain.
9. Host on-line occasions: Plan virtual events like book golf equipment, author talks, or trivia nights to interact with consumers in a brand new way.

By incorporating these techniques into your social media practices, you may increase the effect of your library services and create a dynamic and engaged on line community.

4. Benefits of Integrating Social Media into Library Learning

The integration of social media structures into libraries can provide considerable benefit to both library customers and institutions. It will provide advanced levels of access to library resources and allow the development of on-line groups of users that share common interests leading to increased engagement and interaction between library staff and user. By the same virtue, social media have provided libraries with an effective means of advertising their property: showing off new acquisitions, highlighting digital collections and databases, and raising awareness about research equipment and e-books. They also advertise events, workshops, lectures, and e-book clubs on social media platforms to reach a wider target audience and increase event attendance. Furthermore, social media allows the networking and collaboration of library customers and specialists by connecting them to similar-minded persons, researchers, and experts in different fields.

The encouraging of consumer-generated content associated with library assets can enhance the studying experience and offer valuable insights for others. Social media allows institutions to gather comments in such a way that it offers several opportunities to improve services and facilities to cater more effectively to consumer preferences and needs, and it allows real-time support to be provided, thereby extending academic resources and services to people globally.

Through the use of social media, instructional libraries are able to sell a lifestyle of studies and discovery by sharing findings from research, academic papers, and scholarly resources with customers. Additionally, social media tends to be an affordable method of promoting library services and resources or events, which minimizes the need for traditional advertising and allows the library to reach their respective communities in the delivery of valued resources and guidance through innovative methods.

5. Challenges in Implementing Social Media Learning in Libraries

By embracing social media in library training, there is an possibility for wonderful advantage, however, there are limitations that must be expected:

- The digital divide between those who have get adapted to use technology and those who do not.
- Concerns regarding privacy and information safety.
- The opportunity of facts overload as social media gives a plethora of content
- The allocation of assets to ensure effective use of social media for library studying.
- The best of content shared on social media systems.
- Maintaining cybersecurity even as using social media.
- Moderating the content material shared on social media systems to make sure that it's far appropriate for all audiences.
- The mission of adapting to new social media platforms as they emerge.
- Balancing promotional content with instructional content material.
- Addressing prison and moral issues related to social media use.

- Determining the impact of social media gaining knowledge of programs on library customers.
- Being sensitive to cultural differences even as developing and imposing social media gaining knowledge of applications.
- Managing consumer expectancies regarding the use of social media in library learning.

Despite these challenges, libraries can correctly put in force social media mastering obligations by using developing clean rules, providing digital literacy education, constantly evaluating the effectiveness of their techniques, and prioritizing consumer engagement and education.

6. Future Trends: Social Media and Libraries

The convergence among social media and libraries continues to conform, and there are a plethora of future trends which might be poised to steer how libraries make use of social media to interact customers, sell sources, and decorate their offerings. Here are some critical future developments that warrant interest:

- The emphasis on visible content will stay a critical component in shaping how libraries use social media inside the future. Visual content has been demonstrated to growth person engagement stages and decorate the user enjoy.
- Live streaming and digital activities are probably to emerge as increasingly more famous in the future, as they offer a completely unique opportunity for libraries to connect to their customers in real-time.
- Immersive era, such as virtual and augmented reality, becomes increasingly more standard in libraries, imparting users a more engaging and interactive enjoy.
- Personalization and synthetic intelligence (AI) will hold to form how libraries use social media to have interaction with their users and provide customized offerings.
- Social listening and data evaluation will allow libraries to benefit treasured insights into their users' desires, alternatives, and behaviours, permitting them to tailor their services and sources therefore.
- Collaborative online groups turns into more famous, allowing libraries to create and facilitate groups of exercise which can proportion data, assets, and know-how.
- Short-form video platforms such as TikTok and Instagram Reels will keep growing in popularity, providing libraries with a possibility to engage with their customers in a extra innovative and dynamic way.

- Podcasts and audio content material will continue to be a popular medium for libraries to proportion information, provide get entry to resources, and engage with their customers.
- Accessibility and inclusivity will stay a key attention for libraries when the usage of social media, making sure that everyone users can get entry to and benefit from their offerings and assets.
- Privacy and ethical troubles will stay a key problem for libraries whilst using social media, requiring them to be obvious, responsible, and accountable of their use of consumer information.
- Digital storytelling will become a famous medium for libraries to share tales, hook up with users, and promote their assets and services.
- Global engagement will stay a key attention for libraries, enabling them to connect with users from diverse backgrounds and cultures.
- Collaborations and partnerships with other institutions and companies will become increasingly more critical for libraries, permitting them to pool assets, proportion information, and offer a greater comprehensive variety of services.
- Crisis and emergency communication becomes vital for libraries, requiring them to be organized and responsive inside the face of surprising activities and situations. Libraries will want to increase powerful communication techniques that may reach a extensive audience and provide accurate, well timed, and dependable records.

7. Conclusion

In the field of library education, the impacts of social media have been multidimensional and transformational. The introduction of systems of social media has revolutionized the way in which libraries communicate with their users, promote their resources, and aesthetically enhance educational opportunities. Social media will continue to be an instrumental tool in shaping education and the dissemination of knowledge as libraries continue to evolve within the constantly changing digital environment. Therefore, effective of social media by the libraries extends their reach and impacts. In this way, their reach and impacts can be extended for the improvement of academic experiences of all. Therefore, social media is indeed a game-changer within the subject of library education.

8. Scope for Further Research

Future research directions on the subject of social media in libraries:

- Conducting research on ethical and information privacy issues related to the use of social media in libraries, and making those issues a priority in libraries.

- Assess the interaction between personalization and artificial intelligence in the use of social media through libraries for custom-designed services and increased user involvement.
- The impact of social media implementation in the libraries on reducing the digital gap and ensuring equal access to the library, its services, and resources.
- The quality and credibility of the content material shared by the libraries via social media structures and the level of engagement and dissemination it would cause to happen among its consumers.

References

Ćirić, J., & Ćirić, A. (2021). The impact of the COVID-19 pandemic on digital library usage: a public library case study, 53–68.

du Toit, K., & Mulatiningsih, B. (2013). Social media for libraries. Information & Library Management Africa. https://doi.org/https://doi.org/10.1080/19322909.2021.1913465

Danbaki, C. A., Gado, D. S. M., Mohammed, G. S., & Agbenu, D.. (2020). Library and Social Integration, 14–20.

Green, D. D., Martinez, R., Kadja, A., & Evenson, L. (2018). In a world of social media: A case study analysis of Instagram. https://doi.org/10.21694/2379-1047.18012

Maziriri, E. T., Gapa, P., & Chuchu, T. (2020). Student perceptions towards the use of YouTube as an educational tool for learning and tutorials. 13(2), 119–138. https://doi.org/10.29333/iji.2020.1329a

Mehta, D., & Wang, X. (2020). COVID-19 and digital library services–a case study of a university library. Digital library perspectives, 36(4), 351-363.

Minocha, S.. (2012). How social media is transforming learning, teaching and research dialogues.

Neal, D. (2012). Social media for academics: A practical guide.

Oyetola, S. O., Aderibigbe, N. A., & Oladokun, B. D. (2023). Implications of social media technologies (SMTs) to library services. Library Hi Tech News.

Rafiq, M., Batool, S. H., Ali, A. F., & Ullah, M. (2021). University libraries response to COVID-19 pandemic: A developing country perspective. The Journal of Academic Librarianship, 47(1), 102280.

Rose, M. (2020). Social media for the community college library. 27(1), 1–5. https://doi.org/10.1080/10691316.2020.1737910

Saud, M., Mashud, M. I., & Ida, R. (2020). Usage of social media during the pandemic: Seeking support and awareness about COVID-19 through social media platforms. Journal of Public Affairs, 20(4), e2417.

Shulman, J., Yep, J., & Tomé, D. (2015). Leveraging the power of a Twitter network for library promotion. 41(2), 178–185.

Trucks, E. (2019). Making social media more social: A literature review of academic libraries' engagement and connections through social media platforms. Social media for communication and instruction in academic libraries, 1-16.

Tammaro, A. M. (2020). COVID 19 and libraries in Italy. International Information & Library Review, 52(3), 216-220.

Taylor, F. (2014). Use of social media by the library: current practices and future opportunities.

Xie, I., & Stevenson, J. (2014). Social media application in digital libraries. Online Information Review, 38(4), 502-523.

5

Research Support Services: Insights for the Next Generation Libraries

***Dr. Navkiran Kaur*[1] *and Sukhdeep Kaur*[2]**

[1]*Assistant Professor, Department of Library & Information Science, Punjabi University Patiala-147 002, Punjab, India, E-mail: navkiran1907@gmail.com*

[2]*Librarian, Guru Gobind Singh College for Women, Chandigarh - 160 019, and Researcher, Department of Library & Information Science, Punjabi University Patiala-147 002, Punjab, India, E-mail: sukhdeepkaur05@gmail.com*

Abstract

Societal progress depends on research, often described as the lifeblood of communities, as it is vital for offering solutions to various problems. Libraries have historically played a significant role in supporting research, and now libraries are seen as indispensable part of research process from the inception of research activities to the dissemination and formal publication of research findings. Research support encompasses any activities undertaken by libraries to support and strengthen the scholarly and research-related activities within their respective institutions. This domain of services is relatively new and requires various competencies. Research support services include diverse areas such as collection development (both print and digital), digital preservation and curation, open access services, data management, institutional repositories, citation analysis, plagiarism detection software, and more.Achieving this requires not only the redesigning of library services but also development of new skills among library professionals to proficiently navigate through the flood of information, use advanced information and communication technologies effectively and offer comprehensive research support services to user community.Considering the evolving needs of users and the shifting role of libraries, this study aims to describe emerging research support servicesand propose strategies for libraries to effectively promote and facilitate these services to support research. Effective implementation of research support services and leveraging their full potential require collaboration among authorities, library professionals, faculty, and researchers.The new and innovative research support services are essential for the next generation libraries and can contribute significantly in the research endeavors of the society.

Keywords: *Research support services, Innovative services, Library services, Libraries.*

1. Introduction

For actively participating in today's global information society, having access to information and information resources is crucial. Information serves as a foundational pillar for the success of any nation, and libraries act as the primary gateways to access information. Libraries have been playing a crucial role for decades in supporting educational, learning, and research activities, as well as catering to the diverse informational requirements of individuals. However, changes in the form and formats of documents as well as diverse needs of users, pose new challenges for libraries in meeting these dynamic needs of their users. Libraries must now reconsider their roles and duties, designing a variety of new innovative services that are crucial for providing value-added, needs-based support to the society. Societal progress depends on research, often described as the lifeblood of communities, as it is vital for offering solutions to various problems.Traditionally, libraries have provided their patrons with information resources. However, in addition to these resources, libraries must now offer a range of services to support the research endeavors of scholars and researchers. In essence, while libraries have historically played a significant role in supporting research, they now must take an active role from the inception of research activities to the dissemination and formal publication of research findings. Achieving this requires not only the redesigning of library services but also the development of new skills among library professionals to proficiently navigate through the flood of information, useadvanced information and communication technologies effectively and offer comprehensive research support services (RSS) to user community.

2. Background of the Study

Libraries, guided by their vision and mission, are adapting to the changing landscape ofeducation and research. They are expanding beyond their traditional roles to incorporate new technologies and initiate innovative services, aiming to better support teaching, learning, and research.Libraries are important in the research process as researchers use library resources for commencing and completing research. Now, researchers often seek assistance for tasks such as securing funding, preparing grant proposals, managing research data, and receiving support for publications (Peters & Dryden, 2011). Daland (2013) highlighted the necessity for active involvement of users in order to develop effective research support programs.Universities must restructure their organizational framework to align with library research support services and promote them accordingly (Finlayson & Mithal, 2014). Lightfoot and Sanders (2017); and Ashiq *et al.* (2021) delineated the challenges and opportunities libraries face in providing RSS. There is a need for increased awareness and promotion of guidance in writing technical reports and tools for managing personal content to enhance the utilization of such services (Liji & Devi, 2018). Some nations are particularly attuned to the importance of research support, evidenced by budget allocations and policy development (Xi *et al.*, 2019). Librarians are called upon to provide guidance in developing effective search strategies, which is identified as a significant challenge (Bussell *et al.*, 2020). Borrego and Pinfield (2020) observed that librarians are increasingly engaged with research teams throughout the research process. Their involvement commonly includes reviewing and editing the final article versions. Libraries must cultivate coordination, collaboration, and networks, enhance ICT skills, and demonstrate their value to researchers, students, and faculty to effectively offer new support services (Ducas *et al.*, 2020; Nwabugwu & Godwin, 2020).

Librarians' competencies alone are insufficient; collaboration from authorities, faculty, and researchers is essential to meet their research needs (Daland, 2013; Kersloot *et al.*, 2022). Librarians are proactively initiating and developing new services to bolster research endeavors within their institutions (Searle *et al.*, 2015; Awan *et al.*, 2022; Santharooban, 2022). Librarians have demonstrated their skills and capabilities and are making positive contributions to conducting systematic reviews (Beck, 2015; Knaan & Miller, 2022). Taha (2022) introduced a conceptual model for networked academic library services within a research-intensive university, advocating the exploration of innovative models and strategies in providing research-oriented information services. Schaerfer *et al.* (2023) emphasized the importance of ongoing communication between librarians and requestors during the search process to improve the utilization of bibliographic databases. Payal *et al.* (2019) underscored the importance of data sharing, while Briney (2016), and Gowen and Meier (2020) stressed the necessity of strategic planning in library data services. However, during this evolving phase, librarians have encountered challenges such as time constraints, lack of confidence and skills, and notably, limited or no funding available for their role as research partners (Beck, 2015; Borrego & Pinfield, 2020; Shin, 2020; Knaan & Miller, 2022).

Based on these studies, it can be inferred that libraries are currently transitioning from providing basic information services to emphasizing research-oriented services. Librarians are dedicatedly exploring and innovating new services to address the evolving information needs of their users. They are actively engaging with faculty and researchers to effectively cater to their changing research requirements.No doubt libraries have historically supported research by providing diverse resources such as books, journals, databases, alongside reference and information literacy services, and establishing institutional repositories to manage research outputs. However, the evolving needs of researchers now prompt libraries to offer more robust support services, including bibliometric analysis, research data management, and scholarly publishing services, etc. Considering the evolving needs of users and the shifting role of libraries, this study aims to describe emerging research support services and propose strategies for libraries to effectively promote and facilitate these services to support research.

3. Role of Libraries in Supporting Research

The present digital era offers libraries and librarians with a unique opportunity to either acquire new skills or apply their existing expertise in emerging new areas. Libraries are now expected to embrace new roles and responsibilities to meet the evolving needs of their users. A report by the Research Information Network in 2010 recommended that libraries collaborate with their institution's research offices to establish and promote research support services. Consequently, libraries are fostering partnerships with research offices and other academic departments. Librarians are actively engaging with administrative bodies to shape policies concerning scholarly communication, publishing, and related matters within their institutions. They play a pivotal role in advising faculty and researchers on author rights, navigating publishing agreements, and promoting their institution's research publications through the development of institutional repositories, among other endeavors. While retaining their traditional roles, which remain in demand, libraries are also striving to incorporate new responsibilities to fulfill their core mission. They are innovating and introducing new services to strengthen research efforts. Libraries worldwide are proactively adapting to changes in the research landscape, initiating reforms, and embracing a broader spectrum of research support services. This shift is driven by evolving teaching methods and the rapid advancement of enabling technologies, which have spurred interest in new research support services such as bibliometrics, data management, digital preservation and curation, and open access. Libraries are collectively mobilizing to provide these services through coordinated efforts.These services may include assistance with copyright and licensing, scholarly publishing, citation analysis, generating novel research ideas, and guiding individuals on how to identify funding opportunities.

4. Research Support Services

Research support encompasses the assistance provided by libraries to researchers, aiming to enhance the productivity of their scholarly endeavors. According to Hoffman (2016), research support encompasses any activities undertaken by libraries to support and strengthen the scholarly and research-related activities within their respective institutions. This domain of services is relatively new and requires various competencies. Research support services encompass diverse areas such as collection development (both print and digital), digital preservation and curation, open access services, data management, institutional repositories, citation analysis, plagiarism detection software, and more. Since research is crucial to any educational institution and the value of an institution is judged by its research output, the expectations placed on libraries increase significantly. Consequently, the concept of research librarians is emerging, with higher education institutions appointing dedicated professionals to ensure comprehensive support for research services.Some research support services are discussed below:

4.1. Traditional Research Support Services

- **Open Access (OA) Services**: As per the Budapest Open Access Initiative (BOAI), open access entails the global electronic dissemination of peer-reviewed journal literature, granting unrestricted and free access to it for all scientists, scholars, educators, students, and other interested individuals. Institutions worldwide are leveraging the advantages of open access. With open access, libraries can remotely provide services to strengthen research and educational endeavors. Libraries are offering access to a plethora of electronic resources such as journals and books through their websites. Users can avail these resources both within library premises and from the comfort of their homes or offices.
- **Current Awareness Service (CAS)**: Researchers consistently strive to remain abreast of the latest literature within their research domains. The CAS fulfills this essential requirement by ensuring users are informed about recent advancements, enabling them to stay current with new research discoveries, publications, and opportunities in their respective fields. Through the provision of timely updates and customized alerts, libraries play a crucial role in boosting research productivity and fostering knowledge dissemination within the research community.
- **Selective Dissemination of Information (SDI)**: functions as an alert system, utilizing automatic filtering and distribution mechanisms to deliver pertinent information directly to individual users, tailored to their unique interests, preferences, and profiles. SDI systems actively track new publications, databases, and other information sources, providing personalized updates to users keen on remaining current within their focus areas and who have specified their preferences. Serving as a valuable research aid, it enables researchers to stay informed, streamline their workflow, and make more informed decisions.

- **Document Delivery Services (DDS):** entail granting access to a diverse range of scholarly materials that might not be easily accessible within a researcher's institution, yet are vital for advancing knowledge and conducting top-tier research. Through this service libraries provide documents either physically or electronically upon users' requests. In today's fast-paced environment, timely access to information is paramount. By extending access to resources beyond institutional collections, libraries aid users in meeting their needs promptly, thereby encouraging the success and productivity of researchers while enhancing the research capabilities of academic institutions.
- **Inter Library Loan Services (ILL)**: fosters collaboration between libraries, allowing patrons to borrow books and materials from one library through another. This service is pivotal in aiding research endeavors, granting researchers access to a diverse array of scholarly materials beyond their institution's collection. By facilitating resource sharing among libraries and document suppliers, ILL services bolster research capabilities, foster collaboration, and contribute to the advancement of knowledge.
- **Ask-a-Librarian**: serves as a link between conventional library offerings and contemporary technology. It offers users a versatile platform to seek guidance, assistance, and expert advice in navigating the vast landscape of information resources. This service extends beyond physical library visits, as it also provides virtual reference services. Through email, chat, or video conferencing platforms, users can remotely ask questions and receive assistance from library staff. These virtual reference services ensure convenient access to library assistance for users who may be unable to visit the library in person, thereby enhancing accessibility and support for all patrons.
- **Information Literacy Programmes**: The process of gathering and accessing information to bolster research poses a significant challenge for new researchers. Information literacy programs address this issue by focusing on initiatives designed to equip individuals with the necessary skills and competencies to effectively locate, assess, utilize, and communicate information across diverse contexts. These programs empower researchers and faculty members with the knowledge and abilities required to navigate the complexities of information in the digital era. As part of these initiatives, libraries collaborate with various departments and organizations to host workshops, seminars, webinars, conferences, and similar events aimed at providing training for emerging

researchers. Additionally, they offer access to online resources, tutorials, guides, and instructional materials that support information literacy instructions.

- **Institutional Repositories (IRs):** IRs are digital platform or database developed by academic institutions to gather, preserve, and distribute scholarly works produced by their faculty, researchers, students, & staff, thereby offering open access to academic content. It serves as a centralized archive for a diverse array of materials, encompassing research articles, preprints, conference papers, datasets, theses, dissertations, and other scholarly outputs. As described by Lynch (2003), an institutional repository essentially constitutes a suite of services provided by universities, managing and disseminating digital content created by the institution to its users through IRs, thereby preserving the institution's intellectual content for the long term. These repositories foster wider dissemination of research discoveries and enhance the impact of the institution's scholarly contributions. By offering a platform for showcasing research output, institutional repositories can augment the visibility while also advancing open science principles. Among the primary objectives of libraries is the provision of the most recent and pertinent information to users, keeping them abreast of the latest developments in various fields. And institutional repositories represent the optimal means to achieve this goal, ensuring users remain well-informed.

4.2. New Research Support Services

- **Remote Access to Resources and Services**: Facilitating remote access to resources and services is crucial for supporting research, teaching, and learning endeavors, as it allows users to access scholarly materials and research tools irrespective of their location or the time of day. By ensuring convenient and secure access to electronic resources and services, libraries empower researchers to engage in their research work, retrieve information, and collaborate with peers without the limitations imposed by physical boundaries or time restrictions. Additionally, libraries may offer user support and training through online tutorials, user manuals, troubleshooting advice, and help desk assistance, aiding users in effectively utilizing remote resources and services.
- **Literature Search Assistance**: aimed at aiding researchers in identifying, locating, and accessing relevant scholarly literature pertinent to their research inquiries or topics. Conducting a thorough literature search is a critical step in the research process, enabling researchers to review

existing knowledge, pinpoint gaps in the literature, and position their research within the broader scholarly context. Libraries offer assistance in both fundamental and advanced literature searches, playing a pivotal role in bolstering research activities by assisting researchers in navigating the vast and complex scholarly literature landscape. Through provision of guidance, resources, and training in literature search methodologies, libraries aid researchers to conduct thorough and effective literature searches that enrich their research pursuits.

- **Bibliometric Services:** involve the application of statistical methods to analyze books, research articles, and similar materials. These services focus on analyzing and evaluating the scholarly impact and productivity of researchers, institutions, journals, and research domains using bibliometric indicators and methodologies. It offers quantitative insights into various aspects of scholarly communication and research performance, aiding in the identification of trends, patterns, and citation impact within specific fields or disciplines. It helps to identify trends, patterns, and citation impact within specific research fields or disciplines. Citation analysis, a key component of bibliometric services, helps identify highly cited works, influential authors, and research collaborations that have made significant contributions to their respective fields. By analyzing publication and citation data, bibliometric services enable stakeholders to make informed decisions, assess research impact, and identify opportunities for collaboration and enhancement. Apart from Citation Analysis, services such as evaluating the Impact Factor of Journals or Institutional Research Outputs are commonly included under bibliometric services. Libraries are increasingly stepping forward to provide these services to support research endeavors within their respective institutions.
- **Research Data Management (RDM) Service:** encompass a range of activities and resources designed to facilitate the organization, storage, sharing, preservation, and reuse of research data throughout the data lifecycle. It also addresses concerns regarding data security and confidentiality concerns. Libraries play a key role in guiding researchers on implementing measures to safeguard sensitive or confidential research data against unauthorized access, disclosure, or misuse. These services are essential for ensuring the integrity, accessibility, and long-term sustainability of research data, thereby improving research quality, reproducibility, and impact. Libraries in higher educational institutions are taking the lead in offering these services, assisting researchers in developing Data Management Plans (DMPs) that delineate how

research data will be managed, documented, and shared throughout the research endeavor. To support researchers, libraries provide training sessions, workshops, and resources to enhance their understanding and proficiency in research data management practices. This support includes offering guidance on best practices, tools, and software for data management, as well as ensuring compliance with data management policies and standards.

- **Open Access Publishing**: To support the research, libraries are now offering a novel service that aligns with their mission to facilitate access to scholarly knowledge and promote the dissemination of research findings for the betterment of society. This service entails the practice of freely sharing scholarly research online with the public, removing financial or legal barriers. Libraries are actively engaging in educating researchers about the advantages of open access publishing, which include increased visibility, broader impact, and better collaboration opportunities. Librarians are providing guidance and assistance to researchers in selecting reputable open access journals for publishing their work. Additionally, they are aiding researchers in assessing the quality and reputation of open access journals, understanding various publishing models, and navigating the submission and peer review processes. Furthermore, librarians are offering support to researchers in retaining their rights as authors when publishing in open access journals.
- **Research Tools Recommendations**: The Research Tools Recommendations service, offered by libraries or academic institutions, aids researchers, faculty, and other users in identifying and utilizing suitable tools and software tailored to their research needs, goals, and disciplinary requirement. This service is designed to assist users in discovering and utilizing tools that optimize productivity, streamline research processes, and support tasks such as data management, analysis, visualization, and scholarly communication. Librarians collaborate with various departments and organizations to arrange hands-on training sessions, workshops, and instructional materials, empowering users to effectively utilize the recommended research tools.
- **Training to Use Plagiarism Check Software**: Plagiarism is unethical. Academic institutions offer training sessions on the utilization of plagiarism detection software to assist researchers and faculty members in comprehending the effective use of such tools to safeguard the integrity of their scholarly works. These software applications aid in identifying instances of plagiarism by comparing submitted documents with an

extensive database of academic and online sources. Libraries conduct hands-on training sessions on popular plagiarism detection software like Turnitin, iThenticate, or Similarity Check, acquainting participants with the features, functionalities, and capabilities of these tools in detecting and preventing plagiarism. Participants receive guidance on navigating the user interface of the software, including accessing the platform, uploading documents for analysis, interpreting similarity reports, and comprehending the generated results. Through these training initiatives and support services, users are empowered to maintain the highest standards of integrity and professionalism in their academic work.

- **Guidance on Research Ethics:** Research ethics encapsulate the principles, guidelines, & standards governing the responsible conduct of research, emphasizing integrity, transparency, respect for participants' rights, & compliance with ethical principles and regulatory requirements. Libraries play a pivotal role in providing users with guidance on institutional policies, guidelines, and codes of conduct pertaining to research ethics. These policies address various forms of research misconduct, including plagiarism, fabrication, falsification, and other breaches of research integrity. Libraries also offer support on data management practices and principles to uphold the trustworthiness, reliability, and reproducibility of research outcomes. Users are educated on data acquisition, storage, sharing, and preservation methodologies, alongside best practices for ensuring data security, privacy, and confidentiality. Additionally, guidance is provided on publication ethics and responsible authorship practices within the realm of academic publishing.
- **Guidance Regarding Copyright and Licensing**: refers to offering education, resources, and assistance to comprehend and adhere to copyright laws, licensing agreements, and intellectual property rights in relation to scholarly endeavors. Libraries play a crucial role in educating users about fair use and fair dealing principles, which serve as exceptions to copyright law, permitting limited and transformative use of copyrighted works without explicit permission from the copyright holder. Additionally, libraries offer support in obtaining permissions and licenses for utilizing copyrighted materials in various activities such as research, teaching, publishing, and dissemination. They also advocate for the adoption of Creative Commons licenses and other open licensing frameworks to promote the sharing, reuse, and dissemination of scholarly works. Recognizing the significance of copyright and licensing considerations is imperative for ensuring the legal and ethical

use of intellectual property, spanning written works, images, data, and multimedia content across research, teaching, publishing, and dissemination domains. Libraries actively engage with the research community through advocacy initiatives, workshops, seminars, and educational resources, providing opportunities for researchers to participate in discussions, share experiences, and enhance awareness regarding copyright and licensing issues in research and scholarly communication.

- **Training in Improving Information Retrieval and Management Skills:** Libraries organize training sessions aimed at enhancing the information-seeking capabilities of individuals for efficient research and scholarly activities. These sessions are designed to equip users with the necessary skills and strategies to navigate the extensive array of scholarly resources, databases, and information sources adeptly. They often encompass hands-on workshops, interactive exercises, and practical demonstrations to reinforce learning and skill development. Participants are instructed on effective search methodologies across library databases, academic search engines, and specialized research platforms. They gain proficiency in utilizing advanced search functionalities, such as Boolean operators, truncation, and wildcard symbols, to refine search queries and retrieve pertinent information. Additionally, they learn to select appropriate keywords, synonyms, and search terms tailored to their research topics. By honing these retrieval and management skills, users can leverage library resources, collections, and services effectively to backing their research.
- **Dedicated Research Spaces**: refer to specialized facilities or areas within academic institutions that are specifically designed to support and enhance the research activities of faculty and researchers. These spaces, including collaborative work areas like research lounges, innovation hubs, and co-working spaces, promote collaboration, innovation, and scholarly inquiry among researchers spanning various disciplines and backgrounds. They play a pivotal role in advancing knowledge and fostering research excellence within academic settings. Recognizing the value of discussion and collaboration in research, higher educational institutions, particularly those deeply engaged in research, are increasingly establishing dedicated spaces within their libraries for this purpose.
- **Training to Build Scholarly Networking**: aimed at assisting researchers & faculty members in establishing & nurturing professional

connections and collaborations within their academic and research circles. Cultivating a scholarly network is crucial for promoting collaboration, exchanging knowledge, & propelling research endeavors forward. The training provided in this domain is designed to empower participants with the requisite skills and techniques for adeptly networking with peers, collaborators, and mentors in their respective fields of study. To inspire participants, trainers may share success stories and case studies of effective networking experiences from experienced researchers, faculty members, and professionals.

- **Research Partners**: Librarians serve as invaluable research partners, offering expertise and support throughout the research process. Beyond serving as co-researchers or co-authors, they assist in accessing scholarly resources, contribute to research design by formulating questions and devising effective search strategies, and provide guidance on managing and preserving research data. Librarians also offer technological assistance, information literacy training, and advice on scholarly communication, facilitating collaboration and interdisciplinary dialogue among researchers. Their role extends to assessing research impact and visibility, significantly enhancing the quality and influence of scholarly work across disciplines. Notably, librarians have broadened their role from mere supporters to active research partners.
- **Research Profile**: In order to support research endeavors, libraries are constructing research profiles of the researchers and faculty engaged in scholarly pursuits within their institution, ensuring timely and relevant information delivery in appropriate formats. Services such as CAS and SDI are witnessing increased demand in this context. These profiles enable librarians to select and distribute pertinent information to meet the needs of their users effectively. The creation of research profiles not only aids libraries in fulfilling user demands but also benefits the creators of the information i.e. researchers and faculty as it provide a platform for them for promoting their work, establishing credibility and expertise, fostering collaboration and networking opportunities, and highlighting their contributions to advancing knowledge in their respective fields. Consequently, the development of research profiles proves to be mutually beneficial for both libraries and researchers, constituting a win-win scenario.
- **Research Visibility**: With the exponential increase in publications across disciplines, staying abreast of recent developments & publications within one's area of interest poses a significant challenge. For authors, visibility of their work is paramount to ensure it doesn't

go unnoticed. In this regard, librarians play a crucial role. Libraries are establishing Institutional Repositories (IRs) as an effective means to enhance the visibility of research outputs. Additionally, author or institutional profiles on platforms like Google Scholar can boost visibility and citations. Librarians are actively utilizing social networking sites such as ResearchGate, Academia.edu, and LinkedIn to promote the research outputs of institutional faculty and researchers. They also provide resources, guidance, and support to help researchers disseminate their findings effectively, expand professional networks, and enhance their presence within and beyond the academic community. Workshops and seminars organized by libraries aim to train users on utilizing various social bookmarking sites like Mendeley and Zotero. By implementing strategies to augment research visibility, researchers can amplify the impact and dissemination of their work, contributing to the advancement of knowledge in their respective fields.

- **Awareness Drive Regarding Research Funding Opportunities**: involves efforts aimed at informing and educating researchers about available sources of funding, grant opportunities, and resources to support their research endeavors. The objective is to boost awareness, visibility, and access to funding opportunities, thereby facilitating the pursuit of research projects and scholarly activities. Alongside these campaigns, libraries offer researchers access to various supportive resources and tools for navigating the grant application process. These resources may include guides on grant writing, proposal templates, sample narratives, budget templates, compliance checklists, FAQs, and links to online resources and databases for finding funding opportunities, funding agencies, and grant-making organizations. This awareness campaign plays a crucial role in empowering researchers to identify, pursue, and secure external funding to advance their research endeavors.

5. Proposed Strategies for Libraries

The next generation libraries are believed to have an indispensable role in supporting research and innovation. Libraries have already initiated the designing and provision of innovating services and now the libraries need to effectively promote and facilitate these services to support research.Libraries can develop various strategies and action plans to raise awareness and increase the use of research support services, some of which are as follows:

- **Outreach and Marketing**: Libraries can actively promote their research support services through various channels, including library websites, social media platforms, newsletters, and informational brochures. Additionally, they can collaborate with academic departments and research centers to spread the word about available services.

- **Workshops and Training Sessions**: Organizing workshops, training sessions, and webinars on topics such as bibliographic management tools, research data management, copyright and licensing, and scholarly publishing can help researchers become aware of and proficient in utilizing library resources and services effectively.
- **Personalized Consultations**: Offering one-on-one consultations with librarians or subject specialists can provide researchers with personalized assistance tailored to their specific research needs. Librarians can offer guidance on literature searching strategies, citation management, data management planning, and other aspects of the research process.
- **Collaboration with Faculty**: Collaborating with faculty members to integrate library resources and services into course curriculum can help familiarize students with available research support services from an early stage. Librarians can work with faculty to design assignments that require students to utilize library resources and services for their research projects.
- **Online Guides and Tutorials**: Creating online guides, tutorials, and instructional videos on topics related to research support services can serve as valuable resources for researchers seeking guidance on using library resources effectively. These resources can be made available on the library website for easy access by researchers.
- **Feedback Mechanisms**: Establishing feedback mechanisms such as surveys, suggestion boxes, or focus groups can help libraries gather input from researchers about their experiences with research support services. This feedback can be used to identify areas for improvement and inform future planning and development of services.

Promoting research support services in libraries involves proactive efforts to raise awareness, provide training and assistance, and continuously improve and adapt services to meet the evolving needs of researchers.

6. Conclusion

Research serves as the foundation for global and societal progress. The emergence of new innovative technologies is also an outcome of the research conducted by technologists. Higher education institutions including universities bear the responsibility of furnishing the necessary information, support, and infrastructure to generate and advocate solutions to societal issues through research endeavors and the dissemination of research findings.

Libraries assist higher education institutions in fulfilling their responsibilities. By providing access to a wealth of resources, expert assistance from trained librarians and essential services libraries empowers researchers to pursue their research efficiently and effectively. Libraries continually evolve over time to effectively support changing user demands and promote research.Libraries since inception were supporting research but now the libraries are seen as an indispensable part of the research process. Some of the libraries have already taken steps to introduce advanced research support services, including aiding in literature searches, providing bibliometric services, offering guidance and training on research tools and software, and more. However, to further bolster and encourage research in academic settings, there is a necessity to redefine the roles of libraries in this evolving landscape. The libraries in universities needs to provide advanced research support services to support their researchers in their research endeavors and collaborate with authorities in formulating policies to advance research within their institutions.Effective implementation of research support services and leveraging their full potential require collaboration among authorities, library professionals, faculty, and researchers.The new and innovative research support services are essential for the next generation libraries and can contribute significantly in the research endeavors of the society.

References

Ashiq, M., Rehman, S. U., & Mujtaba, G. (2021). Future challenges and emerging role of academic libraries in Pakistan: A phenomenology approach. Information Development, 37(1), 158–173. https://doi.org/10.1177/0266666919897410

Awan, M. H., Richardson, J., & Ahmed, S. (2022). Current status of research support services in university libraries of Pakistan. Digital Library Perspectives, 38(4), 412-428. https://doi.org/10.1108/DLP-11-2021-0101

Beck, C. (2015). Research partner and search methodology expert: The role of the librarian in systematic reviews. In The Quest for Deeper Meaning of Research Support (pp. 90-104). IFLA. https://doi.org/10.15641/0-7992-2522-8

Borrego, Á., & Anglada, L. (2018). Research support services in Spanish academic libraries: An analysis of their strategic plans and of an opinion survey administered to their directors. Publications, 6(48), 1–10. https://doi.org/10.3390/publications6040048

Borrego, A., & Pinfield, S. (2020). Librarians publishing in partnership with other researchers: Roles, motivations, benefits, and challenges. Libraries and the Academy, 20(4), 655-675.

Brown, S., Alvey, E., Danilova, E., Morgan, H., & Thomas, A. (2018). Evolution of research support services at an academic library: Specialist knowledge linked by core infrastructure. New Review of Academic Librarianship, 24(3–4), 339–350. https://doi.org/10.1080/13614533.2018.1473259

Briney, K. (2016). Strategic planning for research data services. Bulletin of the Association for Information Science and Technology, 42(4), 39-41.https://doi.org/10.1002/bul2.2016.1720420411

Bussell, H., Schnabel, J., & Rinehart, A. K. R. (2020). Meeting graduate student needs: An exploration of disciplinary differences. Public Services Quarterly, 16(4), 213–233. https://doi.org/10.1080/15228959.2020.1818663

Daland, H. (2013). The PhD candidate as an information literate resource: Developing research support and information literacy skills in an informal setting. LIBER Quarterly, 23(2), 134-155.

Ducas, A., Michaud-oystryk, N., & Speare, M. (2020). Reinventing ourselves: New and emerging roles of academic librarians in Canadian research-intensive universities. College & Research Libraries, 81(1), 43–65.

Finlayson, A. J., & Mitha, S. B. (2014). Research support for academic excellence [Conference session]. INASP: The Association of Commonwealth Universities, London.

Gowen, E., & Meier, J. J. (2020). Research data management services and strategic planning in libraries today: A longitudinal study. Journal of Librarianship and Scholarly Communication, 8(1). https://doi.org/10.7710/2162-3309.2336

Haddow, G., & Mamtora, J. (2017). Research support in Australian academic libraries: Services, resources, and relationships. New Review of Academic Librarianship, 23(2–3), 89–109. https://doi.org/10.1080/13614533.2017.1318765

Hoffman, S. (2016). Dynamic research support in academic libraries. London: Facet Publishing.

Knaan, K. B., & Miller, K. (2022). Librarians as research partners on interdisciplinary research teams [Conference session]. Proceedings of the IATUL Conferences.

Kersloot, M. G., Hanna, A. A., Cornet, R., & Arts, D. L. (2022). Perceptions and behavior of clinical researchers and research support staff regarding data FAIRification. Scientific Data, 9(241). https://doi.org/10.1038/s41597-022-01325-2

Larsen, A. V., Dorch, B. F., Nyman, M., Thomsen, K., &Drachen, T. M. (2010). Analysis of research support services at international best practice institutions. CULIS Knowledge Center for Scholarly Communication, August. http://findresearcher.sdu.dk/portal/en/publications/analysis-of-research-support-services-at-international-best-practice-institutions(9730d1da-bd15-447a-8cce-4a8f1a53e545).html

Lightfoot, C., & Sanders, K. (2017). Libraries and research support in small and teaching-led universities: Contextual problems around nascent services in dynamic times. SCONUL Focus, 70. https://www.sconul.ac.uk/sites/default/files/documents/68.Libraries and research support.pdf

Liji, B. J., & Devi, B. M. (2018). Awareness and use of research support services by the research scholars of Kerala university library. Indian Journal of Information Sources and Services, 8(2), 75–80.

NIC & Prajapati, L. (n.d.). Department of higher education, Government of India, ministry of education. https://www.education.gov.in/hi/institutions-national-importance

Nickels, C., & Davis, H. (2020). Understanding researcher needs and raising the profile of library research support. Insights: The UKSG Journal, 33(1), 1–13. https://doi.org/10.1629/uksg.493

Nwabugwu, M. J., & Godwin, L. S. (2020). Research Data Management (RDM) Services in libraries: Lessons for academic libraries in Nigeria. Library Philosophy and Practice (e-Journal), 5–21.

Payal, M., Awasthi, S., & Tripathi, M. (2019). A selective review of literature on research data management in academic libraries. DESIDOC Journal of Library & Information Technology, 39(06), 338-345. https://doi.org/10.14429/djlit.39.06.14451

Peters, C., & Dryden, A. R. (2011). Assessing the academic library's role in campus-wide research data management: A first step at the University of Houston. Science & Technology Libraries, 30(4), 387–403. https://doi.org/10.1080/0194262X.2011.626340

QS World University rankings 2024. (n.d.). Top Universities. https://www.topuniversities.com/world-university-rankings

Research Information Network (2010). Research support services in UK universities. London: Research Information Network.

Santharooban, S. (2022). Research support service model of academic library: Case study at Eastern University, Sri Lanka. Library Philosophy and Practice (e-journal), 7018. https://digitalcommons.unl.edu/libphilprac/7018

Schaefer, N., Morgan-Daniel, J., Struckmeyer, L. R., Myers, C. T., King, L., Jeghers, M., Medhizadah, S., & Beneciuk, J. (2023). Librarian and researcher assessments of search result relevance: How well do they align? Medical Reference Services Quarterly, 42(2), 91-107. https://doi.org/10.1080/02763869.2023.2193122

Searle, S., Wolski, M., Simons, N., & Richardson, J. (2015). Librarians as partners in research data service development at Griffith University. Program: Electronic Library and Information Systems, 49(4). https://doi.org/10.1108/PROG-02-2015-0013

Shin, E. (2020). Embedded librarians as research partners in South Korea. Journal of Librarianship and Information Science, 1-9. https://doi.org/10.1177/0961000620962550

Si, L., Zeng, Y., Guo, S., & Zhuang, X. (2019). Investigation and analysis of research support services in academic libraries. The Electronic Library, 37(2), 281–301. https://doi.org/10.1108/EL-06-2018-0125

Taha, A. (2022). Networked library services in a research-intensive university. The Electronic Library, 30(6), 844-856. https://doi.org/10.1108/02640471211282145

Xi, Q., Wu, W., Ji, J., Zhang, Z., & Ni, F. (2019). Comparing the level of commitment to in-depth reference and research support services in two sets of Chinese universities. Science & Technology Libraries, 1–20. https://doi.org/10.1080/0194262X.2019.1583624

6

The Virtual Shelf: Exploring the Availability and Accessibility of Electronic Resources

Dr. Anuj Kumar Singh

Assistant Professor-Library, Government Girls Degree College, BLW, Varanasi-221001 Uttar Pradesh, India, E-mail: aps.kumbh@gmail.com

Abstract

The virtual shelf has transformed how we access and utilize electronic resources, offering unparalleled convenience and potential for knowledge discovery. However, the sheer volume and diversity of resources pose challenges in ensuring equitable access and effective utilization. This study investigates the availability and accessibility of electronic resources on the virtual shelf, examining factors such as:

- *Resources discovery and visibility*
- *Accessibility features for diverse user needs*
- *Licensing and copyright restrictions*
- *Digital literacy and user training*
- *Infrastructure and technical support*

By exploring these aspects, this research aims to identify best practices and recommendations for optimizing the virtual shelf experience, ensuring that electronic resources are truly accessible and beneficial for all users.

Keywords: *Electronic resources, Virtual library, Virtual shelf, Digital library, Academic library, e-Books, e-Journals.*

1. Introduction

The advent of computers, the Internet and other ICT technologies have enhanced the nature of tasks. No one is immune from technology. Technologies have paved the way for information transformation. Today every section of society is making full use of technologies to fulfil their needs. Electronic devices have affected human life to a great extent. These devices have changed the working styles of human beings. Working with electronic devices is much better than manual work. Therefore, every sector of human working society is updating itself by using or implementing electronic devices. The higher education department is no exception to this. Today every sector of higher education department is using these modern tools and technologies not only for official works but also for teaching. Library is the most important part of Higher Education Institutions and today the entire structure of the library has changed. Today computersare being used in every section of the library and the main work of the library- issue/return is also being done through the use of computers and the Internet. The resources available in the library have also become electronic. Every resource like books, magazines, research journals, dictionaries, glossaries, bibliographies, biographies, who's who etc, is now being published in electronic form and the library has to purchase these resources for its users. The modern library collects these resources, organizes them and then disseminates them as per the demand of the users. Nowadays, information is available only by a simple click on the screen of the computer, laptop or mobile. And all this has been possible only because of the Internet. All information that can be accessed, stored, disseminated, retrieved etc. through electronic devices, is popularly known as electronic resources or *e-resources.* "The resources which are published in digital format and read by a computer are digital resources. Digital resources usually consist of *e-books*, online Journals, Databases, Institutional repositories in electronic form, OPAC and Web OPAC, websites, thesis and dissertation in electronic form, CD/DVDs based resources, etc. This means that the information or resources which are generated in electronic format and applied and enjoyed with the assistance of computer and other electronic instruments is termed as electronic resources"(Ashok Kumar and Meher Singh, 2013,p. 2). "Electronic resources are accessed in the form of digital information through computer terminals not only within the boundary but also beyond the boundary" (Barnes, 1997).

"Library has a collection of printed resources in its arena. The library follows the rules and regulations of acquiring and organizing these resources and providing them to its users in a systematic manner. And now with the advent of new technologies in many forms and updated editions, print resources are now becoming available in digital form. The resources which are available in form of softcopy and can be accessed via any technical devices are known as electronic resources" (Thirumalaimurthuswamy, 1974). There is rapid and constant growth in online versions of print and bibliographical databases and even *e-journals*, *e-books* etc. (Prytherch, 2016).

It was found in several studies that the Department of Higher Education, Govt. of Uttar Pradesh is paying more attention to the richness of electronic resources in the degree college libraries of Uttar Pradesh. Most of the Government Degree College Libraries upgrade their online resources and electronic resources with the help of Govt. of Uttar Pradesh under the project of RUSA (Rashtriya Uchchatar Shiksha Abhiyan). Rs. 32 Lakhs have been received for the upgradation of the library and upgrading of *e*-Resources of each Government degree college of the State and also received an additional budget for the purchase of *e*-Resources under the *e*-Library project (an initiative of the Department of Higher Education, Govt. of Uttar Pradesh). Department of Higher Education, Government of Uttar Pradesh is also running a website (https://heecontent.upsdc.gov.in/) for providing *e-resources* to the students of degree colleges under this project. For this, the Government of Uttar Pradesh has released a total budget of Rs. 10,00,000.00 (Rupees Ten Lakhs) for setting up of *e-learning* park in each government degree college. The aided degree colleges have their management for budget allotment. Aided degree colleges are provided a lump sum amount of money by the government. Then the Management Committee decides the percentage of the budget for the development of each department of the college.Budget is allocated as per requirement for library development.

In this study, e-resources have been evaluated from a critical point of view and you all know that in today's era the study of e-resources has become very important.For better understanding of e-resources, it is necessary to study in depth each form of e-resources like e-books, e-journal, e-thesis, e-dissertation, e-report, online database etc. This study is important so that the researcher can get a clear understanding about the e-resources related to this research work. The study of *e*-Resources is a never-ending process. “And even today progress continues in various aspects of digital resources” (Reddy &Aswath, 2016).

From information and communication technology to mobile technology, all have influenced the sustainable development of library and resources available in it. The print resources have been gradually transformed into electronic resources. The publication agency like Springer, Emerald, J-Gate, OCLC, Frauds Bacons etc. have moved towards digital printing atmosphere. The resources of these reputed publishers are very costly, economically poor libraries could not afford this, therefore UGC has created a platform in the form of consortia to provide electronic resources to the economically poor libraries to enhance the quality of Higher Education status of degree colleges as well as Universities. The most important consortia subscribed by almost all the degree college libraries is N-List consortia. Some others are *e*-SodhSindhu, INDEST consortia; Dyangangotri etc. provide scholarly academic literature related to entire disciplines to enhance the academic development of the faculty members and students of degree colleges. "Information and Communication Technology (ICT) has brought unprecedented changes and transformation to academic library and information services, conventional LIS such as OPAC, users services, reference services, bibliographic service, current awareness services, Document delivery, interlibrary loan, audio visual services and customer relations can be provided more efficiently and effectively using ICT, as they offer convenient time, place, cost-effectiveness, faster and most-up-to-date dissemination and end users involvement in the library and information services process" (Krubu and Osawaru, 2011, p. 2). Kumar (2011) stated that due to the advent of multimedia platforms like MERLOT (Multimedia Educational Resources for Learning and Online Technology), ERIC (Education Resources Information Centres), LESTER (Learning Science and Technology Repositions- a multimedia resource technology), Subject Gateways, DSS (Document Delivery Services), DA (Digital Archives) etc. the nature of services provided by an academic library has been changed. In the present time, electronic resources have become an important component of the academic community to quench the thirst for knowledge. In this way, the resources available in the library are being accessed online not only within the campus but also off campus (Pilgirm and Dolabaille, 2011).

2. *e*-Resources: Concept and Meaning

According to AACR2 (2002, updated 2005) glossary, "An electronic resource is defined as any work encoded and made available for access through the use of a computer. It includes electronic data available by remote access and direct access (fixed media). In other words, Remote access (electronic resources) refers to the use of electronic resources via computer networks and direct access (electronic resources) refers to the use of electronic resources via carriers (e.g. discs, cassettes) designed to be inserted into a computerized device or its auxiliary equipment".

According to IFLA (2015), "Electronic resources are an integral component of library collection building. Electronic resources are those that require personal or mainframe computers and a mobile device to read or access. These resources can be accessed via the Internet by the end users. The resources thatlibraries have are *e*-Books, *e*-Journals, Online databases (Full Text, Indexing, Abstracting), Reference Database such as Dictionaries, Biographies, Encyclopaedias, Directories etc.; Numeric Database, Statistical Databases, *e*-Images, *e*-Audio/Visuals resources".

According to Prytherch (2016), "Electronic resources are those electronic products that provide a collection of data, be it text referring to the full-text basis, *e*-Journals, image collection and other multimedia products and numerical, graphical or time-based, as a commercially available till that has been published with an aim to being marked. This may be delivered on CD-ROM or tape or through the internet and so on".

From the above definitions, it can be concluded that electronic resources is one which can be accessed through Internet only with the help of some electronic gadgets like Computers, Tablet, Mobile, Laptops, Kindle etc.

3. *e*-Resources: It's Evolution

3.1. *e*-Resources: It's Arrival

The development of electronic resources began inmid-1960s. The technology of printing had evolved and a new method of printing had arrived. Now libraries especially research libraries started getting electronic form of resources from some famous publishers. The softcopy version of print resources was gradually taking place in the form of electronic resources i.e. *e-resources*. In the study of Meadow (1988) it was described that "during 1940, the manipulative retrieval system and the post coordinate indexing system was developed and operated manually" (p. 15). Print technology has created a revolution in the development of library resources with a certain vision. The Gutenberg project appeared in 1971 and became the first *e*-book entry in the library. 4 July 2021, it was celebrating its 50 years of *e*-book festival. But in a study "As We May Think" byVennevar Bush, it was found that there he discussed a device named "Memex- a Hypothetical Electro-Mechanical device for reading Microform documents. But it took a long time for these types of documents (*e*-Resources) to come into the hands of tech-savvy users. In 1960, Alan Kay developed a notebook computer which was based on Bush's idea. In 1986, William Morriadge developed a device named Dynamic Book and some more projects and devices were launched for the development of library *e*-resources and are still ongoing (Reddy and Aswath, 2016). This was the period when the MARC (Machine Readable Catalogue) format was developed for *e*-resource entry. At this time, Bibliographic databases were also developed. During the mid-1980s, when microframe computers were invented, libraries started using software (CDS/ISIS) for book catalogue entry and stored data on Diskette and CD-ROM. Full-text databases were also stored on CD-ROM. When www (World Wide Web) was invented in the 1990s by Tim Berners-Lee, the search facilities were also developed. After the invention of www, the cataloguing process has also been developed and now online cataloguing has become very popular. The advent of Mosaic web search browser in 1992 and the development of the Yahoo! Search engine in 1994 brought a revolution in web technology. Such technology had radically changed the functioning, behaviour and conduct of the library. Now (mid-1990s) the reach of resources had become wider and the availability of web-based resources through the Internet was universally accessible to the end-users (Barnes, 1997).

3.2. *e*-Resources: Its Development and Rapid Growth

Change is an eternal truth. Change is the law of nature. It is natural for all things to change humans have discovered. The next generation is much more advanced. Rapid growth in *e-resources* has also occurred. From clay to tablet is an example of this. The main point of development of electronic resources in the view of the researcher has been given below for better understanding:

A) **Indexing and Abstracting Services:** 1907, 1926, 1969, 1973 etc. are considered to be very important periods when indexing and abstracting services started to cover and provide important terminology related to various subjects to their users. for example, American Chemical Society (ACS), Library & Information Science Abstract (LISA), Science Citation Index (SCI), Social Science Citation Index (SSCI) etc.

B) **Cataloguing Services:** The second milestone in the history of the library & Information Science field was the development of the Machine-Readable Catalogue (MARC) in the 1960s by Henriette Avram of the Library of Congress. This was the tremendous transformation of the card catalogue into the MARC. By 1971, the US had accepted the MARC formats as a national standard in the dissemination of bibliographical data. Till now so many versionshave been released for MARC and currently is MARC 21. MARC is based on codes determined for different bibliographic fields. Now, UNIMARC has been accepted by all European countries and developed and maintained by the International Federation of Library Associations and Institutions (IFLA) (Library of Congress, Oct. 2023).

C) **Database Services:** In the mid-1960s, some databases began to be created. The first database which was popularly known as DIALOG software was developed in 1966 under the supervision of Roger K. Summit at Lockheed. This was the first step of resources moving towards an electronic version. After that another database named as ERIC (Educational Resources Information Centres) in 1967 was developed and later it was linked with DIALOG and by 1992, ERIC began its own electronic version of printing. ELHILL (1969) was created for medical subjects with the help of SDC (System Development Corporation) which was run by the National Library of Medicine and later renamed MEDLINE and PubMed (Ardito and Bjorner, 2003).

D) **Online Cataloguing Services:** Other major steps taken by the various organization in the form of a "shared cataloguing system" to move towards online/electronic resources are as follows:

i. **OCLC:** In 1967, the founders (Frederick G. Kilgour and others) met at the Ohio State University campus and decided to form a non-profit cooperative based shared cataloguing system and named it Ohio College Library Centres (OCLC). It was the first of its kind to provide computerized database to the member libraries. The bibliography, Abstract and Full Text information are the core functions of OCLC which provides to its member libraries. Now it has moved towards the OCLC Online Union Catalogue popularly known as 'WorldCat". It is also responsible for the maintenance and publication of DDC (Dewey Decimal Classification). OCLC also runs some community programme like "Web Junction™, Outside the Box™ and Geek the Library™" (Source: website).

ii. **OPAC:** "Online Public Access Catalogue is a catalogue of library's collection that is open and accessible to the public through the Internet. It is an online open-access bibliography of a library holding. A bibliography item can be any information entity (e.g. books, journals, computer files, graphics, maps etc.) that is considered library material. An OPAC is just an electronic version of manual card catalogue. An OPAC makes use of Z39.50 protocol for the search and retrieval of the bibliographical data in online database" (Umeaku and Aghauche, 2016, p.1). OPAC was the advanced version of OCLC, formed in 1975 at Ohio State University. By the 1980s, manual card catalogue were replacing by this system Online Cataloguing System i.e. OPAC (Moore, 1981).

iii. **Web OPAC:** Advanced version of OPAC form to broaden the widely accessible catalogue of libraries by using the Internet (www) beyond the boundaries.

E) **Serials Services:** "Sondak and Schwarz (1973) werethe first to conceptualize electronic scholarly journals. Senders, Anderson and Hecht (1975), Senders (1975, 1977), Roistacher (1978), and Lancaster (1978) were among the first to discuss possible characteristics of an online 'virtual' journal. The first experiment with a true scholarly journal was conducted by EIES (Electronic Information Exchange System) a division of the National Science Foundation in 1979 (TuroffandHiltz, 1982). Shortly afterward, in 1980, the British Library awarded a grant to Loughborough University to establish an experimental online journal in the area of computer-human factors (Shackel, 1991)" (Lancaster, 1995, p. 518-519). During the 1980s and 1990s, a lot of experiments were conducted to give the correct shape and wide accessibility of electronic publishing through newsletters, reports, graphics etc.

F) **WWW Services:** The advent of the World Wide Web (www) was a boon for electronic publishing. "The period of 1990 to 1995 might be called the 'Internet phase' because few academic journals published online during this period and a number of publishers announced to transfer their publishing strategy from offline to online (Hitchcock *et al.* 1996). For electronic publication, the standards related to web publishing were set in World Wide Web Consortia (W3C). For web publishing, there has been number of experiments like Elsevier's Tulip projects (Borguis *et al.* 1996), the American Chemical Society's core programme (Entlich *et al.* 1995), Red Sage projects (Lucier and Brantley, 1995) and CAJUN by Wiley (Smith *et al.* 1993) taken places and after the advent of Mosaic browser and Netscape, these publishing were widely accepted but not regularly. This period might be called 'Web Journals Phase" (Hitchcock *et al.* 1997, p. 286).

G) **Electronic Publishing (Books etc.):** The idea of electronic book publishing was first represented by Vennevar Bush in his article "As We May Think" in the year 1945. In 1969, Alan Kay was the person who came up with the idea of a small personal computer for publishing and recording published work. In 1971, Project Gutenberg was started by Michael Hart at the University of Illinois at Xerox Sigma V Mainframe laboratory. He was the person who created the world's first ebook library. It worked on ASCII (American Standard Code for Information Interchange). By 2001, this project contained 10000 electronic books. This was the early period of the electronic book publishing era (Laura & Robert, 2012, p. 294-296).

H) **CD/ROM Services:** CD-ROM technology was begun in 1989 by Bob Stein's company Voyager. This was the precursor of Voyager's Expanded Book Toolkit software. Electronic books can be created by using this software. By 1996, this technology had become very popular in the public domain (Laura & Robert, 2012, p. 297).

I) **Bibliographical Services:** According to Ulrich's International Periodical Directory, "There are currently 130,000 research journals published in the world". In 1965, Eugene Garfield of ISI (Institute for Scientific Information) started indexing services in the name of SCI (Science Citation Index). Its CD-ROM version was published in 1988.

After that, a number of indexing and abstracting services started and published as CD-ROM versions as well as some of them published online. E.g. SCI with Abstract (1992), SSCI (Social Science Citation Index- 1973), JCR (Journal Citation Report-1974), Current Content, Biological Abstract, Chemical Abstract, Index Medicus, LISA (Library & Information Science Abstract-1969) etc.

J) **Full-Text Data Services:** Full-text databases are now available online through several vendors. The journey began From DIALOG, LEXIX-NEXIS, Dow Jones to Full-text Sources Online (FSO) (https://www.infotoday.com), EBSCO Full text database, INGENTA, CSIR-Science Direct Consortia, J-Gate, INDEST Consortia etc. (*e*-Gyankosh,

K) **Web Interaction Services (Web 3.0):** Darcy Dinucci in 1999 and Tim O'Reilly and Dale Dougherty in 2004 coined the terms Web 1.0 and Web 2.0 in a conference. This technology allows the consumers to participate in the web techniques directly. That's why this technology is known as participatory web or social web. A semantic web was introduced in 2006 as Web 3.0 or Pedagogy 3.0. These technologies enable consumers to use and participate directly on the web and form several social media platforms. For e.g. Podcasting, Blogging, Social media, RSS, Social Bookmarking, Folksonomies, YouTube etc. Text, Video, Images, etc. can be shared without any expertise.

4. *e*-Resources: It's Types

"The resources which can be accessed via computer or mobile with the help of Internet, is called electronic resources or *e*-Resources"(Johnson *et al.* 2012, p. 3). All those resources available in the library which are obtained through computer and other electronic devices are called electronic resources. *e*-Resources are also known as Online Information Resources which covers Full Text Databases, Bibliographical Databases, Digital Data Collection, Search Engines and Electronic Reference Sources (NIOS, LCh.008, 2023). Electronic Resources (*e*-Resources) can be categorized as:

A) **Subscription Access based *e*-Resources**: The resources which can only be accessed through the academic libraries or information centres. These centres or libraries pay huge budget for such resources.

B) **Open Access based *e*-Resources**: The resources which are available on public domain and any one can access them without paying any cost for it. On the basis of this categorization, *e*-Resources can be divided into two types:

a) Online Electronic Resources

b) Offline Electronic Resources

a) Online Electronic Resources

The resources that can be accessed through Internet in any electronic devices such as a computer or mobile and stored, preserved, distributed or provided on the web technology are called online electronic resources. In Online Resources, connection of Internet is consider to be an important factor for preserving, storing, searching, accessing, browsing and publishing of resources on the web. The users are able to access these online resources at their door step from anywhere in the universe by simply putting their finger on computer or mobile keys. Online resources are known by many names such as Digital Resources, Virtual Resources, Internet Information Resources, Web Resources, ICT based Online Resources and so on. The popularity of Online Resources has increased a lot in present era. The whole world is proud of its usefulness and availability. Academic Libraries, renowned publishers and other community libraries prefer online resources because of its easy and round the clock access, cost-effectiveness and timeliness etc. Most of the online resources are originally born on the digital platform whereas few of them are scanned copies of the original printed documents (Verma and Pratima, 2014, p. 245). More than 2.5 million older books are digitized and available for free of cost on public domain by Internet Archives.

b) Offline Electronic Resources

Resources which are available in CD-ROM, DVD, Floppy disk or in some other storage devices and can be accessed through a computer or laptop are called offline electronic resources. Anyone can access them, use them and download and copy them only through the inserting of CDs or DVDs into a computer or laptop or other similar devices. There is no need of Internet connectivity to access them. Users can use them anywhere, anytime and at any place (Ramakrishna and Raj, 2017).

5. *e*-Resources: It's Kind

The electronic resources are available in various forms. Like various types of print resources, the electronic resources are also available in the form of *e*-Books, *e*-Journals, *e*-Reference, *e*-Newspaper, *e*-Thesis/Dissertations, *e*-Encyclopaedia, *e*-Dictionaries, *e*-Biographies, *e*-Bibliographies, OPAC, sound, recording, images etc. (NIOS, LCh. 008, 2023).

5.1. *e*-Books

"*E*-Book is also known as electronic book or digital book in which texts or images are available in digital format and one can read it online with the help of Internet or download and read it offline without the help of Internet" (NIOS, LCh. 008, 2023). The term *e*-Books can be defined as "Any piece of electronic text regardless of size or composition (a digital object) but excluding journal publication, made available electronically (or optically) for any device (hand-held or desk-bound) that includes a screen" (Armstrong, Edward and Lansdale, 2002, p. 217). The books which is available online in electronic version and that is accessed or read with the help of Kindle, Computer, Laptop or Mobile are called *e*-books. "Digital version of a traditional print book designed to be read on a personal computer or any other *e*-book reader" (Loan and Refhat, 2015, p. 305). The most important *e*-books supplier's websites are given below:

- Springer (https://www.link.springer.com)
- *e*-Brary (https://www.ebrary.com/corp/index.jsp)
- EBSCO (https://www.ebscohost.com/ebooks/home)
- MyiLibrary (https://www.myilibrary.com)

Some others the most famous ***e****-book publishing platforms are given below:*

- Smashwords (https://www.smashwords.com)
- Amazon Kindle Direct Publishing (KDP) (https://www.kdp.amazon.com)
- Draft2digital (https://www.draft2digital.com)
- Lulu (https://www.lulu.com)
- Kobo Writing Life (https://www.kobo.com)
- Barnes and Nobel Press (https://www.press.barnesandnobel.com)

As defined in Encyclopaedia of Library and Information Science (Vol. 59, p. 125), three types of *e*-Books are there: a.) Text *e*-Books which contains textual matter, b.) Picture *e*-Books which contains pictorial images, c.) Talking *e*-Books which contains audio and sound effect, d.) Multimedia *e*-Books, e.) Polymedia *e*-Books, f.) Moving picture *e*-Books.

5.2. *e*-Journals

"An electronic journal is a periodical publication published in electronic format, usually on the Internet"(SOAS website, 2023). *E*-Journals or electronic journals contains digital articles/research papers and provide digital content to its users through Internet. Lee and Boyle (2002) opined that "*e*-Journals is an electronic resource published as a serial, often duplicating a journal that already exist in print form, but also occasionally being born digital"(p. 128). Another definition of *e*-Journals is given in the study of Jones & Cook, (2000) that, "an *e*-Journal may not be all that different from a print journal in the fundamental editorial process"(p. 22). Jonson *et al.*(2012) also gave similar definition of *e*-Journals. According to them, "the *e*-journals can be accessed through

Internet, available in digital format and display on the screen of a computer, mobile or other electronic devices"(p. 17). *E*-Journals are distributed directly through publishers and tertiary body known as aggregators (Maxymuk, 2004). The electronic journals are available in the following forms:

i. **Online *e*-Journals**: Annals of Library and Information Studies, Ariadne, D-Lib Magazine, First Monday etc.

ii. **CD-ROM *e*-Journals**: ADONIS, BPO, AHEAD, LISA Plus, DIALOG, Ulrich's Plus, SCI etc.

iii. **Networked *e*-Journals**: Based on Mailing List Software for e.g. LISTSERV, Majordomo, SmartList and Listproc.

e-Journals are accessible in the following modes:

i. **Freely Accessible with Printed edition Journals**: Electronic journals or online journals are provided free of charge with subscription to the print edition of journals.

ii. **Extra charges with Printed edition Journals**: Vendors charged some additional cost for online journals with the subscription of printed edition of journals.

iii. **Charges for selective online journals**: Library or Information centres pays only for selective journals.

iv. **Consortia based subscription of online journals**: Due to higher rates of online journals, Library doesn't afford individual subscription of electronic journals. So consortia model has become very popular for academic libraries. In this way, N-List and *e*-ShodhSindhu of INFLIBNET gain popularity and colleges belong to Arts, Science and Commerce Faculty subscribes these consortia (Sudhier, 2016).

v. **Free of charge (Open Access) based Journals:** The concept of Open Access Journals has gain popularity in the present era. A lot of publishers, vendors, academic institutions, institutional repositories, JSTOR, DOAJ (Directory of Open Access Journals), Google Scholars etc. provide online journals without any charge. Consumers can access these online journals from anywhere at any place without paying any extra charges for it. These online journals work on the basis of no profit no loss (Sahoo, Birtia and Mohanty, 2017; Tanwar, 2017 and Pandita, 2013).

5.3. Website: An Authentic Source of Information

Websites are the core authentic platform for an organization or institution. Websites provide authentic information about the relevant bodies. It is an authentic source of information (Gupta and Walia, 2013). In the present study, most of the information about the colleges has been collected from their websites. Websites provide information about the relevant *e*-resources. For e.g. website of *e*-ShodhSindhu, N-List of INFLIBNET, EMRALD, Govt. of India, Dept. of Higher Education, Govt. of Uttar Pradesh, HESC (Higher Education Service Commission of Uttar Pradesh), MGKVP, Varanasi, VBSPU, Jaunpur, DDUGU, Gorakhpur, JNCU, Ballia, SU, Kapilvastu etc.

5.4. The Search Engines: As Internet Search Tools

The available search engines such as Google, Yahoo, Bing, Ask.com, Duck-Duck go etc. play an important role as an Internet search tool to find relevant information not only from the libraries but also beyond the boundaries of the libraries. "Internet search tools are very important part and parcel part of the Libraries" (Tatnall, 2005, p. 13). In this study, number of respondents prefer Google search engine for searching relevant information.

5.5. Web OPAC: A Bibliographical Tool for Library Users

In Library Automation projects, there is a computer based technique used in the Library to search for bibliographical details of library holdings similar to manual catalogue card, is known as "OPAC (Online Public Access Catalogue)". Users can search for their required information about books or other resources available in the library through OPAC. It is a computer based cataloguing system works on metadata and Z39.50 protocol (Umeaku&Aghauche, 2016). Now OPAC is available on the web and searchable for globally with the help of URL (Universal Resource Locator) and popularly known as Web OPAC. Major Web OPAC is WorldCat, GeoWeb, IndCat, INNOPAC, ALEPH, Voyager (*e*-Gyankosh, Unit 7, 2023, p. 154).

5.6. Subject Gateways

"Subject Gateways work similar to search engines on a specific subject. They have details information about the available resources of a specific subject. In subject gateways, users can search and access databases and indexes through the web-based interface" (*e*-Gyankosh, unit-7, 2023, p. 156). For examples (Cook, 1999 & Lispweb website, 2023):

- Intute https://www.intute.ac.uk/
- BUBL https://www.bubl.ac.uk/
- Alex https://www.infomotion.com/alex/
- AERADE https://aerade.cranfield.ac.uk/
- DMOZ https://dmoz.org/
- INFOMINE https://infomine.ucr.edu/

- Digital Librarian https://www.digital-librarian.com/
- QUEST https://www.re-quest.net/
- BioMedNet https://www.bmn.com/

5.7. Web Interaction (Web 2.0: a Tool for Social Media Application)

Credit for web interaction goes to Tim O'Reilly when he coined the term Web 2.0 with his colleague Dale Daugherty in 2004. Web 2.0 interoperability provides an opportunity to the end users to interact with and generate content on the web. This interaction increases the use of *e*-Resources at large scales. Web 2.0 gave a flurry of 2.0s not only academic arena but also in other marketing and government arena such as Library 2.0, Government 2.0, Classroom 2.0, Publishing 2.0 etc. (Zeinab *et al.*, 2011, p. 1-2). Some examples of Social Media are given below:

- **Blogging**: Library Blogs, Librarian.net, Weblogs, LisNews etc. are the examples of Blogs where academic or social issues are published.
- **Really Simple Syndication (RSS)**: It is a simple feed by which a user can keep themself up-to-date with his/her favourite social media, blogs, websites etc. The notification creates an XML files in computer readable format and add to the RSS feed in the web pages of the website.
- **Social Bookmarking:** "Social Bookmarking is the process of tagging a website page with a browser-based tool so that you can easily visit it again later. Some popular tools are Pinterest, Reddit, Digg etc." (Arens, 2019).
- **Wikis:** Wiki is a kind of database available in the form of a website. It is a hypertext publication where consumers are collaboratively able to add or edit academic notes/contents with the help of a web browser (IGI Global dictionary, 2023). For e.g. Wikipedia, Scholarpedia, Metapedia, Hitchwiki etc. These websites facilitate and provide tremendous information online to its users.
- **Podcasting:** "A podcast is a type of digital media, usually audio, that is available in a series of episodes or parts and is streamed or downloaded by the end user over the Internet. Podcasts can be made available via a release schedule or uploaded to the Web randomly. Podcast delivers not only audio but also other digital media such as video, *e*-Books and radio broadcasting. Podcast are used for everything from new and entertainment to education" (Technopedia, 2023, web page).
- **Tagging:** "Tag can be in the form of words, images or marks. They help search engines identify the content of pages online and using that

information, produce the best results for a given search. Tagging was popularized by websites associated with Web 2.0" (Technopedia, 2023, web page)

- **Social Networking (Facebook, Twitter, WhatsApp etc.):** "The term social networking refers to the use of Internet-based social media sites to stay connected with friends, family, colleagues, or customers. Social networking can have a social purpose, a business purpose or both through sites like Faceboook, Twitter, WhatsApp, Instagram, Pinterest" (Kenton, 2023).

6. *e*-Resources: It's Benefits for the Academic Institutions

Due to the global and round the clock access, usability and availability, the *e*-Resources has become very popular among the users of the world. The popularity of *e*-Resources is for the following reasons (Sharma, 2021):

i. ***Global Accessibility***: *e*-Resources can be accessed from anywhere at any place of the world through Internet. It is universally available for academic purposes.

ii. ***Limitless Speed***: *e*-Resources are available as much fast as they are published from the publishers and put on the website.

iii. ***Abundance of Content***: Vast variety of information about an article is available in *e*-Resource format like text with Video and Audio sometimes animation, doodle etc. User can search by any means.

iv. ***User Friendly Search Interface***: Flexible searching facility is available in *e*-Resources. Users can find their required information through searching by truncation, by its relevancy, by similar terminology, by search history, by thesaurus, by browsing; by BOOLEAN searching facility etc. these are not possible in print resources. Users are free to search/browse by their own way.

v. ***Time Saver***: Due to its fastest speed and universal availability, *e*-Resources are time saver of the users.

vi. ***Space Saver***: *e*-Resources save the physical space of a library. Huge amount of information can be stored in a very little space of a computer. *e*-Resources cannot be lost or misplaced while it is easily searchable.

vii. ***Cost-Effectiveness***: There is no need for serial binding, no need replace missing issues and most importantly single person can run this segment. Hence *e*-Resources are cost-effective rather than others.

viii. ***Hyperlink***: Readers have the hyperlink facility in *e*-Resources by which he can directly move from one valuable article to another without any extra searching efforts. There is no such facility in print resources.

ix. ***Archiving***: Archival facility is available for long term use in *e*-Resources. Content published online can be easily preserve and conserve for future reference in *e*-Resources.

x. ***Current and Nascent Information Facility***: Current and nascent information facility about the content is available in *e*-Resources. Updated information is instantly available on the web for users without any delay through RSS feed and other links.

xi. ***Multimedia Facility***: *e*-Resources have provided the facility of presenting Audio, Video, Animation, 3D images, Bars, Graphs etc. in the content. This technique has enhanced the teaching and learning process of faculty members as well as students.

The list of *e*-Resource benefits is very long. It's hard to mention all of these. But it can be said that *e*-Resource provides the best facilities to its readers.

7. *e*-Resources: Collection Development Policy

Collection development in a library refers to building its collection in a systematic manner with the need of the users. Khan (2010) explained in his study that, "A collection development process is that which permits the library to develop a collection of materials as per information needs and service requirement of users"(p. 16). Collection development also removes the weakness of the library. As Hadagali and Kumbar (2005) stated in his study that, "collection development is a systematic analysis of strength and weakness of the library and then create a strategy to remove the weakness and increase the strength of the library"(p. 73).

For the acquisition and selection of library resources, collection development policy plays an important role in fair selection. If library wants to build strong collection of its resources, there must be an active collection development policy. IFLA has formulated a collection development policy for the acquisition and selection of electronic resources. A report was published in 2012 in the name of "Key Issues for *e*-Resource Collection Development: A Guide for Libraries", in which it has been pointed out that, "policy should be used in conjunction with the more traditional collection development policy and not in isolation" (Johnson, Evensen, Gelfand, Lammers, Sipe&Zilper, 2012, p. 4-5).

7.1. *e*-Resources: It's Selection

Selection of electronic resource depends upon the nature and size of the Library. Library should select electronic resources based on the requirement and need of the users and budget of the library (Budd and Harloe, 1994). College Libraries should take extra precaution in selection of electronic resources because of its limited budget. Relevant and authentic resources which cover the users' need must be preferred (Gregory, 2000). For fair selection of electronic resources at fair price, selection tool must be consulted (Welch, 2002). The following criteria should be adopted for the fair selection of electronic resources (Johnson *et al.* 2012, p. 5-6):

a) Content : peer reviewed and reputed publishers
b) Technical Support
    - Method of Access
    - Authentication: Login IDs, IP filtering etc.
    - Compatibility: hardware and software
    - Browsing: accessed via different web browsers
    - Content format: PDF, HTML, ASCII etc.
c) Reliability and Functionality
    - Interface: User's friendly
    - Search and Retrieval: Flexible in searching viz. Boolean, Truncation etc.
    - Response, Reliability and Availability: Round the clock availability
    - Integration : Reference and Full Text linking
d) Vendor Support
    - User Training support
    - Technical and Customer Support
    - Data Security and Archiving
    - Bibliographic data provision
    - Statistical Report: ICOCL (International Coalition of Library Consortia), COUNTER (Counting Online Usage of Networked Electronic Resour-ces) code, guideline and reporting.
e) Supply
    - Pricing and purchase model
    - Cancellation Rights
    - Invoice
    - Renewal
f) Licensing
    - Termination
    - Refunds
    - Period of Agreement

- Liability of unauthorized use
- Governing Laws

(*Source:* IFLA, 2012- Guideline for selection of e-Resources, Johnson *et al.*)

7.2. *e*-Resources: It's Evaluation

For better quality content and to ascertain the strength and weakness of the current collection status of a library, whether it is print or non-print, should be evaluated regularly (Henry *et al.* 2008). Agee (2007) stated that "an evaluation provides qualitative and quantitative data for evaluating the usefulness and effectiveness of a library collection for successful collection analysis and assessment"(p. 16). Therefore evaluation is very necessary for a library before going to procure electronic resources. The following criteria should be adopted for evaluation of electronic resources:

1. **Costing Factor:** The first factor for evaluation of electronic resources in a library is its cost. The cost of *e*-Resources affects the selection criteria. Sometimes, those *e*-Resources also have to be selected whose cost is relatively high. But it has to take the approval of the higher authority. Davis &Bertot (2004) had also stated that, "selection of an electronic resource which has a high cost should not necessarily be immediately rejected, particularly if there are unique benefits that would proceed for the purchase such as official approval of an academic unit or a part of a particular consortium"(p. 30).
2. **Accessibility:** *e*-Resources accessibility must be there in terms of downloading of data, sharing of data and printing of data (Hua and Borin, 2008). Remote accessibility should also be evaluated while selecting *e*-resources.
3. **Publishers and Vendor determination and Technical and Customer Support:** Budd and Horloe (2008) had stated in this regard that, "reputed and well-known company who provide authentic, real, reliable and timely products and services with ready available user statistics and provide a quick response to the customer queries and give the technical support of the library"(p. 85).
4. **Licensing:** Budd and Horloe (2008) further explained that "the terms and conditions of the agreement between library and vendor or publishers should be clearly explained, straight forward and properly documented"(p. 86). Liability for the vendors should be defined.

The academic committee should thoroughly study about the electronic resources, vendors and publishers before going to subscribe the consortia. The resources available in the print edition in the library should not be subscribed, it increases duplication. Before subscribing the electronic resources for library, higher authority should negotiate with publishers (Joshipura, 2008, p. 48).

7.3. *e*-Resources: It's Acquisition

Acquisition of electronic resources is more complicated than print resources. License agreements, obtaining and verifying IP (Internet Protocol) address and filtering electronic resources are very complex process. This requires more attention from library authority. "Electronic resources are not without their cost. These include the cost of the documents (usually in the form of a license). These services are offered using a variety of pricing models – individual subscriptions, bundled subscriptions, joint print and *e*-Journals subscriptions, maximum number of users, and so on. The negotiations between library purchaser and licensor vendor can be time – consuming and complex"(Pinfield, 2001, p.2).

Acquisition of electronic resources is carried out in the following ways:

i. To Subscribe the *e*-Resources

Electronic resources can be purchased on the basis of subscription. In this process, the vendor or publisher fixes certain subscription fees on an annual basis and after paying the amount, a library or information centre is able to procure and access electronic resources. The electronic resources are available to subscribe in the following forms:

a) The same price is available for both types of journals – Print as well as Electronic. A lot of publishers offer this type of service viz. Wiley-Blackwell, Elsevier etc. The Library has to decide which will good for them.

b) Currently most publishers provide bundle subscription models of electronic resources. In this process, a huge range of electronic journals and other resources are available at affordable price. It is also good for the library. For example Science Direct (Elsevier), IEEE Xplore, IDEAL LIBRARY (Academic Press) etc.

c) Some publishers provide electronic resources in the form of an Aggregator model. They facilitate user interface search system with full text *e*-journals. For example JSTOR, EBSCOHost, LEXIS-NEXIS, Pro-Quest etc.

d) Some publishers provide a big deal on subscription of electronic resources. In this process, publishers provide a complete list of electronic resources with heavy discount but among these lists some are not relevant for the library. But if the Library settles with this deal, it will have to purchase this non-relevant item as well. For example, Science Direct of Elsevier Press.

e) Some publishers provide electronic resources on a "Print-plus-online content" and "Onlin*e*-plus-Print content" basis. The library authority has to choose one (Josipura, 2008).

ii. To purchase "Consortia"

Consortia have become very popular now a day. Each and every library and information centres are purchasing Consortia for their user due its affordable prices. So many organizations and agencies are working on it. In consortia, varieties of electronic resources are available in very reasonable price. For example N-LIST of INFLIBNET provides complete list of electronic resources. The best option is for a library to get consortia to provide electronic resources to its users.

iii. Utilization of Open Access *e*-Resources Available on Internet (Public Domain)

A huge amount of peer-reviewed and scholarly electronic journals and electronic books are available on public domain i.e. Internet. These are free for all. Anyone can access them. The library should provide users with a list of electronic resources available on Internet (Public Domain) along with their relevancy.

7.4. Difficulty Arising During Acquisition

Some problems arise during acquisition of electronic resources. The problems which are faced by the Librarian are as follows:

i. **Lack of Proper Collection Development Policies:** The first and most important problem related to acquisition of electronic resources is its collection development policies. There no well written and well maintain policies available for acquisition of electronic resources. In fact, there is a need of well-written collection development policy as given by IFLA, so that library authority can purchase electronic resources without any hesitation (Sambo *et al.* 2014).

ii. **High Rates and its Policy Issues:** The cost of electronic resources has badly affected the purchasing process. Due to the high price of *e*-Resources, the Library authorities think many times and then reach a conclusion. Library authorities also feel that it is useless to invest budget in *e*-Resources as they do not get anything in physical form (Wilkinson *et al.* 2015). Rhind-Tutt (2003) has stated that, "cost of *e*-Resources is not fixed. It is varies from library to library, publishers to publisher and negotiation to negotiation" (P. 52). Evans (1995) explained that "Negotiation is depend upon the Librarian that how he dealt with". Therefore, the pricing module of electronic resources depends upon the relationship between Librarian and Vendor or Publishers or Aggregators. So there is a need of standard pricing module regarding electronic resources.

iii. **Statistical Report Related Issues**: The vendors and publishers provide the statistical report of electronic resources published so far but are unable to provide electronic resources usage report.

iv. **Licensing Related Issues**: Farb and Riggio (2004) have stated in his study that "A license agreement is a contract between and intellectual property owner (licensor) and a potential user (licensee)" (P. 146). They further explained that, "Licensing becomes a challenging job for the library authorities who are dealing with obtaining a license from the respective publishers to use the *e*-Resources. But there are some burning issues related to internet access and certain classes or user issues associated with some Company License Agreements" (P. 146). "The library has a responsibility to try to negotiate agreements that respect the rights and privileges of users as well as the provider" (P. 147).

8. *e*-Resources: Issues and Challenges in Degree College

As the use of *e*-Resources is increasing across the academicians, it is a challenge before the colleges to serve qualitative content to its users. It is the responsibility of college libraries to provide better service to the faculty members and students of the colleges. They need qualitative content not only in physical format but also in electronic format. So, in this modern technological age, the responsibility of both the Librarian and the Libraries has increased a lot. Managing and organizing of *e*-content and *e*-resources, using Information and Communication Technologies in library, providing ICT enabled atmosphere among the users, helping layman for better utilization of *e*-resources, providing better technical searching facility etc. are the core area which need extra attention.

The following challenges are facing by the Librarian in this modern technological era while maintaining and organising the electronic resources in libraries:

i. **Complexity of *e*-Resources and its Explosion**: The first and most important challenge against to Library personnel is to find out the relevancy of electronic resources. There are explosion of electronic resources on the web. Which one is authentic and scholarly content is very complex for shorting. "It is unbelievable that about 2.5 quintillion bytes of digital data are created every single day (Piccially, Cuomo & Jeon, 2018; Zwitter&Hdfield, 2014; Devakunchari, 2014). Kemp (2021) stated that, "*e*-Resources have seen a huge growth during COVID-19 pandemic. Out of total number of digital content available worldwide today, about 90% of *e*-resources have grown in last two years". McNicol (2005) stated that, "Google Scholar supplies too many hits without any relevance to the researcher's field of interest because the metadata of electronic text such as *e*-Books or article in *e*-Journals has not been annotated by any information specialists"(p. 495). Academic libraries are facing this kind of problem to a great extent.

ii. **Growing Trend Towards *e*-Resources:** The craze of *e*-Resources has increased rapidly. Users are taking interest in reading electronic resources. But the practice of reading electronic resources varies from country to country. In our country, most of the people still prefer offline reading. They first download it then take out print and then read. Their words are "Offline padhane ka jo mazahai, vah online meinkahahai" (Online reading doesn't create interest). But now the trends of people are moving towards online reading whereas academician and others in western countries interested in reading content online. They use *e*-Book Reader, Laptop, Computer, Mobile etc. for reading online content.

iii. **ICT Skill Related Challenges:** Use of Information & Communication Technologies has brought a revolutionary change in the services of Library. ICT application has changed the whole scenario of Library infrastructure. Working nature of information professional has changed now (Ashcroft, 2004). "Information professional are now expected to be aware of and capable of using and demonstrating emerging ICTs" (Nwakanma, 2003). "There is a need for additional training to augment the traditional skills knowledge base with a competency in ICT use. Information professional must be flexible and adapt traditional skills to incorporate the requirements of technological advances" (Biddiscombe, 2001; Sharp, 2001). Quadri (2012) stated that, "Given the current situation, whereby ICTs are being continuously updated or introduced, and traditional formats are being replaced or supplemented by digital formats (such as *e*-Journals and *e*-Books), it seems likely that there will continue to be a need for regular training for information professionals" (p. 2).

iv. **ICT Infrastructure Challenges:** Most of the libraries of India are utilizing the application of ICT and providing the service on it but this is not the case for all libraries. Still they are facing various kinds of problem related to installation of ICT infrastructure. Akinjide and Oyeboade (2007) states that, "Although there are benefits in ICTs, there are challenges, for example, lack of skilled manpower, inadequate funding, poor maintenance of equipment, pwer and equipment failure, regulatory restriction of communication technologies, and expensive or unreliable technologies associated that obstruct the effective utilization of these technologies"(p. 35).

v. **Accessibility Challenges:** Okogwu (2019) states that, "The library faces many challenges regarding accessing the electronic resources in the form of user training programme, technical upgrading, IT skill manpower, financial constraints, perishable citation, URL (Uniform Resource Locator) changes problem, authentication problem etc. (Behera and Singh, 2011; Balangue, 2012). "Bandwidth problem in terms of slow speed of Internet and fluctuation of web in downloading resources"(Dhanavandra and Tamizhchelvan, 2012). Chisenga (2004) indicates that, "lack of fund, lack of ICT skilled personnel and erratic power supply is affected the accessing of electronic resources"(p. 7).

vi. **Technical Searching Capability Related Challenges:** Searching for information on Internet is an art. If users are not able to do so then he cannot access *e*-Resources. Users need training to search for technical phrases. It is like a challenge for most of the students studying in colleges. There are many students in degree colleges who are afraid to sit at the computer and prefer print resources over electronic resources. As mentioned by Asutay *et al.* (2011) "If a user lack searching skills, it means he or she is not be able to access and use of *e*-resources"(p. 153).

9. *e*-Resources: It's Promotional Strategies

Emphasis is now being laid on the use of *e*-Resources in Degree Colleges. Degree Colleges now also make sufficient budget provision for purchasing *e*-Resources. Round the clock availability of *e*-Resources attracts the users. But *e*-Resources are still consumed by a small percentage of students and academician. Widespread publicity is needed for widespread consumption of *e*-Resources. Many studies have concluded that although universities and colleges spending more budget on *e*-Resources, yet its consumption is very less. Most of the *e*-Resources are still unfamiliar to the users. Most of the electronic resources are underused. If the purchased electronic resources are not consumed by the users, understand that this is the wastage of money. Therefore, the colleges should make concrete arrangements for its widespread consumption. Awareness campaign will have to be conducted among users.

Identify good (active) electronic resource users and try to make others aware through them. For optimum utilization of *e*-Resources, Library Professionals should organise regular training, orientation and workshop programme, convey messages through social media, personally visit each department and guide for best use etc. Use of social media platform such as Blogs, Facebook, Wikis, RSS alert etc. is more interactive and effective tools. User's feedback also plays an important role in examining the actual status of the current usage. Traditional libraries are using some traditional methods such as A-Z list of Journals, Access through OPAC, Library Brochures, Library Posters and Banners, Library Newsletters, Exhibitions, Personnel visit, Training, workshops etc. But the existence of new modern library has changed the whole scenario of library, so library have to adapt some new way technologies to boost up the more and more utilization of library resources.

Some of the important modern technical factors that enhance the widespread access of *e*-resources are mentioned below:

i. **Adequate Availability of PC Configuration and Cyber Library**: In order to enhance the better use of electronic resources, the library must install updated computer with strong RAM, Processor Speed and Operating Systems. The library should also set up a Cyber Library to attract more users (Sheble and Jasper, 2005). In Degree College libraries, due to lack of budget, a very few computer is set up. Most of the users are using electronic resources by their own laptop or mobile. Few of them visit library. Therefore, Library should have all the facilities to provide better accessibility of electronic resources.

ii. **Well-developed Infrastructure and Bandwidth**: Slow Internet speed, Lack of adequate computers and printers, sitting facilities, opening hours are some of the factors that affect the use of electronic resources in number of degree college libraries. There should be feasibility of Internet connection with 100 Mbps at each node in Library, office, department, laboratory, hostels etc. so that information can be accessed seamlessly (Jackson *et al.* 2007).

iii. **Training Module for Information Retrieval Skills**: Technical searching required skills. If users do not have technical skills for retrieval of information on the web, they may suffer. In degree colleges, such a situation is often seen with all the students when they are unable to search the relevant information on the basis of exact word phrasing (Ozoemelem, 2009). Gui (2007) stated that a user should have the knowledge about to operate the computer, retrieve the exact data, use the Internet optimally, use the *e*-Mails and other networking application.

Search strategies also provide maximum possibilities for contextual data retrieval (Mabawonku and Ekenna, 2013).

iv. **Well Maintained Library Website**: Each library should have its own website. The resources available in the library should be displayed beautifully on it. The resources of library should be displayed in such a way that it can attract the people and make them feel that their visit will not go in vain. The link of Library website should be displayed on the homepage of main website of the college (Ndungu, 2016).

v. **By Using Web 2.0 Technologies**: Web 2.0 has become very essential tool of modern technologies. It has enhanced the web interaction. It has become the most favourite weapons of Tech-savvy. Library should adapt the following technologies to promote electronic resources.

   a) **Blogs**: In blogs, entries are arranged in reverse chronological orders. very popular web technology in the world. People use it just like a diary. Library should also adapt this technology. Library can use it for the announcement of library events, new arrival as review, special programme, available *e*-resources etc. This is a boon for library professionals, if they use. Although some renowned library of the world are using this technology.

   b) **RSS**: Really Simple Syndication is also a boon for tech-savvy. RSS is used for any changes, any new arrival, any new concepts, any new event or any new resources added to the parent organization. Library can use this technology for providing the Current Awareness Services, new arrival of *e*-Journals and *e*-Books, alert for reference services, favourite electronic resources, Subject information of electronic database etc. Many Publishers, Vendors of world are using this technology and provide RSS feed (Ngundu, 2016). For example Nature, American Chemical Society, LEXIS/NEXIS etc.

vi. **By Using SNS (Social Networking Sites)**: In 21st century, if any technology has become the most popular in the universe, then it is WhatsApp and Facebook. Library should take advantages of these technologies to promote the use of electronic resources.

   a) **WhatsApp Group**: WhatsApp allows a person to share and exchange free Text messages, Audio, Video and Images and chat and share status (Timeline status) with the person who is in his contact. Person has the facility of creating a group in which he can add more than 200 people. Library professionals can utilise the same and make groups of students and faculty members for providing current awareness services and others.

b) **Facebook**: Facebook is a social networking site platform where people join to each other and has feasibility to post their messages for the public. Library should create an account on Facebook make a group in the name of library keep informing users by their resources and new arrival and event.

vii. **By Using Web Technologies:**

a) ***E*-Mails**: it is a SMS alert service of the library. *E*-mail is used for providing latest subscription, reserving of books and others in advanced, providing current status of resources in the library. Library can use this service for promoting the use of electronic resources of the library.

b) **Online Bulletin Board Services**: Library can utilize the website of the parent organization and library website for providing latest updates of library in the form of Bulletin Board.

c) **Online Tutorials**: To train the user for online retrieval and accessing online electronic resources, library can upload audio and video of the tutorials on the website so that user can watch this and use accordingly. Some of the renowned publishers, vendors and libraries of the world are using this technology. For example, Pubmed central, Lexis/Nexis, Proquest etc.

10. *e*-Resources: It's Management and Security Issues

It is the ultimate duty of every Library to provide quick access and friendly search option to its users. Proper data management and data protection is essential for each library for quick access. Every hour thousands of documents are being published in digital format for carrying out academic activities. But a lot of challenges arise before the library management regarding their management and security. As Pradhan (2013) has stated that, "Keeping track of varying access conditions, changing licensing terms and conditions, subscription and renewal dates, authentications, resource sharing limitations, usages data collection etc. are some of the challenges that libraries have to cope up with managing *e*-resources. Therefore ERM (Electronic Resources Management) system is required in library"(Pradhan, 2013, p. 236). He further explained that, "an ERM system based on standards only has the power to integrate all management functions relating to *e*-resources such as subscription product names, descriptions, producer, license start and end dates, renewal alerts, images of contracts, price and payment terms, access URLs, user authentication, interface administrative URLs, verification access methods and restrictions, vendor contracts, contract history usage statistics and links to usage reports etc. into a single package. A good ERM relies increasingly on the effective uses of technology and adopts standardized procedures to sail over interoperability problem. Thus it is very crucial for the library to look for standardized tools when it is *re*-examining the architecture of its existing ILS to incorporate ERM system. For this purpose, a number of projects have been implemented for ERM system like DLF, ERMI, ONIX-PL, KBART, SHUSHI, COUNTER, CORE etc."(p. 237).

NISO (National Information Standard Organisation) is also playing a pioneer role in ERM system "Open URL linking is based on a standard known as ANSI/NISO Z39.88 standard which can improve the open knowledge and metadata"(NISO website, as retrieve on 20 Oct. 2023).

Millman (2010) has stated that, "Data or *e*-Resources can be easily copied, shared and downloaded. This is the violation of copyright act. So security of data is essential in this digital environment. This requires a high security management system"(p. 413). For security purpose, user authentication and authorization is essential in libraries. User authentication can be done in the following ways: By providing Login ID and password, publishers can control over the access of *e*-Resources. Data cannot be misused. Only authentic users can access resources. An Internet Protocol (IP) address is a secure point by which publishers or vendor and library are able to provide secure data to their users. IP address is of two typ*e*- IP V4 and IP V6. IP V4 has four parts separated by a dot (.) for example 203.52.304.35 and IP V6 has eight parts separated by a colon (:) for example 2401:4900:5ae1:265:6:55d3:44cf:4a83. Each electronic machine has an IP address and Internet connection is perform through this. IP address works as a locator (Upshall, 2014,). Some publishers provide cookies on its website for security reason. It is also a secure accessible point. HTTP Cookies use for user authentication. Some publishers use cookies for website access, if user rejects cookies, he will not be able to access resources of the website. Proxy server is used to provide access of electronic resources beyond the premises of the institutions. Users use the proxy server to access *e*-Resources. Proxy server works on combination. EZproxy provides licensed databases for remote access. Some other user authentication tools are used by the publishers and vendor for security reason. For example, Open Athens (http://www.openathens.net), Shibboleth (http://shibboleth.net/), Referring URL, Kerberos (https://web.mit.edu/kerberos/) etc. As Millman (2010) has described in his study that, "Authorization is more challenging than authentication, especially for widely distributed libraries for *e*-resources. Access control is one method for making obligatory authorization"(p. 417). Some policies regarding access control is mentioned below:

a) **MAC (Mandatory Access Control)**: This is the highest level of access control system. In this controlling system a high level security control system works to identify the object and users for data accessing, if object and users match, data can be accessed (Glenn and Millman, 1998). The given figure also explains about MAC (Hoffman, 2023)

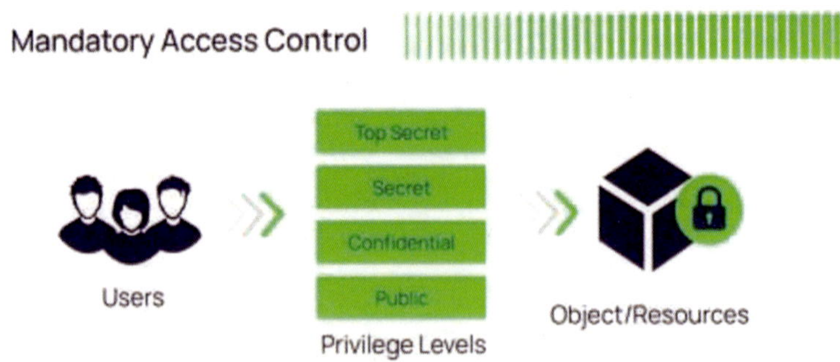

(*Source:* Hoffman, B (2023). https://delinea.com/blog/access-control-model-methods)

b) DAC (Discretionary Access Control): This access control system is based on authorization and identification of users. Users provide identity before accessing the *e*-resources. This can be understood through the figure (Hoffman, 2023).

(*Source:* Hoffman, B. (2023). https://delinea.com/blog/access-control-model-methods)

c) RBAC (Rol*e*-Based Access Control): In this controlling system, roles are assigned to users. Users are able to access *e*-resources on the basis of their roles. The figure explains about RBAC (Hoffman, 2023).

(*Source:* Hoffman (2023). https://delinea.com/blog/access-control-model-methods)

d) **DOI (Digital Object Identifier)**: An identity is created in the form of a character string to identify a particular electronic article available on the Internet, called Digital Object Identifier (DOI). "The Digital Object Identifier (DOI) was conceived as a generic framework for managing identification of content over digital networks, recognising the trend towards digital convergence and multimedia availability. Its key features include persistence, network accessibility, interoperability with other identifiers, shared fault-tolerant infrastructure, and the ability to resolve the identifier in multiple forms. DOI is standardized as ISO 26324. For e.g. 'https://doi.org/10.1109/5.771073' is a DOI link, if anyone click DOI link, it will take to one or more current URLs. If the URLs or services change over time, e.g. the resource moves, this same DOI will continue to resolve to the correct resources or services at their new locations"(DOI website, 2023).

e) **Digital Watermarking**: In this access control system, copyright message is associated with the electronic resources. Sometimes it is invisible. Users have to obtain permission to access these resources.

11. *e*-Resources: It's Usage Statistics

"Usage statistics provide information on the extent to which the *e*-Resources are used as well as, how and where it is being used. This data facilitates libraries to study the usage pattern of *e*-Resources over a period. The usage data also support and help the administrators and other library stakeholders in taking appropriate collection development decisions relating to *e*-Resources"(Sunitha, 2019,p. 47). "SUSHI and COUNTER are the most important tools to measure usage statistics of *e*-Resources. The adoption of COUNTER/SHUSHI compliant statistics helps the libraries in getting a more reliable data in the relatively simple form and these statistics have a significant place in revealing the real value of *e*-Resources in the academic and research libraries"(p. 47). A brief description about SUSHI and COUNTER is given below:

A) ANSI/NISO Z39.39-2014 SUSHI

a) **"SUSHI stands for:** Standard **U**sage **S**tatistics **H**arvesting **I**nitiatives. SUSHI is a protocol which works on ANSI/NISO Z39.39-2014 standard protocol and SOAP (Simple Object Access Protocol). Subsequent and easy harvesting of *e*-Resources is possible only through SUSHI protocol"(NISO website, 2023).

B) COUNTER

a) **COUNTER stands for: C**ounting **O**nline **U**sage of **N**etworked **E**lectronic **R**esources. COUNTER works with SUSHI of NISO (National Information Standard Organisation). It provides usages statistical report of different database like Journals, Books, Reference Books etc. through vendors and publishers (NISO website, 2023).

Both standard protocols is mostly used for university and research library. College libraries are rarely using this protocol. To be honest, most of the librarians of degree colleges are not aware about these protocols. But observant librarians solicit *e*-Resources usage statistical report from vendors and publishers and use it for collection development policy. In this study, when the researcher tries to find out the usage statistical report, no one provides such report. They showed ignorance of such information.

12. *e*-Resources: Role of Library Professionals

The use of modern ICT application in today's era has completely changed the entire landscape of libraries. The role of Librarian and Library professionals has completely changed according to the way the electronic resources are being used by the users. Rising above the orthodox ideas and fantasies, the operations of libraries equipped with modern new technical equipment is now in the hands of new age librarian. "New age library is known by its various names such as automated library, computerized library, electronic library, digital library, cybrary, virtual library, library without walls, transformed library and so on. All these libraries are termed as new generation libraries"(Kaur and Sharma, 2018, p. 131). The way information is being published in digital format by publishers and vendors and students and faculty members are using it online, accordingly the role of librarian has increased a lot. "In this situation, librarian became a predominantly online worker, supporting the citizen/worker by selling services. Finding relevant information faster than the competitors, faster than a non-information – worker can find it, and surviving on the basis of superior knowledge of the networks and digital information resources available through them. In an age of great change in information format, delivery models and technologies, and important new role emerge for the librarian. So, presently a librarian is called as s digital librarian, digital information professionals, cybrarian, and information broker etc."(p.133-134). A new age librarian should possess some important skills such as communication skills, flexibility skills, adaptability skills, innovative skills, decision making skills, presentation skills, time management skills, relationship skills, resource management skills, local and global thinking skills, planning skills, financial management skills, leadership skills, information technology skills (Web based services, content management, Internet use, Database creation, library automation, digitalization, hardware and software networking skills), Information literacy, *e*-resource management skills, knowledge management skills etc. (Janugade, 2016).

13. *e*-Resources: It's Online Readers

Information Seeking Behaviour of users has now been changed due to the emergence of modern tools and technologies applied in the life of each and every person of the universe. The growing demand and universal availability of online resources are making it more interesting. That's why techno savvy users like it more. Techno savvy users download online resources in the form of PDF, Word, Excel and HTML format to their Laptop, Computers, Tablet and Mobile etc. and use it and read it accordingly. Libraries of Universities and Colleges are also managing the same for their faculty members and students. As mention by Shewale (2018), "in India 26% readers read no books, 38% readers read printed books, 28% readers read both printed and digital books and only 6% readers read only digital books". And Rainie, Zickuhr, Purcell, Madden and Brenner (2012) have mentioned that the percentage of online readers is increasing day by day. According to them, "21% online readers read an *e*-book in a day, 42% readers read on Computer, 41% on Kindles and Nooks, 29% on Cell Phones and 23% on Tablets"(p. 5). Online forum has also become very popular nowadays. Google classroom, WhatsApp group, Facebook etc. are the forum platform on which academicians discuss, chat and express their thoughts (Mark and Ahmad, 2016). Kindle, Nook, Kobo etc. are some of the examples of *e*-book reader which are popularly used by most of the academicians.

14. Conclusion

The main objective of the college libraries is to make available the right source of knowledge at the right time to the right students of the college so that it can be utilized within the session. Students of the college must be satisfied. No student should be deprived of the resources available in the library. This is the era of modern tools and technologies. Students have also upgraded themselves towards the use of modern technologies. Library of degree colleges are also moving towards adapting the new way of technologies in sake of modern information provider techniques. Tremendous changes have been made in traditional way of service system of college libraries. The image of only issue and return of books has been changed now. College libraries are also giving the challenges to the University libraries and other special libraries. Most of the college libraries have installed library software and provide electronic resources to their users by taking the membership of electronic Consortia and others. Government is also providing their full support to upgrade the library resources and services by applying ICT tools and techniques in the library. However, awareness regarding use of electronic resources is needed among college students. Library professionals have to be very attentive in this regard. RSS alerts, *e*-Mails, Newsletter, Poster, Banner, Training Programme, workshops, *e*-Circular, *e*-Brochures etc. should circulate among the students so that they become aware of electronic resources. Facebook, Twitter, WhatsApp, Blogs, *etc*. can be used for awareness of electronic resources and create an interactive environment and user-friendly technology for the greatest use of electronic library resources. Students and Faculty members of college can happily move towards the library because of this interactive web technology. Today such technology is required because now publishers and vendors are serving their resource in electronic format as well. Government of India and other countries have also launched a lot of digital platform for the faculty members and students to enhance the academic activities to the world class standard.

References

Akinjide, K. O., & Oyebode, A. S. (2007). Academic staff utilization of information and communication technologies (ICTs) in Nigerian Tertiary institutions: A study of ten selected Universities. Gateway Library Journal, 10(1), 33-42. Retrieve from http://ir.library.ui.edu.ng/bitstream/123456789/5169/1/ui_art_aramide_academic_2007.pdf

Akinola, A.O.; Shorunke, O.A.; Ajayi, S.A.; Odefadehan, O.O.; & Ibikunle, F.L. (2018). Awareness and use of electronic databases by postgraduates in the University of Ibadan. Library Philosophy and Practice (e-journal). Retrieve from http://digitalcommons.unl.edu/libphilprac/2065

Anglo-American Cataloguing Rules. (2nd Ed.). (2002). New York: American Library Association.

Ani, O.E., & Ahiauzu, B. (2008). Towards effective development of electronic information resources in Nigerian University Libraries. Library Management 29(6/7), 504 – 514. Retrieve from http://digitalcommons.unl.edu/libphilprac/2065

Annapurne, S.G. (2019). Use of Electronic Resources by Arts, Commerce, Science Libraries: A Study (Doctoral Thesis). Retrieve from https://shodhganga.inflibnet.ac.in/handle/10603/286564

Arens, E. (2019, February 19). What is Social Bookmarking? (Web logs) Retrieve from https://sproutsocial.com/insights/social-bookmarking/

Armstrong, C., Edwards, L., & Lonsdale, R., (2002). Virtually There? E-books in UK Academic Libraries. Program—Electronic Library and Information Systems 36(4), 216–27. Retrieve from https://crl.acrl.org/index.php/crl/article/view/15842/17288

Arota, J. (2001). Web-Based Digital Resources and Services: Trends and Innovations. Retrieve from https://ir.inflibnet.ac.in/bitstream/1944/105/1/cali_24.pdf

Ashcroft, L. (2004). Developing Competencies, Critical Analysis and Personal Transferable Skills in Future Information Professionals. Library Review, 53 (2), 82-88.

Asutay, A., Efe, K. & Lakhotia, A. (2011). Exploring web neighbours in exploratory search. International Journal of Web Information Systems, 7(2), 152-182. Retrieve from https://www.researchgate.net/publication/220135798

Barnes, J. H. (1997). One giant leap, one small step: continuing the migration to electronic journals.Library Trends, 45 (3). Retrieve from https://link.gale.com/apps/doc/A19192263/AONE?u=googlescholar&sid=googleScholar&xid=3f43d38b

Barnes, J. H. (1997). One giant leap, one small step: continuing the migration to electronic journals. Retrieved from doi:www.ideals.illlinios.edu/bitstream/handle/2142/8106

Biddiscombe, R. (2001). The development of information professionals' needs for Internet and IT skills: Experiences at the University of Birmingham. Program, 35(2), 157-66.

Bjorner, S., & Ardito, S. C. (2003). Online before the Internet, Early Pioneers tell their Stories, Part 2: Growth of the Online Industry. Searcher, 11(6). Retrieve from https://www.infotoday.com/searcher/jun03/ardito_bjorner.shtml

Budd, J.M., & Harloe, B. (1994). Collection Development and Scholarly Communication in the Era of Electronic Access. The Journal of AcademicLibrarianship, 20(2), 83-87.

Cook, A. (1999). A Guide to Finding Quality Information on the Internet. Library Association Publications

Devakunchari, R. (2014). Analysis on Big Data over the Years. International Journal of Scientific and Research Publications, 4(1), 383-389. Retrieve from http://citeseerx.ist.psu.edu/viewdoc/download?doi=10.1.1.429.3175&rep=rep1&type= pdf#page=384

DOAB (2023). Retrieve from https://www.doabooks.org/

DOAJ (2023). Retrieve from https://doaj.org/

DOAR (2023). Retrieve from https://v2.sherpa.ac.uk/opendoar/

DOI (2023). Digital Object Identifier. Retrieve from https://www.doi.org/

e-Content, U.P. (2023). Retrieve from https://heecontent.upsdc.gov.in/

Egyankosh (2023). Bibliographic Full Text Services, Unit-9. Retrieve from https://www.egyankosh.ac.in/bitstream/123456789/76919/1/Unit-9.pdf

e-Gyankosh (2023). Retrieve from https://egyankosh.ac.in/

eGyankosh (2023). Retrieve from https://egyankosh.ac.in/bitstream/123456789/34888/1/Unit-7.pdf

e-PG PATHSHALA (2023). Retrieve from https://epgp.inflibnet.ac.in/

ERIC (2023). Retrieve from https://eric.ed.gov/pdf/ERIC_Retrospective.pdf

Evans, G. E. (1995). Developing library and information center collections. Library science text series. California: Libraries Unlimited, Inc. Englewood.

Farb, S. E., & Riggio, A. (2004). Medium or message? A New Look at Standards, Structures And Schemata For Managing Electronic Resources. Library HiTech, 22(2), 144-152. Retrieve from doi:10.1108/07378830410524576/

Ganaie, S. A., Jan, S., Fayaz, A. L., & Refhat-un-nisa. (2014). Current Trends of the Open Access Digital Repositories in Library and Information Science. International Journal of Information Dissemination and Technology, 4(4), 278-282. Retrieved from https://www.researchgate.net/publication/275189947

Glenn, A., & Millman, D. (1998). Access Management of Web-bases Services: An Incremental Approach to Cross-organizational Authentication and Authorization. D-Lib Magazine. Retrieve from https://www.dlib.org/dlib/september98/millman/09millman.html

Gregory, V. L. (2000). Selecting and Managing Electronic Resources. New York: NealSchuman Publishers.

Gui, M. (2007). Formal and Substantial Internet Information Skills: The role of socio–demographic differences on the possession of different components of digital literacy. First Monday, 12(9). Retrieve from https://:boa.unimib.it/handle/10281/10863

Gupta, M., & Walia, P.K. (2013). Usability analysis of homepage of websites of national libraries in Asia. Retrieved from https://:digitalcommons.unl.edu/cgi/article=2335/

Hadagali, G.S., & Kumbar, B. D. (2005). Collection Development in the Electronic Environment: Challenges before Library Professionals. Retrieved from https://www.researchgate.net/profile/Basavantappa_Kumbar/publication/228718554_

Haneefa, K., & Parameswaran, A. (2018). Library consortia in India. In Bavakutty, B. and Abdul Azeez, T.A. (Ed.), ICT Application in Academic Library Managament, (pp. 127-50). Retrieve from https://www.researchgate.net/publication/322599485

Hitchcock, S., Carr, L., & Hall, W. (1996). A Survey of STM Online Journals: the Calm before the Storm. In D. Mogge (Ed.), Directory of Electronic Journals, Newsletters and Academic Discussion Lists, (pp. 7-32). Washington, D.C.: Association of Research Libraries

Hitchcock, S., Carr, L., & Hall, W. (1997). Web Journals Publishing: A UK Perspective. Serials, 10(3). Retrieve from https://serials.uksg.org/articles/10.1629/10285/galley/341/download/

Hoffman, B. (2023, October 20). Access control: Model and methods (Web log message). Retrieve from https://delinea.com/blog/access-control-model-methods

Howtharne, D., & Watson, J. (2019). Electronic resource management systems: Alternative solutions. Retrieve from https://cdn.ifla.org/wp-content/uploads/2019/05/assets/serials-and-continuing-resources/Conferences/hawthorne.pdf

IFLA. (2015). Freedom of access to information and freedom of expression, libraries and intellectual freedom. Retrieved from http://www.ifla.org/faife/faife/present.htm>

IGI Global Dictionary (2023). Wikis. Retrieve from https://www.igi-global.com/dictionary/cybercells-integration-actual-virtual-groups/32586

Indexing and Abstracting Periodicals (Unit-4). Retrieve from http://egyankosh.ac.in/bitstream/123456789/33138/1/Unit-4.pdf

Jackson, M., Gider, A., Feather, C., Smith, K., Fry, A., Brooks-Kieffer, J., & Nelson, R. (2007). Electronic resources & libraries, 2nd annual conference 2007. Library Hi Tech News, 24(4), 4-17. doi:10.1108/07419050710778473/

Janugade, A.V. (2016). Changing role of librarian in ICT Era: A Study. Knowledge Librarian" An International Peer Reviewed Bilingual E-Journal of Library and Information Science, 3(5), 01-05. Retrieve from http://www.klibjlis.com/3.5.1.pdf

Johnson, S., Evensen, O. G., Gelfand, J., Lammers, G., Sipe, L., & Zilper, N. (2012). Key issues for e-resource collection development: a guide for libraries.IFLA.Retrieve from https://www.ifla.org/files/assets/acquisition-collection-development/publications/IFLA_ELECTRONIC_RESOURCE_GUIDE_FINAL_May2nd%5B1%5D.pdf

Jones, S. L., & Cook, C. B. (2000). Electronic journals: are they a paradigm shift. Online Journal of Issues in Nursing, 5(1), 1. Retrieved from http://ojin.nursingworld.org/Contents/Volume52000

Joshipura, S. (2008). Selecting, acquiring and renewing electronic resources. In Electronic resource management in libraries: Research and Practice (pp. 48-70). IGI Global.

Kaur, R., & Sharma, A.K. (2018). 21st Century Library Professionals in Dynamic Role in Digital Era. Library Progress (International), 38(1), 129-139. Retrieve from https://bpasjournals.com/admin/upload/dynamic2/13-Lib-67-2018%20Review%20P_129-139_.pdf

Kemp, S. (2021). Digital 2021: Global overview report. Retrieved from https://datareportal.com/reports/digital-2021-global-overview-report

Kenton, W. (2023, September 15). Social networking. (Web logs). Retrieve from https://www.investopedia.com/terms/s/social-networking.asp#

Khan, A. M. (2010). Managing collection development and organization in globalizing Indian university libraries. Collection Building, 29(1), 15-21. Retrieve from Doi:10.1108/01604951011015259

Krubu, D.E., & Osawaru, K.E. (2011). The Impact of Information and Communication Technology (ICT) in Nigerian University Libraries. Library Philosophy and Practices (e-journal), 583.Retrieve from https://digitalcommons.unl.edu/libphilprac/583

Kumar, K. P. (2011, Oct- Dec.). Information Communication Technology and E-Resources: Impact on Academic Libraries. PEARL- A Journal of Library and Information Science, 5(4), p. 18-26. Retrieved from www.IndianJournals.com

Lancaster, F. W. (1978). Toward Paperless Information Systems. New York: Academic Press.

Lancaster, F.W. (1995). Evolution of Electronic Publishing. Library Trends, 43(4), 518-27. Retrieve from https://core.ac.uk/download/pdf/4817317.pdf

Laura Manley, & Robert P. Holley (2012) History of the E-book: The Changing Face of Books. Technical Services Quarterly, 29(4), 292-311. DOI: 10.1080/07317131.2012.705731

Lee, S. D., & Boyle, F. (2002). Building an electronic resource collection: a practical guide (pp. 41-61). London: Library Association Publishing.

Loan, F. A., & Refhat-un-nisa. (2015). Open Access E-books in Science and Technology: A Case Study of Directory of Open Access Books. DESIDOC Journal of Library and Information Technology, 35(4), 304-309. Retrieved from www.desidocjounals.com

Mabawonku, I. & Ekenna, M. (2013). Information Retrieval Skills and Use of Library Electronic Resources by University Undergraduates in Nigeria. In Information and Knowledge Management 3(9), 6-14. Retrieve from doi:10.1.1.856.4155/

Mark, B., & Ahmad, P. (2016). E-book User Behaviour in Academic Libraries: The Role of User Agents in Perception and Satisfaction. Malasian Journal of Library and Information Science, 21(3), 95-109. Retrieved from www.nlistinflibnet.ac.in.

Maxymuk, J. (2004). Electronic journals redux. The Bottom Line,17(2),72-74. doi:10.1108/08880450410536116

McNicol, S. (2005). The challenges of strategic planning in academic libraries. New Library World, 106 (11/12), 496-509. Retrieve from doi:10.1108/03074800510634982/

Millman, D. (2010). Authentication and authorization. Encyclopaedia of library and information science. Marcia, B. & Niles, M., M. (Eds.) (3rd Ed.).Vol. I. Baco Raton: CRC Press.

Naqvi, T.H. (2012). Use of Electronic Databases by Postgraduage Students and Research Scholars at GBPUAT Library, India. Library Philosophy and Practice (e-journal). https://digitalcommons.unl.edu/libphilprac/809

NDL (National Digital Library, 2023). Retrieve from https://ndl.iitkgp.ac.in/

Ndungu, M. W. (2016). Promotion of Electronic Resources in Academic Libraries on a Minimal Budget. International Information & Library Review, 48(2), 94-101. Retrieve from doi: 10.1080/10572317.2016.1176449

NIOS, LCh. 008 (2023). Retrieve from https://nios.ac.in/media/documents/SrSecLibrary/LCh-008.pdf

NISO (2023). ANSI/NISO standard. Retrieve from https://www.niso.org/

NISO, (2023). SUSHI and COUNTER. Retrieve from https://www.niso.org/niso-io/2015/09/brief-history-counter-and-sushi

N-List (2023). Retrieve from https://nlist.inflibnet.ac.in/veresources.php

N-LIST (2023). Retrieve from https://nlist.inflibnet.ac.in/vstatus.php

NROER (2023). Retrieve from http://scert.cg.gov.in/ict2014/pdf/NROER.pdf

NSDL (2023). Retrieve from http://nsdl.niscair.res.in/

Nwakanma, C.D. (2003). Information technology competencies: Identifying knowledge depths for library and information science education. Canadian Journal of Information and Library Science, 27(4), 75-6.

OCLC (2023). Retrieve from https://www.oclc.org/en/home.html

OhoLINK (2023). Retrieve from https://www.ohiolink.edu/

Okogwu, F. I. (2019). Difficulties of accessing electronic resources among postgraduate library users in university libraries in Southeast Nigeria. Library Philosophy and Practice (e-journal). 2425. Retrieve from https://digitalcommons.unl.edu/libphilprac/2425

Ozoemelem, O. A. (2009). Use of electronic resources by postgraduate students of the department of library and information science of Delta State University, Abraka, Nigeria. Library Philosophy and Practice. Retrieve from doi:digitalcommons.unl.edu/libphilprac/301/

Pandita, R. (2013). Open Access Publishing in India: An analysis of Directory of Open Access Journals (DOAJ). International Journal of Information Dissemination and Technology, 3(3), 176-183. Retrieved from www.IndianJournals.com

Piccialli, F., Cuomo, S., & Jeon, G. (2018). Parallel approaches for data mining in the Internet of Things Realm. International Journal of Parallel Programming, 46, 807–811. Retrieve from https://doi.org/10.1007/s10766-018-0565-y

Pilgrim, M., & Dolabaille, A. (2011). Our journey from print to electronic resources: An acquisitions Perspective at a Caribbean Academic Library. The Serials Librarian, 61, p. 90-104. Retrieve from doi:10.1080/0361526X.2011.582618

Pinfield, S. (2001). Managing electronic library services: Current issues in UK higher education institutions, Ariadne 29. Retrieve from http://www.ariadne.ac.uk/issue29/pinfield/

Pradhan, S. (2013). Developing standards for electronic resource management. International Journal of Information Dissemination and Technology, 3(4), 236-238. Retrieve from http://ijidt.com/index.php/ijidt/article/download/3.4.2/pdf

Prytherch, R. (2016). Retrieve form https://www.routledge.com/Harrods-Librarians-Glossary-and-Reference-Book-A-Directory-of-Over10200/Prytherch/p/book/978081 5399636

Quadri, G. O. (2012). Impact of ICT skills on the use of e-resources by information professionals: A review of related literature. Library Philosophy and Practice (e-journal). Retrieve from https://www.researchgate.net/publication/280049138

Rainie, L.; Zickuhr, K.; Purcell, K.; Madden, M., & Brenner, J. (2018). The rise of e-Reading. A project for Pew research centre. Retrieve from http://libraries.pewinternet.org/2012/04/04/the-rise-of-e-reading/

Ramakrishna, & Raj, R. R. (2017). Use of E-resources and services by the post graduate students of various disciplines in Bangalore: A cross sectional study. Library Progress (International), 37(2), 326-335. Retrieve from doi:10.5958/2320-317X.2017.00031.9.

Reddy, N. A., & Aswath, L. (2016). E-books: A paradigm shift and legal issues. PEARL-A Journal of Library and Information Science, 10(4), p. 261-265. Retrieve from doi:10.5958/0975-6922.2016.00036.X

Reitz, J.M. (2023). Database. ODLIS (Online Dictionary of Library and Information Science. Retrieve from https://products.abc-clio.com/ODLIS/odlis_d

Rhind-Tutt, S. (2003). Pricing models for electronic products–as tangled as ever? The Charleston Advisor, 4(4), 52-56. Retrieved from https://docserver.ingentaconnect.com152

Roistacher, R. C. (1978). TheVirtual Journal. Computer Networks, 2(1), 18-24.

Sahoo, D.R., & Sharma, D, (2015). Impact of Internet on library and information services. International Journals of Innovative Science, Engineering and Technology, 2(4), 515-522. Retrieve from https://ijiset.com/vol2/v2s4/IJISET_V2_I4_74.pdf

Sahoo, J., Birtia, T., & Mohanty, B. (2017). Open Access Journals in Library and Information Science: A Study on DOAJ. International Journal of Information Dissemination and Technology, 7(2), 116-119. Retrieved from www.IndianJournals.com

Saini, A. (2017). Library consortia: An overview. International Journals of Digital Library Services, 7(4), p. 119-123. Retrieve from http://www.ijodls.in/uploads/3/6/0/3/3603729/12ijdosl174.pdf

Sambo, A. S., Abu-udenyi, H., Urhefe, E. A., &Yakubu, M. D. (2014). Collection development policy of e-Resources in Nigeria libraries: Certified librarians perception. Information and Knowledge Management, 4(9), 25-34. Retrieved from https://www.researchgate.net/publication/304681991_

Senders, J. W.; Anderson, C. M. B., & Hecht, C. D. (1975). Scientific publications systems: An analysis of past, present and future methods of scientific communication. Toronto, Ontario, Canada: Toronto University. (NTIS No. PB 242 259)

Senders, J. W. (1977). An on-line scientific journal. Information Scientist, 11(1), 3-9.

Shackel, B. (1991). BLEND-9: Overview and appraisal. Boston Spa, England: British Library

Sharma, N. (2021). Use of e-Resources by faculty members and students of Kurukshetra University and Maharshi Dayanand University in Social Sciences. (Unpublished Master's Thesis). University of Delhi, New Delhi, India.

Sharp, K. (2001). Internet librarianship: Traditional roles in a new environment. IFLA Journal, 27 (2), 78-81.

Sheble, L., & Jasper, R.P. (2005). Evolutionary approach to managing e-Resources. The Serials Librarian, 47(4), 55-70. doi: 10.1300/J123v47n04_07

Shewale, N. (2018). Reading habits: Current trends. Retrieve from https://dspace.gipe.ac.in/xmlui/bitstream/handle/10973/47456/2017-01-20-BGC-Presentation.pdf?sequence=1&isAllowed=y

Shukla, P. (2020). Open educational resources: Initiatives in India. Library Philosophy and Practices (e-journals), 3876. Retrieve from https://digitalcommons.unl.edu/libphilprac/3876

Singh, M., & Kumar, A. (2013, April-June). Digital resources and preservation: An overview. International Journal of Information Dissemination and Technology, 3(2), p. 132-135. Retrieved from www.IndianJournals.com

Sinha, M. K. (2017). ICT and Internet literacy skills for accessing to E-Resources available under N-LIST Programme: A case study of college library users of Barak Valley, South Assam. Retrieved from www.inflibnetnlist/ebsco host

SOAS, University of London, (2023). Retrieve from https://www.soas.ac.uk/research/library/electronic-resources/electronic-journals/what-electronic-journal

Sondak, N. E., & Schwartz, R. J. (1973). The paperless journal. Chemical Engineering Progress, 69(1), 82-83.

Subject Gateways (2023). Retrieve from https://lispweb.wordpress.com/subject-gateways/

Sudhier, K. (2016). Awareness and use of E-Journals by the students of TKM College of Engineering, Kollam, Kerala: A case study. Journal of Knowledge and Communication Management, 6(1), 43-58. doi:10.5958/2277-7946.2016.00004.8

Sunitha, T. (2019). E-Resources usage statistics as an indicator and measure of knowledge consumption in higher learning academic institutions. Asian Journal of Information Science and Technology, 9(S1), 46-50. Retrieve from https://doi.org/10.51983/ajist-2019.9.S1.223

SWYAM (2023). Retrieve from https://swayam.gov.in/about

Tanwar, S. S., & Tanwar, N. (2017). Electronic journals: Changing face of research and scholarly communication. PEARL- A Journal of Library and Information Science, 11(3), 287-294. doi:10.5958/0975-6922.2017.00038.9

Tatnall, A. (Ed.). (2005). Web portals: the new gateways to internet information and services. IGI Global.

Technopedia (2023). Podcast. Retrieve from https://www.techopedia.com/definition/5546/podcast

Trawick, B.W., & McEntyre, J.R. (n.d.). Bibliographic databases. Retrieve from https://global.oup.com/booksites/content/019963887x/13183238.

Turoff, M., & Hiltz, S. R. (1982). The electronic journal: A progress report. Journal of the American Society for Information Science, 33(4), 195-202.

UGC-INFONET (2023). Retrieve from https://www.inflibnet.ac.in/

Umeaku, C.F., & Aghauche, E.E. (2016). Online Public Access Catalogue (OPAC). Retrieve from https://www.researchgate.net/publication/317345059

Understanding MARC, (Oct. 2023). Retrieve from https://www.loc.gov/marc/umb/

Upshall, M. (2014). Content licensing: buying and selling digital resources. Retrieve from https://www.researchgate.net/publication/308590038

Uzuegbu C.P., Chukwu, C.O., & Ibegwam, A. (2012). Creating universal resource locator links desktop: A panacea for students' utilization of subscribed electronic databases inacademic institutions in Nigeria. Annals of Library and Information Studies 59, 97-105. Retrieve from http://digitalcommons.unl.edu/libphilprac/2065

Varma, M. K., & Rajiv, P. (2014). Shifting paradigm from books to E-book in academic library: A boon for users and librarians. International Journal of Information Dissemination and Technology, 4(3), 245-251. Retrieved from www.IndianJournals.com

Walsleben, M. (2019). Specialized database. Retrieve from https://www.ttuhsc.edu/libraries/documents/information-literacy/specialized-databases.pdf

Welch, J. M. (2002). Hey! What about us?! Changing roles of subject specialists and reference librarians in the age of electronic resources. Serials Review, 28(4), 283-286. Retrieve from doi:10.1080/00987913.2002.10764760/

Wikipedia (2023). List of library consortia. Retrieve from https://en.wikipedia.org/wiki/List_of_library_consortia

Wilkinson, F. C., Lewis, L. K., & Lubas, R. L. (2015). The complete guide to acquisitions management. California: ABC-CLIO.

Zeinab H., Nicolas T., Nelly V., Vassilis C., & C´edric D. M., (2011). Everything you would like to know about RSS feeds and you are afraid to ask. p. 1-20. Retrieve from https://www.researchgate.net/publication/254200530

Zwitter, A. J., & Hadfield, A. (2014). Governing big data. Politics and Governance, 2(1), 1-3. Retrieve from DOI: https://doi.org/10.17645/pag.v2i1.2

7

Best Practices Used in Nehru Library, CCS Haryana Agricultural University Hisar: A Case Study

Dr. Bhanu Partap

Assistant Librarian, CCS Haryana Agricultural, University, Hisar - 125 004 Haryana, India, E-mail: bpartaps2005@gmail.com

Abstract

The main focus of this paper is to share some of the best practices followed by the Nehru Library, Chaudhary Charan Singh Haryana Agricultural University, Hisar (CCSHAU, Hisar). The Nehru Library always tries to serve its users' community on their desired level by adopting traditional as well as digital information resources and services. This paper discusses in detail about the best practices usage and technological impact as well as the users changing information needs and enhancing the quality of teaching learning process in the Nehru Library. It is stated that most of the best practices parameters have been followed and maintained by the library and tried to serve its users in the best way. In this paper, it has also tried to include traditional or general best services; ICT based best services and some other best practices adopted by the Nehru Library.

Keywords: *Best practices, Library services, Physical facilities, Nehru Library, CCSHAU.*

1. Introduction

The Online Dictionary of Library and Information Science (ODLIS) defines best practice as, "in the application of theory to real-life situations, procedures that, when properly, applied consistently yield superior results and are therefore use as reference points in evaluation of the effectiveness of alternative methods of accomplishing the same task. Best practices are identified by examining empirical evidence of success". Commonly, best practices are those services, which satisfy the respective users' community on their desired level. Earlier, best practice term was used in the manufacturing industries for implementation of benchmarking methods so that ultimate quality results may be achieved. But later on the term "best practice" has also been used in academia and other fields as well. The mandate of any library is to work according to the informational needs of the user's and other facilities, which are desired by the users. The service delivery pattern has totally been changed gradually due to the technological impacts and ever changing needs of the users. Now library users' wants quality information services, which satisfy their needs so it becomes necessary for any library to provide the satisfactory services in the best and desired way. Best practices can play a major role in the comprehensive development of any academic institute and the agricultural university libraries are not exception to it.

Mantha (2019) discussed in his study that "the father of Library Science in India Dr. S. R. Ranganathan says that 'the kind of education given to the library profession is normally reflected in the service given by the profession'. Therefore, it is a great responsibility of the academic library personnel to identify and adopt best and active practices to meet the hassles of the end-users. Furthermore, when we talk about best practices, what we are referring to is a method or a technique that is generally accepted as better, then alternatives due to the fact that it produces results that are superior to those that might be achieved via other means, and/or practices. Over the time, these best practices have become crystallized and turned into standards that are accepted across the industry. Eventually, best practices might even evolve to be absorbed into standard requirements in terms of legal or ethical industry standards. Best Practices may be used to ensure sustained quality output as an alternative to mandatory legalized standards and may also be based on self-assessment or benchmarking".

An agricultural university library is considered to be a central place or a hub of teaching, research and extension activities. Best practices must be adopted and followed by the library of higher academic institutions so that the institute's vision and mission or intellectual goals can be promoted and achieved effectively. The Nehru Library of CCS Haryana Agricultural University, Hisar has a very good reputation in agricultural universities in India as well as in Asia. Therefore, a study has been planned to explore the best practices used in Nehru Library, CCS Haryana Agricultural University, Hisar.

2. Meaning and Definition

Merriam Webster Dictionary defined best practice as "a procedure that has been shown by research and experience to produce optimal results and that is established or proposed as a standard suitable for widespread adoption".

Similarly, Oxford Advanced Learner's Dictionary explained best practice as "a way of doing something that is seen as a very good example of how it should be done and can be copied by other companies or organizations".

Therefore, it can be said that all the possible ways, which can improve the service delivery and satisfy modern-day users' needs can be termed as best practice.

"National Accreditation and Assessment Council (NAAC) developed a set of best practices followed in academic libraries and presented under the following four broad areas:

1. Management and Administration of Library
2. Collection and Services
3. Extent of User Services
4. Use of Technology"

"For college and university libraries, NAAC has developed some of the best practices guidelines given below that can enhance the academic information environment and usability:

- Computerization of library with standard software
- Inclusion of sufficient information about the library in the college/ university prospectus
- Compiling student/teacher attendance statistics and locating the same on the notice board
- Displaying newspaper clippings on the notice board periodically
- Career/Employment Information/Services
- Internet facilities to different user groups
- Information literacy programs
- Suggestion box and timely response
- Displaying new arrivals and circulating a list of those to academic departments
- Conducting book exhibitions on different occasions
- Organizing book talks
- Instituting Annual Best User award for students
- Organizing competitions annually
- Conducting user surveys periodically."

3. CCS Haryana Agricultural University, Hisar

"After the division of Punjab State, the first established university in Haryana State in 1970 is Haryana Agricultural University, which is one of Asia's biggest agricultural universities. In 1991, it was renamed after India's seventh Prime Minister Chaudhary Charan Singh. Thus, now it is known as Chaudhary Charan Singh Haryana Agricultural University, Hisar. The university covers an area of 8645 Acres, out of which, 7219 acres at main campus and 1426 acres at sub-campuses. The University has six colleges in its main campus with different departments, *i.e.*, College of Agriculture (COA), College of Home Science (COHS), College of Agriculture Engineering & Technology (COAE&T), College of Basic Science & Humanities (COBS&H), and College of Fisheries Science (COFS), and College of Biotechnology (COBT). The University has two more colleges in its outstation campuses other than these six colleges situated in main campus, *i.e.*, College of Agriculture, Kaul (COAK) and College of Agriculture, Bawal (COAB). The University is affiliated to Indian Council of Agricultural Research (ICAR)" (CCS HAU, 2024).

4. Nehru Library– A Profile

"Though the library was established in 1948 with the establishment of the College of Veterinary Sciences, in 1975, it was named after Jawaharlal Nehru, the first Prime Minister of India. The covered area of the building is 9076 m^2 with a capacity to accommodate more than 3.5 lacs books and bound journals (CCSHAU, 2022). The library possesses a rich collection of more than 3.8 lacs books, bound journals and other reading material to meet the informational requirements of students, teachers, research scholars, extension specialists and other staff of the Chaudhary Charan Singh Haryana Agricultural University (CCSHAU) as well as Lala Lajpat Rai University of Veterinary and Animal Sciences (LUVAS) located at Hisar and sub-campuses. The library is also maintaining its collections at College of Home Science, Hisar, College of Agriculture, Kaul, College of Agriculture, Bawal, Campus School, 19 Krishi Vigyan Kendras and 7 Regional Research Stations. The library is successfully exploiting the power of information and communication technologies in various activities for retrieval and dissemination of information" (Nehru Library, 2024).

Table 1: Current Status of the Nehru Library, CCS HAU, Hisar

Information Resources	**Quantity**
Books	2,53,643
Bound journals	1,05,952
Thesis	14,574
CD-ROM databases	155
Books CDs	2,555
Theses CDs	4,179
E-books	2,318
EBSCO E-Books Academic Collection	2.4 Million
NIPA Computer Examination Package	4,30,000

Source: Annual Report, 2022-23, Nehru Library, CCS HAU, Hisar

5. Best Practices Used at Nehru Library, CCS HAU, Hisar

The Nehru Library, CCS HAU, Hisar is providing quality services to its users under best practices adopted and broadly can be classified into two parts as follows:

- Traditional/General best practices
- ICT based best practices

5.1. Traditional/General Best Practices

5.1.1. Library Orientation Programme

To create awareness and interest about the library and its resources and services among the students, orientation is one of the best practices followed by the Nehru Library. At the beginning of the academic session every year, faculty wise orientation programmes have been conducted such as agriculture, community science, agricultural engineering, fisheries science, and veterinary sciences. Library orientation helps the students to be aware and understands the library use facilities, library membership, reading facilities, importance of library and its various library services so that the users can become information literate.

5.1.2. Library Orientation for Teachers

The Nehru Library has also conducted library orientations separately for those teachers and scientists who have been newly recruited in the institution. It helps them to be aware the library facilities, services and information resources, library membership and other digital resources.

5.1.3. Library Visit

Every year, the Nehru Library has conducting library visit for the visitors of various categories such as school students, college students, university students, newly recruited teachers and scientists, Government officials, foreign dignitaries, and other dignitaries of academia and industries from India and abroad, which helps them to aware and provide proper exposure about the library facilities, resources, services and other associated activities.

5.1.4. Book Exhibition

Every year, the Nehru Library, CCS HAU, Hisar have been organising a book exhibition on the University's foundation day, *i.e.*, 2nd February. All the empanelled publishers have participated in this activity and displayed their latest titles. Thousands of books have been selected and recommended by the faculty members of various departments and even students can purchase the books from the exhibition on their own and it saves time and will also get the maximum discount. Apart from this, the Nehru Library has also organized book exhibitions on different occasions and students can visit, see and explore the books for their knowledge purposes.

5.1.5. New Arrivals

The Nehru Library has purchased and added thousands of books and other documents to its collection every year for the use of students, teachers, scientists and other users. To inform the users about the newly purchased books, the Nehru Library has displayed the jackets of the newly added books in the display location. The jacket's of books contained brief information about the book, location of the documents in the shelves of the library.

5.1.6. Library Facility for Outside Readers as an Extension Services

The Nehru Library offered permission to use the library facility to the sons, daughter and spouse of University staff. Apart from this, outsiders are also permitted to use the library facilities with nominal fee and refundable security amount. This is one of the best practices of the library as recognized highly by various categories of users in the Hisar region.

5.1.7. Library Trainings

Every year, a good number of students who are freshers after completing their diploma/bachelor/master degree in library and information science have approached the Nehru Library to get the training or sufficient practical knowledge about the working and various professional activities. It is the most admirable practice of the Nehru Library among the young Library and Information Science (LIS) professionals.

5.1.8. Library Course

The Nehru Library teaches one compulsory paper "Library and Information Services" to postgraduate and Ph.D. students of the University in every semester. It is made mandatory by the Indian Council of Agricultural Research (ICAR), the apex body, and by the University to learn and pass the examination to be entitled to get the degree. This is a unique feature of the library in which the faculty or LIS professionals of Nehru Library teaches all the PG and Ph.D. students so that the students can able to learn about the basics of library, working pattern, resources and services and other facilities, which can be useful in their education, research and extension activities as well as for lifelong learning purposes.

5.1.9. Training on Awareness and use of e-Resources

The Nehru Library has conducted training programmes on awareness and use of e-resources for the students and teachers of the University on a frequent basis. It helps them to get awareness and use knowledge about the existing e-resources and newly added e-resources collection in the Nehru Library. This is one of the best practices of the Nehru Library and highly demanded among the library users because practical sessions are there so that the users can easily do hands-on-practice and can learn about the use of e-resources.

5.1.10. Informative Notice Board

The Nehru Library has maintained its notice boards with useful information such as rules and regulations of Nehru Library, exam related information, career guidance, recruitments, social information, current or upcoming events, *etc.* so that users can be benefitted and keep themselves updated.

5.1.11. Reprographic Services

The Nehru Library provides a basic facility of reprographic services to its users as well as to the different offices of the University. The services included photocopy, enhancing or reducing the paper size, spiral binding, lamination, colour Photostat and printout, thesis printing and binding, *etc.*

5.1.12. Resource Management

The Nehru Library has huge print and electronic collection of information resources such as books, bound volumes of journals, thesis, dissertations, pamphlets, bibliographies, magazines, newspapers, e-books, e-journals, e-databases, e-repository, *etc.*, for the benefit of the library users. In this digital era, a lot of information has been generated everyday and published in print as well as in electronic form; hence, handling of information resource collections is termed as resource management.

5.1.13. User Surveys

The Nehru Library has conducting user surveys on a periodic basis to know the current status of the library and their satisfaction whether they are satisfied with the library staff, collection, and services. The survey reports also have published in the form of research papers in journals. Users' surveys are very useful for libraries to know whether their users are satisfied with the library and its resources and services or not and further take the initiative to improve the overall library services.

5.1.14. Book Bank Facility

The Nehru Library has established a Book Bank for the benefit of those students, who are entitled to borrow textbooks from the Book Bank on nominal charges. "Book Bank is actually a collection of multiple copies of textbooks and these books can be borrowed by the students for the semester/year against nominal rental charges (10% of the cost of the book if the cost is up to Rs. 100/- or Rs. 10/- + 5% of the cost of the book exceeding Rs. 100/-). There is also a Social Welfare Section (SWS) in the Book Bank in which books are kept for the students of scheduled castes and scheduled tribes for issuing free of any charge". The Book Bank has a collection of 14,333 volumes as per the annual report of Nehru Library for the period 2022-2023.

5.1.15. HAU Bookshop

For the University community, the Nehru Library has offered a very unique service called as HAU Bookshop because "no other library in the country offers this type of facility. The Nehru Library has also a Bookshop facility for the purchase of its reading material. The University staff and students are also entitled to purchase personal books through the Bookshop and avail of discount facility. The University community is passed on the entire amount discount which the Bookshop gets on cash sales but retains a part of discount of credit sales".

5.1.16. Reference Service

Reference service is a very important and adopted as a best practice in every type of library and has gained reputation amongst its users. The Nehru Library has also offered reference service to its member users as well as to the other visitors of the library. In Nehru Library, reference desk is headed by a senior library professional (Reference Librarian) and assisted by other staff guides/ helps the library users to explore the information and its location, access the print and e-collections, in the use of OPAC, and to procure information or resource from other libraries if needed by the user concerned.

5.2. ICT Based Best Practices

5.2.1. Automated Library

The Nehru Library implies Koha library management software to automate its activities such as acquisition of books, subscription of printed and e-journals, cataloguing of current acquisitions, retro-conversion of the library catalogue, bar-coding and RFID of documents, circulation of books and other reading material, linkage of library online catalogue to the University website and AgriCat union catalogue, digital library, *etc*. It saves the time of the users as well as of the library staff.

5.2.2. Library Portal

The Nehru Library has its own portal, which is linked to the University's website. All about the Nehru Library including digital library, OPAC, and links to access to other e-contents are represented on the library portal so that the users could be able to get all the information and access to information resources from a single platform. This is the best practice or better to say highly demanded service among the agricultural teachers, scientists and students of the University.

5.2.3. Internet Surfing

The Nehru Library provides Internet surfing facility to the students and faculty of the University for their academic and research use. For this purpose, the Library has a well equipped computer lab consisting of 32 latest PCs with Internet facility, whereas 32 Internet points have been provided for laptop users. This facility is offered to the bonafide library users free of cost; however, if printout is required, the users have to pay Rs. 1.00 per page.

5.2.4. Document Delivery Service

A highly demanded service amongst the students and teachers/scientists of agricultural, veterinary, fisheries, dairies, and allied sciences of state agricultural universities, deemed-to-be-university, institutions, directorate, bureaux under Indian Council of Agricultural Research (ICAR). This is a resource sharing initiative under Consortium for e-Resources in Agriculture (CeRA) wherein all the CeRA member institutes provide copy articles available in their libraries to those member institutes who do not have the print version of a particular scientific article. Being one of the members of CeRA, users of CCS HAU, Hisar also availed this facility and received their required articles either in print or in electronic form.

5.2.5. Automated Circulation

The Nehru Library has started its check-out and check-in service a long back with the implementation of LibSys library management software. Later on, the library management software-Koha was implemented during 2009-2014 under the e-granth project of NAIP to follow the standards; maintain the uniformity for compilation of union catalogue of agricultural libraries. Now, all the documents are issued to the users with the help of barcode technology and from September 2019, after implementation of RFID technology in the library, RFID based circulation services have been started.

5.2.6. Self Issue/Return

The Nehru Library offered self issue/return facility to its users with the help of RFID enabled self check-out/check-in kiosk. A user can himself/herself get issued the documents using this facility at any time. The user must be carrying his/her smart ID Card and desired documents for this, which he/she has to place on the RFID machine/kiosk for issuance of documents. After he/she gets the document issued, a slip is generated just like on ATM machine, which he/she has to show on the exit gate as gate pass.

5.2.7. OPAC/WebOPAC

The Nehru Library offered the Online Public Access Catalogue (OPAC) facility to its users including scientists, teachers, extension specialists, students, and other users. Anyone can search and access the status of documents and other related information by using OPAC as all the Personal Computers (Desktop Computers) have linkage to the Campus wide network and locations all over the University campus. Similarly, the provision of Web-OPAC is also there and by using Web-OPAC, anyone can access the library catalogue from anywhere in the world as the OPAC of Nehru Library is also put into global IP, which works through Internet. Online catalogue impacts the users' opinion about the library as online catalogue is termed as face of the library.

5.2.8. Video-conferencing Facility

The Nehru Library is providing video conferencing facilities to the teachers, scientists and students of the University for different academic purposes. It is one of the best practices adopted by the Nehru Library, which has gained great popularity. The scientists of CCSHAU using this facility and students can have live interactions with their fraternity at remote locations. In fact, this is a multipurpose facility, which can also be used for delivering a PowerPoint Presentation to a group of the audience.

5.2.9. CCTV Security System

The CCTV security system consisting of high resolution 69 HDD cameras with highly advanced night vision technology has been installed for strengthening the security of reading material and monitoring proper discipline in the Nehru library. This service is highly used by the students by checking CCTV footage for missing or stolen things.

5.2.10. Smart Class Room

The Nehru Library has a smart class room for the purpose of teaching the course, *i.e.*, PGS-501- Library and Information Services to the postgraduate and Ph.D. students. It is located on the 2nd floor of the Nehru Library. Library course has been taught to postgraduate and Ph.D. students of different colleges/ faculties of the CCS HAU, LUVAS and MHU using the audio visual facilities of this smart class room. Apart from academic purposes, this smart classroom is also used for delivering lectures/interaction, presentations, talks, *etc.* on some other occasions.

5.2.11. Centralized Air Conditioned

A completely centralized air conditioning system supported with advanced technology has been installed in the Nehru Library, which creates a healthy and cosy environment in the library premises. This service is the state-of-the-art service and after installation, the use of library has increased.

5.2.12. Personalized Service

Information needs of teachers, scientists, students, non-academic staff and other special members of the University or sometime outside of the University can also be fulfilled by the Nehru Library especially by the reference librarian, the head of Readers Service Division. The information demands have been fulfilled by using WhatsApp, e-Mail, or telephonically.

5.2.13. Access to Electronic Resources

The Nehru Library provides access to its subscribed and purchased electronic/ digital resources by giving links on its web portal under "digital library" tab. All the acquisitioned e-books, e-journals, e-databases, e-repositories, and other e-resources can be accessed by its respective users 24x4 as per their need and time convenience. All the e-resources have been acquisitioned and can only be accessed on IP addresses of the University.

5.2.14. Discovery Service/Remote Access to Access e-Contents

The Nehru Library also provides remote access to its e-resources through discovery platform "CCSHAU e-Library". Any user is able to access all the subscribed and purchased electronic information resources such as e-journals, e-books, OPAC, e-databases, e-repositories, e-theses, and other e-resources remotely from anywhere in the world. This is one of the best practices adopted by the Nehru Library to serve its member users.

5.2.15. Mobile App

A mobile app named "CCSHAU e-Library" has also been developed by the Nehru Library, which enables the University community to access the e-contents remotely through proxy server. This application is available and can be downloadable from Google Play Store. However, only registered users of Nehru Library are able to use and access the mobile application.

5.2.16. Krishikosh- e-Repository

A repository or an institutional repository is an online archive for collecting, preserving, and disseminating digital copies of the intellectual output of an institution. Various kinds of information resources can be made available to the users' community by the library such as question papers, copy of syllabus, research papers, various notes, material related to career guidance, minutes of meetings, electronic theses, *etc*. The Nehru Library has provided access to its M.Sc. and Ph.D. theses through its participation in Krishikosh, which is an e-repository maintained centrally by Indian Council of Agricultural Research (ICAR).

5.2.17. Inter Library Loan (ILL)

Under inter library loan services, a user of one library can borrow books or receive photocopies of documents that are owned by another library with some terms and conditions. The concerned user has to make a written request with their parent library, which will act as a mediator, identifies the owner or library of the desired item, places the request, receives the item, makes it available to the user, and arranges for its return as well. This all process will occur under Memorandum of Understanding (MoU) between two or more libraries/institutions. The Nehru Library is a user member of DELNET and provides ILL services majorly through this platform to its users, if any demand is raised by any users.

5.2.18. Document Delivery Service (DDS)

Document delivery service is one of the crucial services of the Nehru Library, which undertakes under the able guidance of a reference librarian. Under DDS, the Nehru Library delivers copies of research articles published in journals and book chapters and other important reading material to the users of other libraries/institutions electronically or manually. This service is mostly free of any charges and provides to all the users throughout India. Under the Consortium of Electronic Resources in Agriculture (CeRA), the Nehru Library provides document delivery services to the users, who submitted their demands electronically from all over India.

5.3. Other Best Practices at Nehru Library, CCSHAU, Hisar

- Compiling annual reports of Nehru Library
- Binding of books/theses/periodical volumes
- Newspapers and magazines
- Common spaces
- Space for Laptop users
- Information literacy programs
- Teachers/scientists awareness and training programs
- Special awareness and training programs for SC/ST students
- Separate faculty collection
- Reading facility for more than 700 users
- Flexible timings for Night Reading Hall (NRH)
- Special drives for neat and clean library premises on various occasions

- Power backup
- Drinking water and washroom facility at all floors of Nehru Library
- Conducting user surveys periodically about their opinion on services of Nehru Library.

6. Users' Feedback Analysis

A small survey had been conducted in the premises of Nehru Library in the first week of August 2024 to know the effectiveness of library services and satisfaction level of users. A total of 200 self-made questionnaires distributed randomly among the postgraduate and Ph.D. students. Out of a total 200 questionnaires, 187 valid filled-in questionnaires received back with overall response rate @93.50%. The collected data were than tabulated and evaluated accordingly as following:

Table 2: Demographic profile of library users

Gender wise distribution	**Number of respondents**	**Percentage**
Male	107	57.22
Female	80	42.78
Total	**187**	**100**
Academic status		
M.Sc.	126	67.38
Ph.D.	61	32.62
Total	**187**	**100**
Frequency of library visit		
Everyday	59	31.55
Once or twice in a week	66	35.30
Once in a fortnight	47	25.13
Once in a month	15	08.02
Total	**187**	**100**
Time spent in the library		
<2 hours	32	17.11
3-4 hours	55	29.41
5-6 hours	76	40.64
>6 hours	24	12.84
Total	**187**	**100**

Table 2 highlights the demographic profile of the library users who participated in the survey and responded positively. Out of a total 187 respondents, 107 were male and 80 were female users. On the other hand, out of 187 respondents, 126 of those users responded who were studying in M.Sc. courses and rest of the respondents (61) were studying in various PhD. programs. With regard to the frequency of library visit among the users, 35.30 percent of the respondents were preferred to visit the library once or twice in a week, whereas, 31.55 percent respondents were also there who visiting the library every day, followed by 'once in a fortnight' (25.13 percent) and 'once in a month' (08.02 percent) respectively. On asking the question of how much time they prefer to spent in the library, 40.64 percent of the respondents spent 5-6 hours once they visited, while 29.41 percent of the respondents spend 3-4 hours in the library, followed by <2 hours (17.11 percent) and >6 hours (12.84 percent) respectively.

Table 3: Satisfaction level among library users towards library services/best practices

Satisfaction level	Number of respondents	Percentage
Fully satisfied	151	80.75
Partially satisfied	28	14.97
Not satisfied	08	04.28
Not at all satisfied	00	00
Total	**187**	**100**

The data given in Table 3 illustrates the satisfaction level of library users towards library services or best practices followed by the Nehru Library. It was found during the study that a majority of the respondents (80.75 percent) were fully satisfied with the library services and facilities offered to them on priority basis, while 14.97 percent respondents were partially satisfied. On the other hand, only 4.28 percent of respondents were not satisfied with the best practices of the Nehru Library. Interestingly, no respondent has given the answer to the question 'not at all satisfied'. It shows that more than 96 percent of the respondents (library users) have been satisfied with the services, facilities and other best practices being followed by the Nehru Library.

Table 4: Problems faced by library users towards accessing library services and facilities

Problems faced	Number of respondents (N= 187)	Percentage
Less reading space	00	00
Library timings	00	00
Library physical facilities	12	06.42
Lack of relevant print documents	00	00
Inability to retrieve documents from library stacks and databases	29	15.50
Lack of e-resources	07	03.74
Library staff behaviour and competencies	20	10.70
Less number of computers	39	20.85
Lack of Wi-Fi facility	111	59.36

The presented data in Table 4 depicts the various problems faced by library users towards accessing library services and facilities. It has been found during the study that out of 187 respondents, 59.36 percent of the respondents replied that they are facing the problem of non-availability of Wi-Fi facility in the library premises, whereas, 20.85 percent respondents reported about the availability of less number of computers in the computer/Internet lab. On the other hand, 15.50 percent of the respondents reported that they have not proper ability to retrieve and access the documents of their requirement from library stacks and databases, whereas, 10.70 percent of respondents faced the behaviour of library staff and competency level not up to the mark. Lack of some physical facilities in the library premises is the problem for 6.42 percent respondents, while lack of e-resources which are required on priority basis by the library users is the problem for 3.74 percent respondents. Interestingly, it was noticed during the study that no respondents have replied to the questions of 'less reading space', 'library timings' and 'lack of relevant print documents'.

Table 5: Impact of best practices of Nehru Library on library users

Impact analysis	Number of respondents (N= 187)	Percentage
Ease and convenience of accessing the library services	97	51.87
Helpful towards research and academic activities	109	58.28
Comfortable environment	161	86.09
Resource rich place	118	63.10
Flexible timings during examinations	43	22.99
Helpful for lifelong learning	68	36.36
Best in class services among all users' categories	51	27.27
Helpful in preparing for exams, seminars, and other academic activities	95	50.80
Achieve utmost satisfaction	86	45.98

Table 5 shows the various impacts on library users in view of best practices of Nehru Library. Out of 187 respondents, 86.09 percent of respondents felt comfortable with the library's physical and academic environment, while 63.10 percent of the respondents replied that the Nehru Library is a resource rich place where they can find and access the desired information freely and quickly. On the other hand, 58.28 percent of respondents feel that the Nehru Library is always proven to be helpful towards their research and academic affairs, whereas, 51.87 percent respondents were overwhelming with the ease and convenience of accessing the library services. 50.80 percent respondents replied that Nehru Library is always helpful in preparing for examinations, seminars, and other activities, while 45.98 percent respondents were happy to achieve utmost satisfaction from the library services and best practices, followed by 'helpful for lifelong learning' (36.36 percent), 'best in class services among all users' categories' (27.27 percent), and 'flexible timings during examinations' (22.99 percent) respectively.

7. Issues and Challenges Encountered in Implementing the Best Practices

Some of the issues and challenges encountered in implementing the best practices in Nehru Library, CCS HAU, Hisar have been identified as given below:

- Budgetary issues
- Shortage of professional library staff
- Non appointment of University Librarian on regular basis
- Unnecessary audit objections
- Additional responsibilities
- Non-cooperative behaviour of Para-professional library staff and other non-professional staff
- Some other issues and challenges also encountered such as non attending routine maintenance work, lighting issues, and users' behaviour, *etc.*

8. Conclusion

This is the digital era and we all are widely affected by Information and Communication Technology (ICT). Due to digital effects and rapidly changing user's demands, a library has to re-look in its activities and services. Now users need only those services, which satisfy their information requirements on their desired level. So, whatever services are provided by the libraries in today's digital scenario, that must be according to the users' current need and in a best practice way. Shukla and Khajuria (2021) elaborate in their study that "to provide the best services to the users, library professionals should go on with the ever-changing environment and serve the best of its kind. The 'best practice' is an approach, a loom or an attitude based on the need for continuous learning and improvement. It helps to determine and secure performance gaps, resolve a problem, generate new opportunities, and positively impact the institution". Libraries need to be more vibrant and must adopt state-of-the-art resources and services apart from the traditional services to serve their users' community. It can be concluded that the Nehru Library, Chaudhary Charan Singh Haryana Agricultural University, Hisar is providing best services in the best way to its users on their utmost satisfaction level. Best practices only indicate the changes in service delivery mode; however, flexibility must be there to learn new things and try to implement them with the help of technologies and curiosity and the Nehru Library always adopted up to date policies and best practices, which cope up with the ever changing environment. Furthermore, it has been suggested that proper budgetary provision must be given priority in terms of library services and resources, proper qualified library professional staff must be appointed in sufficient number, and professional training sessions must be organized by the library authority to train the library staff for their enhancing professional competencies and overall personal grooming.

References

Annual Report of Nehru Library, CCS HAU, Hisar for 2022-23

Best practice (2024). https://www.merriam-webster.com/dictionary/best%20practice Retrieved June 15, 2024

Best practice (2024). https://www.oxfordlearnersdictionaries.com/definition/american_ english/ and?q=and Retrieved June15, 2024

Best Practices College Libraries (2024). http://naac.gov.in/docs/Guidelines/Guidelines-Colleges-Library.pdf

CCS Haryana Agricultural University, Hisar (2024). http://hau.ernet.in/about/ history.php.

Mantha, Raghu (March 4, 2019). Best practices in academic libraries. SSRN: https://ssrn.com/abstract=3696300 or http://dx.doi.org/10.2139/ssrn.3696300

NAAC (2007, April). Best Practices in Library and Information Services. Bangalore, Karnataka, India.

National Assessment and Accreditation Council- NAAC (2024). http://www.naac.gov.in

Nehru Library, CCS Haryana Agricultural University, Hisar (2024). https://hau.ac.in/college/nehru-library.

Prabhakar, S.V.R., & Manjula Rani (2017). Best practices adopted in academic libraries and information centres: At a glance. International Journal of Librarianship and Administration, 8(1), 7-15.

Reitz, J. M. (2021). Online Dictionary for Library and Information Science. Retrieved June 15, 2024 from http://www.abc-clio.com/ODLIS/searchODLIS.aspx

Shukla, Ravi, & Khajuria, Riya (2021). Central library best practices in Mizoram University: A case study. In Krishna Devi, Sur Chandra Singha and Manoj Kumar Verma (Eds.), Web based services in library and information science (pp. 110-120). New Delhi: Shree Publishers & Distributors.

8

Role of OER Initiatives in Achieving the Sustainable Development Goals

Dr. Madhu Midha

Deputy Librarian & Head, (KRC), IKG Punjab Technical University, Jalandhar Kapurthala Highway, Near Science City, Jalandhar - 144 603, Punjab, India
E-mail: ptu_madhu@yahoo.com

Abstract

Open Educational resources are seen as major contributor to Sustainable development goal 4 open education. OERs are catalysts of micro learning, nano learning and as a support in continuing professional education by lifelong learners. OERs have the potential to achieve the SDG 4 quality education as it provides access to quality material to lifelong learners, has coordinated partnerships and collaborations to establish and promote social and educational equality. Many authors have claimed that OER initiatives at global level have contributed majorly in achieving SDG 4. This paper sets out the scale and scope of the SDGs, explains the role of OERs in quality educations and provides insights on potential contribution of OER initiatives at global level in relation to SDG 4.

Keywords: *Open educational resources, Open education, Quality education Sustainable Development Goals, OER initiatives.*

1. Introduction

Universal development through Education for all had been a prominent feature earlier during Millennium Development Goals and recently it has gained a fresh momentum when United Nations General Assembly set 17 inspirational Sustainable development Goals. Quality Education is the focus of SDG 4, "Ensure inclusive and equitable quality education and promote lifelong learning opportunities for all" (Sustainable Development Goals, 2023). Role of education in achieving these sustainable development goals in developing countries is well recognized by OECD. OECD in its report on 'Education at a Glance 2013' states that "Educational attainment is frequently used as a measure of human capital and the level of an Individual's skills. The level of educational attainment is the percentage of population that has reached a certain level of education. Higher levels of educational attainment are strongly associated with higher employment rates" (OECD, 2013). Furthermore, OECD in its report published in 2017 explains that "Making SDG 4 a reality will transform lives around the globe. Education is so central to the achievement of a sustainable, prosperous, and equitable planet that failure to achieve this particular SDG puts at risk the achievement of the 17 SDGs as a Whole" (OECD, 2017).

There had always been a difference of opinion among authors in the context of SDGs. Some authors believe that education and particularly higher education plays a critical role to SDGs while some other advocates Distance education as a key role player. Uvalić-Trumbić & Daniel (2016) in his research on sustainable development begins with education has emphasized that Distance education plays critical role in SDGs. Many authors find open educational resources as prominent factor in achieving SDGs (Mackinnon, 2016). The SDGs have large scope and complexity so having different perspectives among authors is inevitable.

2. Education and Sustainable Development Goals

On 25th September, 2015 United Nations General Assembly and 193 countries adopted 17 Sustainable development Goals for its 2030 development agenda. Out of these 17 Goals 16 SGDs are sector based, intergovernmental and equally applicable to all countries of the world. The 17th SDG, "Strengthening the means of implementation and revitalize the global partnership for sustainable development" deals with the agenda on how to achieve the aforesaid 16 SDGs through partnerships and collaborations". There are total 179 targets for 17 SDGs and further there are 10 separate targets within SDG 4, "Quality Education".

The Sustainable Development Goals Report 2022 provides "a global overview of progress on the implementation of the 2030 Agenda for Sustainable Development, using the latest available data and estimates. It tracks the global and regional progress towards the 17 Goals with in-depth analyses of selected indicators for each Goal" (United Nations, 2022).

As per SDG 2022 Report, "cascading and interlinked crises are putting the 2030 Agenda for Sustainable Development in grave danger, along with humanity's very own survival. The Report highlights the severity and magnitude of the challenges before us. The confluence of crises, dominated by COVID-19, climate change, and conflicts, are creating spin-off impacts on food and nutrition, health, education, the environment, and peace and security, and affecting all the Sustainable Development Goals (SDGs)" (Korngold, 2023).

"The Report details the reversal of years of progress in eradicating poverty and hunger, improving health and education, providing basic services, and much more. It also points out areas that need urgent action in order to rescue the SDGs and deliver meaningful progress for people and the planet by 2030" (United Nations, 2022).

United Nations Department of Economic and Social Affairs in its "Overview of SDG 4 Quality Education states that the world is falling far behind in achieving the quality education, without additional measures by 2030 around 84 million Children and youth will be out of schools, 300 million will lack basic numeracy/ literacy skills, only 1 in 6 will achieve universal secondary school completion target. The pandemic caused learning losses in 4 in 5 countries out of 104 countries studied. Primary and secondary school completion rates are rising but the pace is slow and uneven. Low and lower middle-income countries face a $ 100 billion annual financial gap to reach their education targets" (Sustainable Development Goals, 2023).

3. Role of OER in SDGs Quality Education

"MOOCs and Open Educational Resources both are becoming prominent factors in achieving SDGs. After the `Open Access and Open-Source Software movements` success, Open Educational Resources is an extended version of these globally accepted higher education movements towards openness" (Midha & Kumar, 2022).

"The two most important aspects of openness have to do with free availability over the Internet and as few restrictions as possible on the use of the resource. There should be no technical barriers (undisclosed source code), no price barriers (subscriptions, licensing fees, pay-per-view fees), and as few legal permission barriers as possible (copyright and licensing restrictions) for the end-user. The end-user should be able not only to use or read the resource but also to adapt it, build upon it and thereby reuse it, publish it given that the original creator is attributed for his/her work. In broad terms, this is what is meant with "open" in all three movements" (Hylen, 2015).

"Open educational resources movement is a part of open access movement that particularly caters to the course curriculum related educational content requirements of teaching and learning community" (Midha & Kaur, 2022). Smith & Casserly (2006) have explained that open access is not the only feature of open educational resources that distinguishes it from the other content available over the internet. OERs available with open license also permits the users not only to use the resources but also to modify, share and reuse it. Ed Walker (2009) has defined the term 'open' as "convenient, effective, affordable, and sustainable and available to every learner and teacher worldwide".

"It is reasonable to argue that OER can greatly benefit developing countries, even more so than those that are considered developed. As such, they can play an important transformative role in supporting the SDG4 goal" (McGreal, 2017).

4. Open Educational Resources Initiatives at Global Level

There are many key role players in the open educational resources movement that has stimulated and facilitated this noble movement to reach at global level from local level. These key role players have supported the open educational resources movement in many ways such as by providing technical support, financial aid, policy making, standardization and in promotion and awareness etc. In the coming section few of such key role players are listed along with their activities and roles in open educational resources movement. Many philanthropic organizations like Mellon Foundation, UNESCO, Hewlett Foundation and Commonwealth of Learning etc have provided financial support to propel the OER movement worldwide (Midha, 2022).

a) UNESCO's Contribution in OER Movements

UNESCO is the sole agency of the United Nations that has a specialized program focused on free educational resources. UNESCO is credited with coining the term "open educational resources" during its forum on open courseware in higher education in 2002. According to UNESCO, the provision of high-quality education that ensures everyone has access to information is essential for promoting peace, sustainable social and economic development, and fostering intercultural communication. Open Educational Resources (OER) present a valuable chance to enhance the caliber of education and the dissemination of knowledge, while also fostering global policy discussions, knowledge exchange, and capacity development, as stated by UNESCO in 2012. UNESCO has achieved another significant accomplishment in the open educational resources movement by adopting the OER recommendations (UNESCO, 2019; UNESCO & CoL, 2019).

The UNESCO General Conference, during its 40th session in November 2019, unanimously accepted the Recommendation on OER. This recommendation promotes the development, utilization, and customization of inclusive and high-quality Open Educational Resources (OER). It also aims to enhance international collaboration in this domain (Midha, 2023). The UNESCO OER recommendations are the sole globally recognized framework in the field of Open Educational Resources (OER). The recommendation is the sole internationally recognized vehicle for setting standards on Open Educational Resources (OER). It is the result of more than ten years of work to unite a broad range of stakeholders.

The Recommendation identifies five specific Areas of Action, which are as follows:

1. Enhancing the ability of stakeholders to generate, obtain, modify, and distribute OER.
2. Establishing policies that support the use of OER.
3. Promoting the availability of high-quality OER that is accessible to all.
4. Fostering the development of sustainable models for OER.
5. Encouraging and strengthening international collaboration in the field of OER.

b) Effective Role Played by Commonwealth of Learning in OER Movement

The Commonwealth of Learning was founded as an international organization with the objective of advancing and facilitating the development and dissemination of open learning and remote education, as well as the resources and technologies associated with them (Commonwealth of Learning, 2012). The organization COL is headquartered in British Columbia, Canada, with a regional office located in New Delhi. The Consortium for Open Learning (COL) is the sole international and intergovernmental organization dedicated to advancing the progress of remote education and open learning. A total of 53 countries are members of the Commonwealth. COL has achieved significant accomplishments in promoting the utilization and advancement of Open Educational Resources (OER) on a global scale. Outlined below are some of the notable accomplishments of COL: In 2012, UNESCO and Col collaborated to organize a consultation series, during which government ministries were invited to share information regarding policy pertaining to Open Educational Resources (OER).

The College of Law (COL), in collaboration with the Government of Slovenia, undertook two surveys on a global scale. The survey results revealed numerous gaps and obstacles in the development and implementation of OER. The results of the poll were released in the form of the Global OER report 2017. During the 2016-2017 year, CoL organized six regional forums worldwide to discuss OER studies. The results of this study were condensed and presented in a book titled "Open Educational Resources: From Commitment to Action."

The 2nd World OER Congress was organized by COL and UNESCO in partnership and took place in Slovenia in 2017. The event resulted in the formulation of a "OER action plan" with the objective of integrating OER into the education system to promote equal access to high-quality education, enhance capacity building, facilitate knowledge sharing, and ensure universal access to quality learning and teaching resources by 2030 (Commonwealth of Learning - Learning for Sustainable Development, n.d.).

The COL hosted the inaugural World Open Educational Resources Congress in Paris. The Paris OER declaration, a significant outcome of this event, advocates that all educational materials created using public funds should be made accessible under a public license (Midha, 2023).

In 2019, COL and UNESCO collaborated to publish the "Guidelines on the development of open educational resources policies." This tool is designed specifically to analyze the present status of policies and assist in comprehending copyright and licensing alignment policies (UNESCO & CoL, 2019).

In 2020, COL and UNESCO collaborated to organize a series of workshops focused on enhancing capacity building. The workshop aimed to raise awareness about OER policy, enhance OER capacity building, assist in the formulation of OER policy, and monitor the impact of implementing OER recommendations (Commonwealth of Learning, 2012).

c) Open Learn Platform

Open Learn is a cost-free educational portal. The Open University established it in 2006. Open Learn offers a wide range of free educational resources, including whole courses, as well as articles, videos, games, and other open learning materials. Open Learn offers a wide range of subject categories, including arts, forensic, and many more, for giving free comprehensive courses (Open Learn, 2019).

Open Learn offers a wide range of academic information, including over 950 short courses and a variety of audio/video materials covering topics from Greek history to the current political landscape and all in between. Additionally, it encompasses a multitude of puzzles, essays, quizzes, and interactive activities. Open Learn is a collection of printed materials created by Open Learn and BBC that accompany various TV and Radio programs. Open Learn is a highly renowned provider of open educational resources, with numerous notable honors in the field of open education. The educational content offered by Open Learn is accompanied by Creative Commons licensing, allowing for its reuse, remixing, and redistribution.

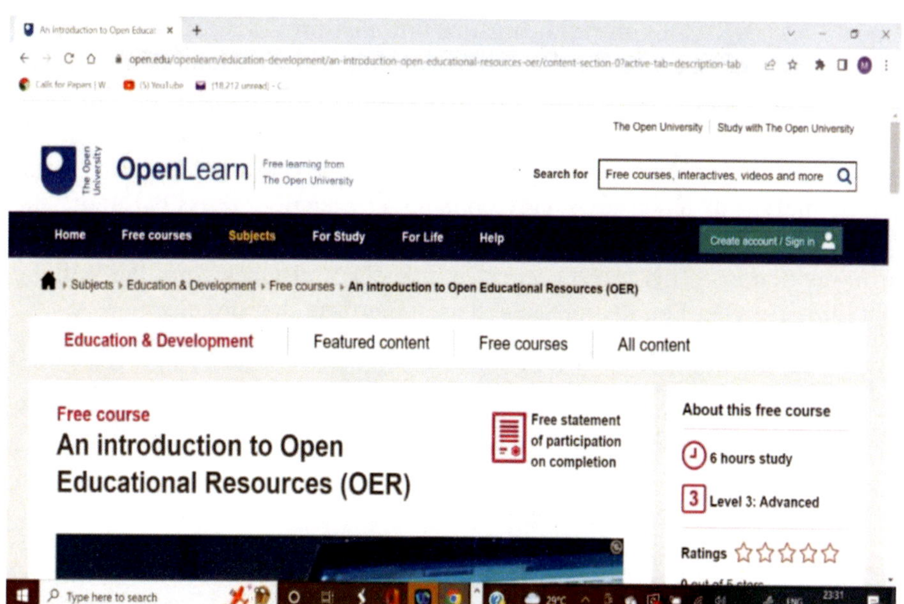

Fig. 1.1: Open Learn Homepage (Source: www.open.edu) (Open Learn edu., n.d.)

d) Rice University's OER Journey from Connexions to OpenStax

Connexions is a digital education system that includes an open-source content management system. This system helps in the distribution of open educational resources and provides a repository for open access learning objects. Connexions is a highly renowned open educational materials web platform globally. This resource has an extensive assortment of over 17,000 learning objects, as well as numerous open textbooks, research articles, and other materials. The Connexions website is the most extensively utilized open educational resources website globally, with over 2 million monthly visitors accessing and utilizing the Connexions library. The instructional content caters to learners of all age groups, ranging from primary to postgraduate level. It covers a wide range of topics, including Mathematics, Science, History, English, Psychology, and more.

The instructional information on Connexions is accessible in a readily downloadable format and compatible with any mobile device, in addition to online open educational resources. Connexions also assists schools by offering physical copies of textbooks at a really affordable price.

Dr. Richard Baraniuk established Connexions in 1999 at Rice University in the United States. Subsequently, it underwent a name change to Open Stax and is presently recognized as Open Stax CNX (Baraniuk, 2008).

Open Stax CNX is a non-profit organization that provides a dynamic digital learning system, offering free educational information to millions of users every month. Open Stax is intentionally designed to promote the sharing and reuse of educational resources. All the information of OpenStax is developed using semantic HTML format, making it very versatile and flexible in any context. Furthermore, all of the open stax contents are legally accessible under the Creative Commons license, allowing authors to use and reuse them without any legal restrictions. This project receives financial assistance from the William and Flora Hewlett Foundation as well as numerous other organizations.

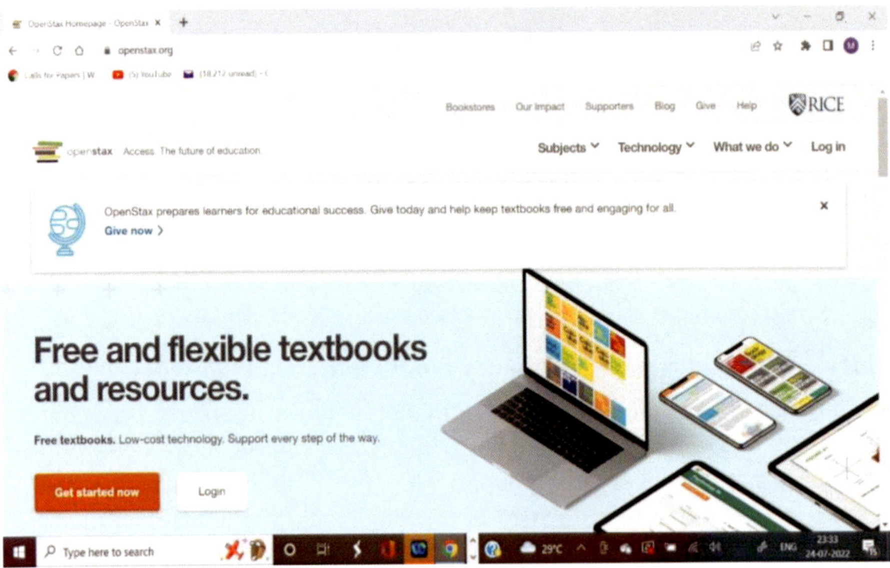

Fig. 1.2: Open Stax Homepage (Source: www.openstax.org) (OPen Stax, n.d.)

e) California State University's MERLOT Project

The project commenced at California State University in 1997. The abbreviation MERLOT stands for Multimedia Educational Resources for Learning and Teaching. The Merlot collection comprises 91,000 educational resources available in 22 distinct languages and in 22 different sorts of materials. That encompasses case studies, online courses, a learning object repository, assignments, and more. Merlot is an invaluable online platform for all needs pertaining to open educational resources. There is no cost associated with being a member of Merlot. Merlot promotes the evaluation and ranking of open educational resources by its members. This allows other users to explore and choose open educational materials based on these ratings and reviews (Rutledge *et al.*, 2001).

Merlot also offers advanced search features that assist customers in refining their own search criteria. The resources can be searched using various criteria such as keywords, languages, title, material kind, material format, peer-reviewed status, copyrighted status, and other options. Merlot features a powerful search mechanism that allows users to search by the ISBN number and discover related books or material.

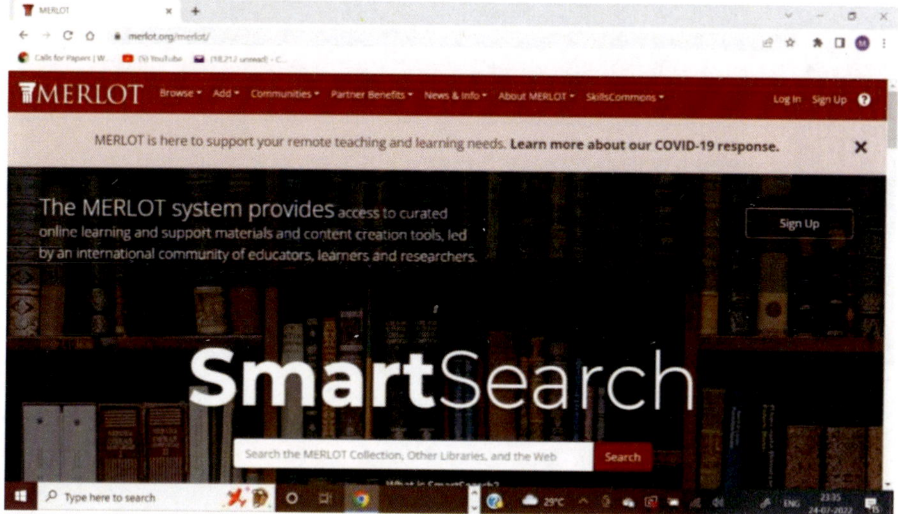

Fig. 1.3: Merlot Homepage (Source: www.merlot.org) (Merlot , n.d.)

f) OER Commons: an Initiative by IKSME

ISKME initiated the launch of OER Commons in 2007. OER Commons is an online library that offers excellent Open Educational Resources (OER) and also provides extensive infrastructure and experienced academics that work together to adapt, evaluate, and use OERs. They also cater to the needs of instructors and students (ISKME, 2007). OER Commons is a comprehensive search tool for all open educational resources. It offers a vast collection of over 50,000 high-quality resources, including university-level full courses, open textbooks, K-12 lessons, activities, worksheets, interactive mini lessons, and simulations. OER Commons is a platform that provides a wide range of educational resources in several subject areas, including applied science (7964), life science (7787), arts and humanities (6193), physical science (6130), education (4907), and social science (4709).

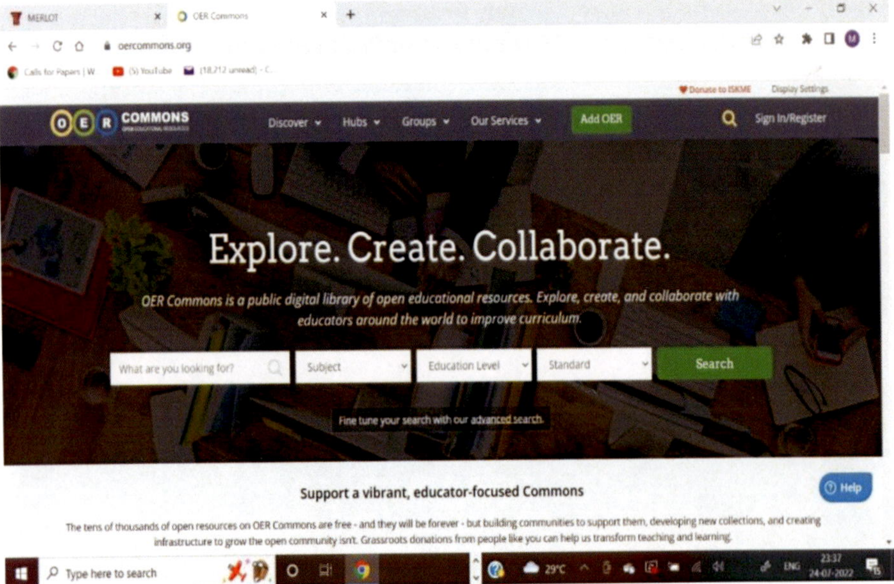

Fig. 1.4: OER Commons Homepage (*Source*: www.oercommons.org) (OER Common: Open Educational Resources, n.d.)

g) Digital Commons Network

The digital commons network serves as a portal that grants unrestricted entry to a vast collection of scholarly publications and open educational resources from numerous global institutions of higher learning. The digital commons collection comprises 3,952,975 works from 642 colleges and universities. The Digital Commons, available through their institutional repository, offers authors, faculty, researchers, and librarians the chance to contribute and store their intellectual work in a freely accessible worldwide repository (Bepress, 2015).

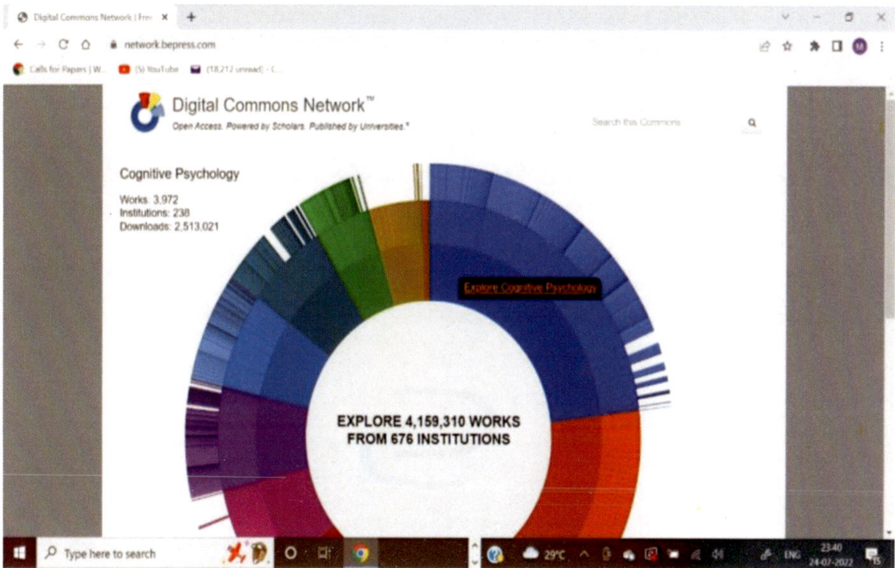

Fig. 1.5: Digital Commons Homepage

(*Source*: www.network.bepress.com) (Be Press, n.d.)

h) European Commission's Open Edu Framework

The "Open Edu Framework" was developed by the European Commission to serve as a support framework for higher education institutions. The Open Edu initiative is included as one of the six priorities of the European Commission, specifically specified within the E&T 2020 priority list (queromi, 2016). This objective emphasizes the importance of fully embracing the digital world in order to promote open and innovative education and training. This focus emphasizes the expansion of education by utilizing open education materials and new technology to facilitate innovative teaching and learning for all. The Joint Research Centre (JRC), a component of the European Commission, provides assistance for this project aimed at expanding access to education. The JRC examined how criteria including as quality, efficiency, equity, and innovation are impacting the adoption of the Open EduProject. Therefore, JRC has published the "Open Education framework for higher education." This framework was specifically developed to assist European higher education institutions in making strategic decisions pertaining to open education. The framework is a practical instrument that encompasses 10 aspects of open education. The initial six elements are the fundamental dimensions, encompassing access, content, pedagogy, recognition, collaboration, technology, and research. The latter four dimensions are transversal, comprising strategy, leadership, technology, and quality.

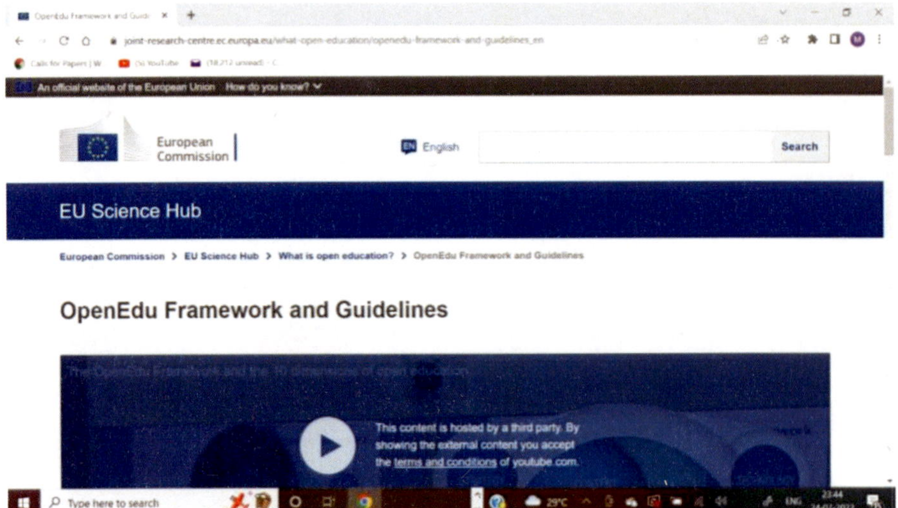

Fig. 1.6: Open Edu Framework Homepage (European Commission, n.d.)

5. Conclusion

"Any implementation of OER becomes an opportunity that presents challenges. OER can save time and money, but as has been noted above, the implementation does not come without costs both one-time and recurring. One-time costs include those of searching, transferring, adapting, assembling, etc., while the recurring costs are for the infrastructure and updating. Note, that there are similar costs when implementing commercial content. Many teachers and learners still do not understand the difference between freely accessible content on the internet and OER, believing that all web resources are OER" (Chen & Panda, 2013).

"We must rise higher to rescue the Sustainable Development Goals – and stay true to our promise of a world of peace, dignity and prosperity on a healthy planet." António Guterres Secretary-General, United Nations. (United Nations, 2022).

References

Baraniuk, R.G. (2008). Challenges and oppurtunities for the Open Education Movement: A Connexions case study. In Opening Up Education: The collective advancement of education through open technology, open content, and open knowledge (pp. 229–246). Cambridge University Press.

Bepress. (n.d.). Retrieved May 10, 2024, from https://bepress.com/

Bepress (2015). Education | Open Access Articles | Digital Commons Network™. https://network.bepress.com/education/

Chen, Q., & Panda, S. (2013). Needs for and utilization of OER in distance education: A Chinese survey. Educational Media International, 50(2), 77–92. https://doi.org/10.1080/09523987.2013.795324

Commonwealth of Learning. (2012). A Government Policy Development Template to Progress Effective Implementation of Open Educational Resources (OER) An Argument for Including OER in Country Policies. http://www.educause.edu/library/resources/dramatically-bringing-

Commonwealth of Learning - Learning for Sustainable Development. (n.d.). Retrieved December 14, 2021, from https://www.col.org/

McGreal, Rory (2017). Special report on the role of open educational resources in supporting the Sustainable Development Goal 4: Quality Education Challenges and Opportunities. International Review of Research in Open and Distributed Learning, 18(7).

Ed Walker (2009). OpenEd2005-A Reality Check for Open Education : Ed Walker : Free Download, Borrow, and Streaming : Internet Archive. https://archive.org/details/OpenEd2005ARealityCheckforOpenEducation

European Commission. (n.d.). Open Education Framework. Retrieved May 10, 2024, from https://joint-research-centre.ec.europa.eu/what-open-education/openedu-framework-and-guidelines_en

Hylen, J. (n.d.). Open educational resources: Opportunities and challenges. http://www.okfn.org

ISKME (2007). Browse OER Materials | OER Commons. https://www.oercommons.org/oer

Korngold, A. (2023). A better World, Inc. In A Better World, Inc. Springer International Publishing. https://doi.org/10.1007/978-3-031-31553-4

Mackinnon, T. P.-N. s. mannas. (2016). A Meta analysis of Open Educational Communities of practice and sustainablity in higher educational policy. 19(1). https://doi.org/10.4000/alsic.2983

Merlot . (n.d.). Retrieved May 15, 2024, from https://www.merlot.org/merlot/

Midha, M. (2022). Open Educational Resources: An insight into Indian initiatives. In B. Partap (Ed.), Cutting Edge Technology and Libraries (pp. 57–66). New Delhi: NIPA.

Midha, M. (2023). Role of international organisations in awareness and promotion of open educational resources. Library Herald, 61(1), 39–47. https://doi.org/10.5958/0976-2469.2023.00004.0

Midha, M., & Kaur, B. (2022). Exploring the awareness of Open Educational Resources among users of Central University Libraries of North India. IASLIC Bulletin, 67(2), 110–122.

Midha, M., & Kumar, J. (2022). Users' awareness and usage of open educational resources in central universities of North India. DESIDOC Journal of Library and Information Technology, 42(1), 47–56. https://doi.org/10.14429/DJLIT.42.1.17304

OECD (2013). Education at a Glance. https://www.oecd-ilibrary.org/education/education-at-a-glance-2017_eag-2017-en

OECD (2017). Education at a Glance. https://www.oecd-ilibrary.org/education/education-at-a-glance-2017_eag-2017-en

OER Commons: Open Educational Resources (n.d.). Retrieved May 15, 2024, from https://oercommons.org/

Open Learn (2019). An introduction to Open Educational Resources (OER) - OpenLearn - Open University - H800_1. https://www.open.edu/openlearn/education-development/introduction-open-educational-resources-oer/content-section-0?active-tab=description-tab

Open Learn edu. (n.d.). The Open University. Retrieved May 15, 2024, from https://www.open.edu/openlearn/

OPen Stax (n.d.). Open Staxs: Access the future of education. Retrieved May 15, 2024, from https://openstax.org/

Queromi (2016). Opening up Education: the OpenEdu Framework Andreia Inamorato dos Santos @aisantos. http://bit.ly/2aFm9x2

Rutledge, J., Petersburg, S., & College, J. (2001). MERLOT Multimedia Educational Resource for Learning and Online Teaching. www.merlot.org

Smith, M. S., & Casserly, C. M. (2006). The promise of open educational resources. Change: The Magazine of Higher Learning, 38(5), 8–17. https://doi.org/10.3200/CHNG.38.5.8-17

Sustainable Development Goals. (2023). United Nations; Department of Economic Affairs. United Nations; Department of Economic Affairs

UNESCO (2012). Open Educational Resources (OER). https://en.unesco.org/themes/building-knowledge-societies/oer

UNESCO (2019). UNESCO Recommendation on Open Educational Resources (OER). https://en.unesco.org/news/unesco-recommendation-open-educational-resources-oer

UNESCO, & CoL. (2019). Guidelines on the development of open educational resources policies. http://oasis.col.org

United Nations (2022). The Sustainable Development Goals Report. https://unstats.un.org/sdgs/report/2022/

Uvalić-Trumbić, S., & Daniel, J. (2016). Sustainable Development Begins with Education. Journal of Learning for Development, 3(3). https://doi.org/10.56059/jl4d.v3i3.164

9

Designing the Future Libraries: Sustainable Strategies at Modern Era

Dr. Sucheta Bhattacharjee

Assistant Librarian, Techno India University, Agartala - 799 004, Tripura, India

E-mail: sucheta.aus@gmail.com

Abstract

Eco-friendly technologies are so essential in modern society, and libraries are leading the way in this regard. Nations all over the world are creating affordable solutions to deal with environmental issues. Acknowledging the advantages of environmentally sustainable methods, the computer sector highlights their significance for cost-effectiveness and public relations. Libraries use cutting-edge technologies in the field of information and communication technology to effectively serve information consumers. Driven by environmental awareness initiatives, libraries are incorporating eco-friendly procedures into their everyday operations, bolstered by resources that advise institutions on energy efficiency and paper reduction. The placement of libraries must be carefully considered as more universities open up in India in an effort to improve the quality of education. In order to promote environmental awareness, this study suggests choosing locations for new libraries while utilizing eco-friendly architectural design practices. Libraries provide a compelling example of the way public institutions may support long-term sustainability and environmental protection while also enhancing the reading experiences of future generations by integrating green concepts into their physical environments.

Keywords: *Eco-friendly library, Sustainable library, Recycling materials.*

1. Introduction

The term "sustainable development" was originally used in a 1953 article by Joseph L. that was published in Land Economics. The notion was employed to depict the degree or pace at which an environmentally friendly resource is employed for a particular objective while guaranteeing its long-term availability. Organizational dynamics, social interactions, and cultural norms all change with the times. These changes are making environmental problems worse and bringing up questions about pollution and the requirement for a safe natural environment. They are being caused by changing human behavior and population expansion (Kamińska, 2021; Gaffar *et al.*, 2021). Organizations have been forced to review their approaches to dealing with these urgent environmental issues as a result of the increased awareness. Academic libraries have been particularly well-positioned to implement eco-friendly practices because they are service-oriented establishments to adopt eco-friendly practices in their operations (Pangail, 2017).

Improving user services, increasing productivity, and offering various forms of relaxation are the main objectives of technology. But just as the Industrial Revolution has created problems with the environment, the technology revolution has also brought forth problems with trash and its effects on the environment (Gaffar *et al.*, 2021). There has been a discernible movement from home concerns toward business boardrooms where environmentally conscious and sustainable living are prioritized, as customers as well as companies appreciate the need of environmental stewardship. Growing social and federal regulations that promote sustainability are supporting this movement, which is a reflection of a larger cultural shift towards eco-conscious behaviors (Mohammed, 2021). This changing viewpoint emphasizes how important it is to incorporate eco-friendly procedures into every facet of corporate operations and technology.

Global climate change is one of the 21st century's most urgent problems. Buildings in the European Union contribute 47% of the total emissions of carbon dioxide, making it imperative to address how development and operation affect the environment. It is imperative to design, construct, and manage buildings sustainably in order to lessen the negative consequences of climate change. Institutions of higher learning, such as universities & colleges are essential to this endeavor. These institutions provide scientific research and train staff, but they also act as leaders and examples in environmental sustainability (Liu *et al.*, 2019). Academic institutions may drastically reduce their environmental footprint and set a good example for others to follow by adopting environmentally friendly practices through their operations and infrastructure (Pangail, 2017).

2. Environment Friendly Library Design

With their dual roles as communal hubs and educational institutions, libraries have a special place in society. It is essential to rethink library building design and construction in order to minimize environmental effect and maximize efficiency, as the demand for sustainable living practices grows. Libraries can benefit from sustainable architectural design by using a comprehensive strategy that takes into account the building's entire ecological footprint, waste management, material selection, and energy use. Libraries have used state-of-the-art technology to meet the needs of fervent information seekers (Kamińska, 2021).

Consequently, a novel movement dubbed "green libraries" is starting to surface. This revolution is not only thought-provoking but also necessary for the long-term sustainability of the library industry. James and Suzanne proposed in 1991 that green building principles, such as energy, water, and material conservation, should be maximized in library architecture. Since then, Lewis has suggested ways to lower building energy consumption through the use of energy-saving technologies (Liu *et al.* 2019). Library sustainability literature has expanded since 1990, with an emphasis on green library operations and practices. Sustainability is now a key factor in library design, and there are numerous instances of sustainable libraries worldwide (Afacan, 2017). Preferred to as "green building" or "sustainable architecture," this architectural approach prioritizes the interaction between buildings and their local and global environments. The primary objective is to create buildings that are energy-efficient and environmentally conscious.

The rapidly growing field of environmentally conscious technology is focused on novel scientific and technological methods that are environmentally helpful. The goal of this discipline, sometimes referred to as "green technology," is to preserve our natural resources. This involves both the development of new technologies and the improvement of existing ones (Tundele, 2015). Environmental protection and sustainability are essential elements of green management. Going green with infrastructure, operations, and goods may result in major changes to a company's operations. As a result, their sustainability business might benefit more. Companies and institutions that use green innovation preserve a hygienic and safe environment. This can be accomplished by making use of products, services, and technology that don't deplete the natural world's resources (Afacan, 2017).

When every aspect is taken into account an environmentally friendly library architecture not only benefits the local community by offering a rich atmosphere full of resources for study and inquiry, but it also shows a dedication to environmental care and sets an excellent precedent for other both governmental and nonprofit structures.

3. Factors Effecting Green Library Building

The foundation of sustainable library design is careful site selection and planning. By determining a location with optimal natural light and ventilation, one can reduce the need for artificial lighting and mechanical heating or cooling.

Additionally, through the use of energy-efficient construction techniques, such as LED lighting, high-performance HVAC systems, and alternative energy sources like solar panels, one can significantly lower the building's energy consumption. Finally, the layout of an environmentally responsible library should foster energy efficiency and user comfort.Designing flexible spaces that can be adapted to various functions and using natural daylight to illuminate reading areas can enhance user experience while reducing the reliance on artificial lighting. Moreover, incorporating features like passive solar design and thermal mass can help maintain stable indoor temperatures. Green library buildings are influenced by several factors that contribute to their sustainability and environmental impact. The choice of construction materials, energy-efficient designs, and renewable energy sources significantly affect a library's carbon footprint (Bhattacharjee, 2019).

Sustainable landscaping and rainwater harvesting contribute to water conservation, while proper waste management systems minimize environmental impact. Location and site selection play a crucial role, influencing accessibility, transportation, and the overall ecological context. Interior features like efficient lighting, ventilation, and eco-friendly furniture contribute to energy efficiency and occupant well-being. Additionally, community engagement and awareness programs promote sustainable practices among library users, fostering a culture of environmental responsibility. Overall, a holistic approach to these factors ensures green libraries serve as eco-friendly, resource-efficient, and socially responsible structures (Liu *et al.* 2019).

In India, several initiatives focus on promoting green and sustainable practices in library buildings. The Indian Green Building Council (IGBC), a part of the Confederation of Indian Industry (CII), offers the Green Library Building rating system. This system assesses libraries based on factors such as energy efficiency, water conservation, use of sustainable materials, and indoor environmental quality. Many Indian libraries, particularly in academic institutions, are adopting green building principles. The LEED-India green building rating system is a widely recognized national and international standard for the design, construction, and operation of high-performance green buildings. To make one library sustainable some of the key areas have to follow:

a) **Site Location:** Strategic site selection considers climate change resilience, anticipating potential impacts and implementing necessary measures. Harmonizing with local aesthetics and infrastructure compatibility further enhance the library's cultural relevance and operational efficiency. In essence, the site's thoughtful selection is fundamental, shaping the library's environmental impact and community integration. It takes a sustainable site plan to make the environmentally friendly library project a reality. A sustainable site plan for a green library project is one that meets the project's objectives and minimizes environmental impact while meeting users' diverse information needs (Biswas, 2021). The site plan needs to be modified to accommodate project specifics without compromising environmental concerns. Numerous guidelines for the site selection process have been provided by LEED (Ragheb, 2016).

- The library should be easily accessible to all kind of library user by public transportation and situated in a densely populated area.
- To lessen the impact of heat, there should be environment friendly green surroundings with natural shading.
- Site should have environment friendly modes of transportation, walking and biking, allow libraries to create welcoming and cozy pathways for patrons to enter the building.(Bangar, 2018)

b) **Water Conservation:** Water conservation in green libraries is essential for sustainability. Efficient plumbing systems, rainwater harvesting, and landscaping practices reduce water consumption. Implementing water-efficient fixtures and promoting user awareness ensures responsible water usage. Libraries should design effective water-saving measures, such as

- Collecting and storing rainwater for use in landscaping around buildings.
- If there are lakes and pond near the library building those should be conserved which will help to collect the water naturally.
- It is possible to waste water and reuse it for irrigation (Gaffar, 2021).

c) **Energy Efficiency:** A library should have an adequate number of windows, including glass windows and skylights, to let natural light the space. This way, during the day, no artificial lighting is needed.

- It is better to use LED Lights to reduce power consumption.
- Renewable energy sources like solar energy can be used for light the library during evening or night time. A solar panel might be installed on the library building to generate solar energy that would then be used inside the building. (Wang, et al. 2021).

d) **Materials:** By enhancing the indoor environment quality and reducing the need for mechanical systems, the library's durability can be achieved through the use of recycled materials in its construction. The selection of library materials is the primary responsibility; waste is minimized.

- Utilize locally sourced store components, saving on fuel and transportation expenses.
- The new library building should use much of the modern technology that has emerged in green building. India has a lot of rules and regulations, so a green building outside of the country recycles and uses environmentally friendly materials.
- Naturally obtained material like stone, bamboo can be used as a material. (Hartini, 2020)

e) **Indoor Air Quality:** Implementing proper ventilation systems, using low-emission materials, and regular air quality monitoring are essential. This ensures a healthy and comfortable environment for library users while aligning with the sustainability goals of green building design. To keep the air circulation good big windows and door can be added. Indoor plants may be planted inside the library.(Bangar, 2018)

f) **Green Roofs:** A library building can benefit from a green roof in a number of ways, including rainwater absorption, insulation, wildlife habitat creation, benevolence enhancement and stress reduction for those nearby by creating a more aesthetically pleasing landscape, and lowering urban air temperatures and reducing the heat island effect (Vandermeulen, 2011).

g) **Green Walls:** Adding plants to the façade of a building is also referred to as vertical greenery. Where skyscrapers are the primary building type in the built environment, green walls can cover more exposed hard surfaces than green roofs can (Ragheb, 2016).

4. Discussion and Suggestion

Building a green library with eco-friendly initiatives while remaining cost-effective involves strategic planning and innovation. Begin by selecting a site that optimizes natural resources like sunlight and wind. Design energy-efficient structures using sustainable materials to minimize environmental impact. Integrate rainwater harvesting systems for water conservation, and implement efficient waste management practices. Utilize energy-efficient lighting and heating, ventilation, and air conditioning (HVAC) systems to reduce operational costs. It is essential to understand the advantages of this innovation. The advantages it offers for people's lives and businesses should be known. It can lower people's expenses and maintain the health of the planet. They ought to be aware that buying green products can save them money and lead to better health. Government grants and financial support for the upkeep of these kinds of libraries should be used as a means of promoting green libraries.

To guarantee sustainability, the library can set up sessions for strategic planning and thought. Whenever feasible, governments and universities should work to transform libraries into green structures (Kamińska, 2021).

In order to meet community needs and handle today's environmental issues, sustainable library design is essential. Libraries have the potential to drastically lower their carbon footprint by giving energy efficiency first priority. Lowering operating expenses and environmental effect can be achieved by putting measures like energy-efficient lighting, passive solar design, and renewable energy sources like solar panels into practice. Furthermore, utilizing sustainable materials reduces the environmental effect of building and promotes waste reduction. Examples of these materials include recycled or quickly renewable resources. Conserving water is still another essential element(Tundele,2015). Water consumption can be decreased and stormwater can be efficiently managed by including low-flow fixtures and rainwater harvesting devices. In addition to improving aesthetics, living walls and green roofs help sustainability objectives by promoting biodiversity and thermal insulation (Gupta, 2021).

To maximize the benefits of sustainable design, libraries should focus on flexible, adaptive spaces that optimize natural light and ventilation, reducing the need for artificial lighting and mechanical systems. Prioritizing indoor air quality through the use of low-VOC materials ensures a healthier environment for users.

5. Conclusion

One of the most important global concerns is protecting and maintaining life on Earth. Green libraries have the power to encourage people to conserve the environment by taking proactive measures. As a result, government backing for the creation of green libraries is crucial. Given that many Indian libraries have successfully launched green programs, librarians ought to take an active role in the push for green libraries. As a long-standing institution, the Green Library enhances the environment and fosters wellness. Using environmentally friendly practices not only supports sustainability objectives but also strengthens the library's environmental responsibilities by giving stakeholders and patrons a pleasant, green impression.

Involve the community in the process to promote environmental responsibility and a sense of ownership. Look for joint ventures with nearby companies to provide affordable, eco-friendly products. Grants and incentives from the government might also help with start-up expenditures. For efficiency, take into account prefabricated materials and modular construction methods. By lowering energy use and maintenance, putting these steps into practice not only ensures long-term cost savings but also conforms to eco-friendly principles. In conclusion, a green library can be constructed by striking a balance between economical and environmentally beneficial measures, resulting in a resource that is both accessible and long-lasting for the neighbourhood.

References

Afacan, Y. (2017). Sustainable library buildings: Green design needs and interior architecture students' ideas for special collection rooms. The Journal of Academic Librarianship, 43(5), 375–383. https://doi.org/10.1016/j.acalib.2017.07.002

Bangar, M. S. (2018). Green libraries in India: An overview. Knowledge Librarian: An International Peer Reviewed Bilingual E-Journal of Library and Information Science, 16(2), 98.

Biswas, A. (2021). Site selection and planning for green library. 10.13140/RG.2.2.13438.31049.

Gaffar, S. A., N, Sindhu P., & Kumar, S. (2021). The green library initiative in Indian perspective: A study. Library Philosophy and Practice (e-journal), 6189.

Gupta, S. (2021). Green library: A strategic approach to environmental sustainability. International Journal of Information Studies and Libraries, 5(2), 82-92.

Gupta, Saumya (2020). Green library: A strategic approach to environmental sustainability. International Journal of Information Studies and Libraries, 5(2), 82-92. SSRN: https://ssrn.com/abstract=3851100

Hartini, L., Wibawa, B., Situmorang, R., & Raissa, F. (2020). Interior design of national library with environmentally sustainability materials. IOP Conference Series: Materials Science and Engineering, 1007(1), 012004. https://doi.org/10.1088/1757-899X/1007/1/012004

Kamińska, A.M., Opaliński, Ł., & Wyciślik, Ł. (2021). The landscapes of sustainability in the library and information science: Systematic literature review. Sustainability, 14,441.

Liu, Q., Chen, J., Wu, K., & Du, S. (2019). Sustainable design practice of university library building: Take the guangdong university of finance & economics library as an example. IOP Conference Series: Earth and Environmental Science, 242, 052046. https://doi.org/10.1088/1755-1315/242/5/052046

Meher, P. and Parabhoi, R. (2017). Green library: An overview, issues with special reference to Indian libraries. International Journal of Digital Library Services, 7(2).

Mohammed, S.I. (2021). Advantages of green technology to mitigate the environment problems. IOP Conference Series: Earth and Environmental Science, 779(1), 012133. https://doi.org/10.1088/1755-1315/779/1/012133

Pangail, R.K. (2015). Green libraries: Meaning, standards and practices. Episteme: an online interdisciplinary, multidisciplinary & multi-cultural journal, 4(3).

Rabidas, S. (2016). Green library buildings: A sustainable process. IJARIIE, 2(6).

Ragheb, A., Elshimy, H., & Ragheb, G. (2016). Green architecture: A concept of sustainability. Procedia- Social and Behavioral Sciences, 216. DOI: 10.1016/j.sbspro.2015.12.075

Tundele, S. (2015). Eco-friendly technology-key for sustainable development. International Journal of Research Granthaalayah, 3(9), 1–3. https://doi.org/10.29121/granthaalayah.v3.i9SE.2015.3171

Vandermeulen, V., Verspecht, A., Vermeire, B., Van Huylenbroeck, G., & Gellynck, X. (2011). The use of economic valuation to create public support for green infrastructure investments in urban areas. Landscape and Urban Planning, 103(2), 198–206. https://doi.org/10.1016/j.landurbplan.2011.07.010

Wang, Z., Wang, W., Zhang, Y., Song, J., & Zhu, L. (2021). Research on energy-saving design strategies of libraries in severe cold regions: Taking a university in Xinjiang as an example. Energy Reports, 7, 1101–1113. https://doi.org/10.1016/j.egyr.2021.09.160

10

Fostering Sustainable Library Services in the Era of AI: Insights from Bir Chandra State Central Library, Agartala

Dr. Sudip Bhattacharjee

Assistant Professor, Department of Library and Information Science, Maharaja Bir Bikram University, Agartala-799 004, Tripura, India, E-mail: sudipbht12@gmail.com

Abstract

In an era where artificial intelligence is reshaping information access and management, fostering future library services is critical. Libraries must adapt to include AI technologies to improve resource accessibility and streamline operations while maintaining their primary objective of community assistance. The study sought to examine how the Birchandra State Central Library provides different community information services to its users. This study dives into the common function of public libraries as important community information distributors, which are frequently acknowledged as public institutions. Public libraries play an important role in providing individuals with the information they need to face their daily challenges. Furthermore, this study emphasizes the importance of Birchandra State Central Library in sustainable growth by providing a variety of services. The paper defines community information service (CIS) as a systematic approach for preserving and disseminating a diverse range of information within a community. The data demonstrate that the Birchandra State Central Library considerably contributes to community information services, constituting an integral element of its user.

Keywords: *Birchandra State Central Library, Library services, Sustainable development, Public library, Community service.*

1. Introduction

A public library is a repository of diversified material that serves as the primary hub for community information the spread. It provides fair access to a wide range of information and data, meeting the diverse demands of users. Services must be inclusive, with information appropriate for all age groups and the incorporation of both classic materials and contemporary technologies across many mediums. In 1987, the ILA defined the purpose of public libraries, stating that they should serve as independent training centers for community activities. Public libraries play a pivotal role in fostering community engagement, welcoming all members to participate in various programs. Initially established by the public, these libraries later received government funding for expansion and other requirements. The systematic development of India's public library system ensued after the establishment of the Raja Rammohan Roy Library Foundation (RRRLF) in 1972.

The enactment of library legislation in several states further facilitated well-planned library services, benefiting both those states and the entire nation. This evolution underscores the vital role of public libraries as dynamic sources of knowledge and community development.Community Information Service (CIS) is defined as information that addresses problems and aids individuals in making the best decisions when necessary. A committee formed by the Library Association in 1980 defined CIS as services that help people and groups resolve everyday concerns and participate in the democratic process. These services are designed primarily for people who have limited access to other forms of support and address essential issues such as housing, employment, and rights. These services contain community information, which is required for daily functioning (Cobigo, 2016; Das, 2018).

Effectively utilizing information resources is integral to the seamless functioning of a community development system, offering support across all aspects of community life. Public libraries, by providing informational support, play a crucial role in raising awareness and motivating communities to actively participate in their own development. To effectively serve their communities, public libraries must assess how, with the available resources, they can contribute (MacQueen,2004). Community involvement is paramount for the development of a locality, with public libraries serving as vital connection points between information and society. These libraries offer diverse services that contribute to societal growth, and in turn, communities actively contribute to shaping the library's future. This reciprocal relationship underscores the significance of public libraries as dynamic agents fostering community development and knowledge dissemination (Chatterjee, 1970; Sarkar, 2019).

Public libraries, strategically positioned within communities and equipped with resources for human development, are considered ideal partners in community development initiatives. Their influence extends beyond the realm of knowledge dissemination, impacting social and economic development as a driving force for progress (Sarkar, 2019; Chidozle, 2022). Community information serves as a problem-solving tool, aiding individuals in making informed decisions when needed. Often referred to as "survival information," it encompasses knowledge essential for day-to-day community survival. Despite the critical importance of information, public libraries are essential for offering free access to a diverse audience, ensuring that valuable knowledge is available to everyone in the community (Joselin, 2015). The diverse contributions of public libraries highlight their fundamental role in enhancing community well-being and fostering development.

2. Birchandra State Central Library

The Birchandra State Central Library, formerly known as the public library of Tripura, has a rich history dating back to its establishment in 1896 during the Maharaja's Period. Originally named Bir Chandra Library, it was dedicated by King Maharaja Radhakishore Manikya in honor of his late father, Maharaja Bir Chandra Manikya. Functioning as a social institution and information center, Birchandra State Central Library houses a comprehensive collection of reading and information materials in Bengali, English, Hindi, and various regional languages such as Kokborok, Manipuri, Chakma, and more. On June 27, 2013, Birchandra State Central Library expanded its services by inaugurating the digital library of Tripura, a project executed in collaboration with CDAC under the Digital Library of India scheme by the Government of India. This initiative, equipped with four computer machines, facilitates users and scholars in accessing information. The library aims to extend this digital scope to other public libraries in the future. Additionally, Birchandra State Central Library stands out by providing cybercafé services, further contributing to the accessibility and utilization of information in the modern era (Bhattacharjee, 2019).

The Bir Chandra State Central Library in Agartala provides a wide range of services and facilities, including a large collection of books, journals, and digital resources including e-books and online databases. Users can take advantage of peaceful reading rooms, reference services for research assistance, and community programming such as workshops and lectures. The library has a separate children's area, public computers with internet access, and interlibrary loan services to increase resource availability. It also holds archives and special collections, which preserve rare documents and serve as vital resources for historical research. These aspects highlight the library's function as an important educational and cultural hub.

Table-1 Name of library services and facilities of Birchandra State Central Library

Sl. No	Services/ Facilities
1	Lending service
2	General reading facilities
3	Career guidance service
4	Reference service
5	Free news paper reading service
6	Senior citizen section
7	Wi-Fi facility
8	Cyber café service
9	Reprography service
10	Children corner section
11	Women's section
12	Journal or Periodical section
13	Ramaprasad Govesanagar (for Researcher)
14	Braille section

Table-1 shows that, Birchandra State Central Library which services or facilities library provided for their users or community peoples and library play most important role in sustainable development in community which is reduced the problems of people who faced daily life, which are listed below:

Lending Service: Lending service also known as circulation service. The main function of lending service is to issue and return the books to the library users and Government Departments. The service time 07:30 AM to 08:30 AM for central library and 12:30 PM for others libraries. Only those books are lent out which is printed books.

General Reading Room Facility: The reading room are daily open at 8a.m to 08:00 PM except Monday. Users without library card do not entrance their room and do not take their own books.

Career Guidance Service: This service is most important for unemployment peoples and students. The service time 07:30 AM to 08:30 PM. This section holds various types of books like competitive exam books, entrance exam books, general knowledge books etc. Many students in our society belong in below poverty line who cannot afford this types of books but in this way library provide this service which is more helpful to the students for their study.

Reference Service: This service is to get quick information of a specific subject. The library staffs always help their users for quickly collected their information. The service time 08:00 AM to 8:30 PM.

Free News Paper Reading Service: This service is most important for community peoples. This service totally free for all peoples including students, old peoples etc. the service time 07:30 AM to 08:30 PM.

Wi-Fi Facility: Birchandra State Central Library provides this facility only for Bonafide members or users with computer basic knowledge. For this facility they also show computer operating card and, in this way, they also deposited INR 300.00 as security amount, which is refundable. In our community there are many people belongs below poverty level who cannot cost available money to buy a computer. In these regard Birchandra State Central Library play most important role in sustainable development, which reduced poverty in community and give better life.

Senior Citizen Service: This section was established in 2011. This section holds senior persons related books. This section also systematically arrange their documents and they easily find their information from the rack as per their needs. This section also available various daily newspapers.

Cyber Café Service: Only Birchandra state central library provide this service. The service time 12.30 PM to 07:00 PM. This section was established in 2011, 14th September. The users, research scholar daily come and collected their information at a time.

Reprography Service: Reprography service hour of Birchandra State central Library 12.00 PM to 07:00 PM. Nominal cost of per A4 size page of Birchandra State Central Library is 1rs.

Children Corner Section: This section is most important for children's. This section holds various story books with English, Bengali, Hindi and Kokborok, toys, video games, computer game. About 12,243 nos. books and 650 children members of Birchandra State Central Library. In our society, most of the families belong poverty line who cannot fulfil their children demands but the library provide this service to the children for increase their interest and reading habits, *etc*.

Braille Section: Currently, there are three Braille libraries in Tripura, namely Birchandra State Central Library, Nazrul Granthagar, and Unakati District Library. Birchandra State Central Library inaugurated this section on May 3, 2012. The service is provided free of charge, operating from 10:00 AM to 05:30 AM. Presently, Birchandra State Central Library has 48 members, and it offers the Jaws software to facilitate computer operation for users. To effectively support residents, public libraries must adapt their offerings to align with the needs and strengths of local communities. This necessitates thorough research and analysis to tailor services for maximum community impact.

3. Sustainable and Modern Library Service Provided by Bir Chandra State Central Library

Public libraries must engage in sustainable community service to meet pressing needs and improve the social, economic, and environmental well-being of the community. Libraries play a crucial role in guaranteeing fair access to knowledge, which is essential for advancing literacy and aiding in a range of educational sector. Serving as the focal points of community development, they offer priceless tools for enhancing one's skill set, locating employment, and encouraging civic engagement. Libraries need to change with the times to stay relevant and useful in their communities. This entails adopting creative ideas and incorporating new technologies to satisfy a range of user needs. Libraries provide venues for learning and personal development through a variety of programs and seminars in addition to providing access to digital and physical resources.

A fundamental aim of long-term community service in libraries is inclusivity and same can be observed in the case of Bir Chandra State Central Library, Agartala. Libraries establish relationships by close collaboration with local institutions and groups, which improves their capacity to successfully serve the community. Through these partnerships, libraries are able to provide a broad range of programs that target a variety of community needs, including digital literacy, language proficiency, and job development, and that appeal to a range of demographics, from young children to senior citizens. Additionally, public libraries contribute significantly to the development of a strong, cohesive community. Their dedication goes beyond just supplying resources; it also involves encouraging environmental stewardship via sustainable practices including trash reduction programs and energy-efficient building operations. Additionally, libraries promote social justice by guaranteeing that all members of the community have equitable access to their resources and chances for lifelong learning.Bir Chandra State Central Library also provide various kinds of Community Information Services which are given below:

a) **Educational Guidance:** Bir Chandra State Central Library is provided various information about education and they give to the people proper way to choose decision. Library takes good initiative to developed sustainability which is reduced lack of literacy level in our community.

b) **Career Guidance:** Naturally, having the right education is crucial to landing the best job in your field. Naturally, the prerequisites or stepping stones for selecting one's own ideal career are the credentials needed to succeed in it. Top jobs are available to those with the necessary credentials, and choice would be abundant. Career counselling can help students choose the classes that are right for them. Bir Chandra State Central Library Give this service to the students and unemployed peoples of the community. This service is a ground breaking opportunity and it show the way to take the right path to their career. This leads them to choose the proper decision to grow their career.

c) **Competitive Examination Guidance:** Competitive exam are the first stepping stone for those who are seeking for government jobs. This kind of exam shows the ability and knowledge of the candidates from different field. Bir Chandra State central Library is providing the materials, weekly classes and proper books to those candidates who are preparing for the competitive exam. They also give the individual attention to the candidate not only that they also make sure the availability of the books which ishelpful the candidates to wants for their preparation.

d) **Job Notification and Guidance:** Bir Bircharndra Library is giving this service for past 4 month, Head librarian Dilip Kumar Das, Ramkrishna Bhattacharjee and Samir Das has taken this initiation to provide this community service and more then hundreds are taking this service. They have invited various TCS Officers, TPS officers, SP, SDM, Ex- Principal, Professors and other renowned people. They provided the classes and guidance to the candidates. Among the students some of the candidates already have cleared various exams and joined in their jobs.Library provide Group discussion which is helpful the candidates can share their thoughts and Knowledge, prepare for different interviews which is reduced unemployment and increase income level in our community.

4. Discussion and Conclusion

Libraries are changing from being seen of as knowledge stores to becoming active participants in the development of communities. The Bir Chandra State Central Library is a prime example of this change, having gone from being a basic book lending facility to a vital hub for a variety of community services. The library is dedicated to sustainability and runs a number of programs that deal with the interrelated issues of social, economic, and environmental sustainability. Its wide approach, which goes way beyond typical library services, reflects its passion. The library works in a variety of fields, including health, education, women's empowerment, self-employment, and career training. Through user orientation programs, expert discussions, and health awareness campaigns, it actively responds to community needs. Health initiatives are designed not just to tackle immediate issues but to promote long-term sustainability. For instance, the provision of nutritious food charts highlights a preventive approach to healthcare, reinforcing the library's role in fostering overall community wellness.

Additionally, the library plays a crucial role in women's empowerment by offering skill development workshops and raising awareness about legal rights. These efforts are instrumental in combating social issues such as child marriage, with targeted campaigns aimed at educating and empowering women. By integrating these comprehensive services, Bir Chandra State Central Library demonstrates its commitment to enhancing community life and contributing to sustainable development.In promoting economic sustainability, the library plays a pivotal role in disseminating information about vocational courses, empowering individuals with skills that enhance self-confidence and reduce economic challenges and unemployment. This aligns with the library's broader objective of achieving sustainable development within the community.

Public libraries, such as Birchandra State Central Library, function as social institutions, bridging gaps and actively addressing community needs. They have evolved into dynamic hubs for community information services, extending their outreach beyond students to encompass individuals of all backgrounds without discrimination. The library's proactive stance involves organizing user orientation programs and awareness campaigns to better understand and address community needs. Community information services are pivotal in enhancing knowledge, decision-making capacity, and self-confidence within a community. By providing a vast array of information on education, health, self-employment, and career guidance, public libraries serve as channels for effective communication.

The library's role as a repository of historical and cultural knowledge further enriches community life. While Birchandra State Central Library has made commendable strides in offering community information services, this study highlights opportunities for expansion. Recommendations include incorporating additional services such as information on agriculture, horticulture, animal husbandry, and financial assistance. By broadening the scope, the library can further contribute to the comprehensive development of the community, addressing a more extensive array of needs and fostering sustainability in all aspects of community life. Birchandra State Central Library transcend traditional roles, actively engaging with diverse community needs, and contribute to social, economic, and environmental sustainability. The ongoing commitment to expand services ensures that libraries remain dynamic agents of positive change, fostering a resilient and empowered community for the future.

References

Agustín-Lacruz, C., & Saurin-Parra, J. (2020). Library services to diverse communities in Europe: The case of the Roma Community in Spain. The International Journal of Information, Diversity, & Inclusion, 4(3/4), 20–35. https://www.jstor.org/stable/48645283

Akhter, Md. Yeosuf, Deb, Prasenjit, and Biswas, Jahar (2021). Role of public libraries with special reference to women empowerment through Kanyashree Prakalpa in West Bengal of India. Library Philosophy and Practice (e-journal), 6099. https://digitalcommons.unl.edu/libphilprac/6099 .

Chatterjee, P., & Koleski, R. A. (1970). The concepts of community and community organization: A review. Social Work, 15(3), 82–92. https://doi.org/10.1093/sw/15.3.82

Chatterjee, S., Samanta, M., & Dey, S. (2020). Role of public libraries in providing community information services. https://ssrn.com/abstract=3535051 or http://dx.doi.org/10.2139/ssrn.3535051

Chidozle, O. (2022). The roles of public library in community development in the 21stCentury. https://www.researchgate.net/publication/365489157

Cobigo, V., Martin, L., & Mcheimech, R. (2016). Understanding community. Canadian Journal of Disability Studies, 5(4), 181. https://doi.org/10.15353/cjds.v5i4.318

Hazarika, Sanjoy Kumar (2020). Public library as community information service provider: A study in North East India. Library Philosophy and Practice (e-journal). https://digitalcommons.unl.edu/libphilprac/3843

Joselin, J. Jaba, & Panneerselvam, P. (2015). Community information services in public libraries– A solution to information poverty. https://www.researchgate.net/publication/280244522

MacQueen, K.M., McLellan, E., Metzger, D.S., Kegeles, S., Strauss, R.P., Scotti, R., Blanchard, L., & Trotter, R.T. 2nd. (2001). What is community? An evidence-based definition for participatory public health. Am J Public Health, 91(12), 1929-38. DOI: 10.2105/ajph.91.12.1929. PMID: 11726368; PMCID: PMC1446907.

Ndungu, J., & Otike, J. (n.d.). Community information services in public libraries: A case study of KNLS Nairobi County. https://ssrn.com/abstract=3189375 or http://dx.doi.org/10.2139/ssrn.3189375

Rakshikar, Nidhi. (2012). Career guidance services in academic libraries: A study of college libraries in Mumbai. Entire Research, 4. 107-112.

Sarkar, A. (2019). Public libraries and community information services: An experience in Blocks of Sundarban Region, South 24 Parganas. Journal of Information and Knowledge, 56(1), 32–35. https://doi.org/10.17821/srels/2019/v56i1/131264

11

Managing Anger in the Library Workplace: Strategies for a Safe and Productive Environment

Rohan Makwana

Library Associate, INFLIBNET Centre (An IUC of UGC), Infocity Gandhinagar - 382 007 Gujarat, India, E-mail: raviraj.rohan@yahoo.com

Abstract

Anger management is a critical issue in the library workplace, where tensions can run high and conflicts can escalate quickly. Unmanaged anger can lead to decreased productivity, strained relationships, and even violence. This article explores the importance of anger management in libraries, highlighting the warning signs of anger and the cultural and organizational factors that can contribute to its escalation. Drawing on research and best practices, the article provides practical strategies for library managers and employees to recognize, address, and diffuse anger in the workplace. From proactive communication and conflict resolution techniques to stress management and employee wellness initiatives, this article offers a comprehensive approach to creating a safe and respectful library environment. By prioritizing anger management, libraries can promote a positive and productive work culture, ensuring the well-being of employees and the success of the organization.

***Keywords**: Anger, Anger management, Library, Workplace, Strategies, Preventive measures*

1. Introduction

The library, often perceived as a tranquil oasis, can be a hotbed of emotions. Behind the scenes, library employees and patrons alike struggle to manage their emotions, and anger can be a particularly pervasive and destructive force. The pressure to meet deadlines, the stress of dealing with difficult patrons, and the frustration of working with outdated technology can all contribute to a simmering sense of anger and resentment. But anger is not just a personal problem, it can have far-reaching consequences for the library as a whole. Unchecked anger can lead to decreased productivity, strained relationships, and even violence. It can also create a toxic work environment, driving away talented employees and eroding the trust of the community. Despite the importance of anger management, it is often overlooked in library training and professional development. Library employees are frequently expected to manage their emotions without being given the tools or support to do so. This can lead to a culture of silence, where employees suffer in silence, afraid to speak out about their struggles with anger.

This chapter aims to break that silence. By exploring the warning signs of anger, the cultural and organizational factors that contribute to its escalation, and practical strategies for managing anger, we hope to provide library managers and employees with the tools they need to create a safe and respectful work environment. We will examine the ways in which anger can manifest in the library workplace, from subtle changes in behavior to overt outbursts, and explore the impact that anger can have on individuals, teams, and the organization as a whole. Through real-life examples, case studies, and expert insights, this chapter will provide a comprehensive guide to managing anger in the library workplace. We will explore the importance of proactive communication, conflict resolution, stress management, and employee wellness in preventing and addressing anger. We will also examine the role of cultural competence and diversity in creating a more inclusive and respectful work environment.

By prioritizing anger management, libraries can promote a positive and productive work culture, ensuring the well-being of employees and the success of the organization. It is our hope that this chapter will provide a valuable resource for library professionals, helping them to create a workplace where everyone can thrive.

2. Understanding Anger: Defining Anger, its Causes and Effects, and its Impact on the Library Workplace

Anger is a universal human emotion that can manifest in various ways, from mild irritation to intense rage. In the library workplace, anger can be a significant challenge, affecting not only individual employees but also the overall work environment and the quality of service provided to patrons.

2.1. Defining Anger

Anger is a complex emotion characterized by antagonism toward someone or something you feel has deliberately done you wrong. It is a natural response to a perceived threat, injustice, or frustration, and can manifest as a range of emotions, including frustration, irritation, resentment, and hostility.

2.2. Causes of Anger

Anger can be triggered by a variety of factors, including:

- **Frustration**: Feeling blocked or thwarted in achieving a goal or meeting a need
- **Injustice**: Perceiving unfairness or inequality in the workplace or in interactions with patrons
- **Fear**: Feeling threatened or vulnerable, either physically or emotionally
- **Stress**: Experiencing chronic stress or feeling overwhelmed by workload or responsibilities
- **Lack of control**: Feeling powerless or unable to influence events or outcomes

2.3. Effects of Anger

Unmanaged anger can have serious consequences for individuals and organizations, including:

- Decreased productivity and performance
- Strained relationships and conflict with colleagues and patrons
- Decreased job satisfaction and increased turnover Negative impact on physical and mental health
- Damage to the library's reputation and relationships with the community

2.4. Impact on the Library Workplace

Anger can manifest in various ways in the library workplace, including:

- **Verbal aggression**: Yelling, shouting, or using aggressive language with colleagues or patrons
- **Passive-aggressive behavior**: Expressing anger indirectly through sulking, procrastination, or sabotage
- **Withdrawal**: Withdrawing from interactions or responsibilities due to feelings of anger or frustration
- **Burnout**: Feeling exhausted, demotivated, and disconnected from work due to chronic anger and stress

By understanding the causes and effects of anger, library employees and managers can take steps to prevent and manage anger in the workplace, creating a more positive and productive work environment.

3. Recognizing the Warning Signs: Identifying Physical, Verbal, and Behavioral Cues of Anger in Oneself and Others

Recognizing the warning signs of anger is crucial for effective anger management. By identifying physical, verbal, and behavioral cues of anger in oneself and others, library employees can take steps to prevent escalation and resolve conflicts more effectively.

3.1. Physical Cues of Anger

Physical cues of anger can include:

- **Facial expressions**: Tightened jaw, narrowed eyes, furrowed brow, or a red face
- **Body language**: Crossed arms, clenched fists, or a rigid posture
- **Physiological responses**: Increased heart rate, sweating, or trembling
- **Proximity**: Invading personal space or standing too close to others

3.2. Verbal Cues of Anger

Verbal cues of anger can include:

- **Tone and pitch**: Raised voice, sarcastic tone, or a condescending pitch
- **Language**: Using aggressive or confrontational language, such as insults or threats
- **Volume**: Speaking loudly or shouting
- **Pace**: Speaking quickly or rapidly

3.3. Behavioral Cues of Anger

Behavioral cues of anger can include:

- **Defensiveness**: Becoming overly defensive or dismissive of others' concerns
- **Aggression**: Engaging in aggressive behavior, such as pushing or shoving
- **Avoidance**: Avoiding interactions or conversations due to feelings of anger or frustration
- **Passive-aggressive behavior**: Expressing anger indirectly through sulking, procrastination, or sabotage

3.4. Recognizing Anger in Oneself

Recognizing anger in oneself is just as important as recognizing it in others. Library employees can take steps to identify their own physical, verbal, and behavioral causes of anger, such as:

- **Self-reflection**: Taking time to reflect on one's emotions and behaviors
- **Self-monitoring**: Paying attention to physical sensations, such as a racing heart or tightened muscles
- **Journaling**: Writing down thoughts and feelings to identify patterns and triggers of anger

3.5. Recognizing Anger in Others

- Recognizing anger in others can help library employees respond more effectively to conflicts and difficult situations. By paying attention to physical, verbal, and behavioral cues of anger, library employees can:
- **Stay calm**: Remain calm and composed in the face of anger or aggression
- **Empathize**: Show understanding and empathy towards the angry person
- **De-escalate**: Use active listening and problem-solving skills to de-escalate the situation

By recognizing the warning signs of anger, library employees can take steps to prevent conflicts, improve communication, and create a more positive and respectful work environment.

4. De-escalation Techniques: Strategies for Calming Down Angry Patrons and Colleagues, Including Active Listening, Empathy, and Problem-solving

De-escalation techniques are essential for managing anger and conflict in libraries. By using effective de-escalation strategies, library employees can calm down angry patrons and colleagues, resolve conflicts, and create a more positive and respectful work environment.

4.1. Active Listening

Active listening is a crucial de-escalation technique that involves fully concentrating on what the angry person is saying, understanding their perspective, and responding in a non-judgmental way. Active listening can help to:

- **Reduce tension**: By acknowledging the person's feelings and concerns, active listening can help to reduce tension and calm them down.
- **Build trust**: Active listening can help to build trust and rapport with the angry person, making them more likely to listen to and respond to solutions.
- **Identify solutions**: Active listening can help to identify the root cause of the problem and potential solutions.

4.2. Empathy

Empathy is another essential de-escalation technique that involves understanding and sharing the feelings of the angry person. Empathy can help to:

- **Validate feelings**: Empathy can help to validate the person's feelings and concerns, making them feel heard and understood.
- **Reduce anger**: Empathy can help to reduce anger and aggression by acknowledging the person's emotional experience.
- **Build rapport**: Empathy can help to build rapport and trust with the angry person, making them more likely to respond to solutions.

4.3. Problem-Solving

Problem-solving is a de-escalation technique that involves working with the angry person to identify and solve the underlying problem. Problem-solving can help to:

- **Identify solutions**: Problem-solving can help to identify potential solutions to the problem, reducing anger and frustration.
- **Build trust**: Problem-solving can help to build trust and rapport with the angry person, making them more likely to respond to solutions.
- **Reduce conflict**: Problem-solving can help to reduce conflict and improve communication, leading to a more positive and respectful work environment.

4.4. De-escalation Techniques

Other de-escalation techniques that can be effective in calming down angry patrons and colleagues include:

- **Remaining calm**: Remaining calm and composed can help to de-escalate the situation and reduce tension.
- **Using open-ended questions**: Using open-ended questions can help to encourage the person to talk and express their feelings, reducing anger and frustration.
- **Paraphrasing and summarizing**: Paraphrasing and summarizing what the person has said can help to ensure understanding and show that you are actively listening.
- **Offering choices**: Offering choices can help to give the person a sense of control and agency, reducing anger and frustration.

5. Conflict Resolution: Approaches to Resolving Conflicts in the Library, Including Mediation, Negotiation, and Communication

Conflict resolution is an essential skill for library employees, as conflicts can arise between patrons, between patrons and staff, or among staff members. Effective conflict resolution can help to resolve issues, improve relationships and create a more positive and respectful work environment.

5.1. Mediation

Mediation is a conflict resolution approach that involves a neutral third party facilitating a conversation between the parties in conflict (Moore, 2014). Mediation can be an effective way to resolve conflicts in the library, as it:

- **Encourages communication**: Mediation encourages open and honest communication between the parties in conflict, helping to identify the root cause of the issue.
- **Fosters understanding**: Mediation helps to foster understanding and empathy between the parties, reducing tension and improving relationships.
- **Promotes mutually beneficial solutions**: Mediation encourages the parties to work together to find a mutually beneficial solution, rather than relying on a winner-takes-all approach.

5.2. Negotiation

Negotiation is a conflict resolution approach that involves finding a mutually acceptable solution through compromise and concession (Fisher &Ury, 1981). Negotiation can be an effective way to resolve conflicts in the library, as it:

- **Encourages flexibility**: Negotiation encourages flexibility and creativity in finding a solution, rather than relying on a fixed position.
- **Builds trust**: Negotiation can help to build trust and rapport between the parties, improving relationships and reducing conflict.
- **Finds mutually beneficial solutions**: Negotiation encourages the parties to find a solution that meets the needs of both parties, rather than relying on a winner-takes-all approach.

5.3. Communication

Effective communication is essential for conflict resolution in the library Communication can help to:

- Prevent conflicts: Clear and respectful communication can help to prevent conflicts from arising in the first place.

- Resolve conflicts: Effective communication can help to resolve conflicts quickly and efficiently, reducing tension and improving relationships.
- Improve relationships: Communication can help to build trust and rapport between the parties, improving relationships and reducing conflict.

5.4. Conflict Resolution Strategies

Other conflict resolution strategies that can be effective in the library include:

- **Active listening**: Active listening can help to ensure that all parties are heard and understood, reducing tension and improving relationships
- **Empathy**: Empathy can help to build understanding and rapport between the parties, reducing conflict and improving relationships
- **Problem-solving**: Problem-solving can help to identify and resolve the underlying issue, reducing conflict and improving relationships

6. Stress Management: Techniques for Managing Stress & Preventing Anger, Including Mindfulness, Self-Care, and Time Management

As library professionals, we often find ourselves in high-pressure situations, dealing with demanding patrons, tight deadlines, and limited resources. If left unchecked, these stressors can escalate into anger, compromising our well-being and the overall work environment. In this section, we'll explore effective stress management techniques to help you cope with stress, prevent anger, and maintain a safe and productive workplace.

6.1. Mindfulness: The Power of Presence

Mindfulness is the practice of being fully present in the moment, without judgment. By cultivating mindfulness, you can reduce stress and anxiety, leading to a decrease in anger and frustration. Here are some mindfulness techniques to try:

- **Deep Breathing Exercises**: Take slow, deliberate breaths, focusing on the sensation of the air moving in and out of your body.
- **Body Scan**: Pay attention to physical sensations in your body, starting from your toes and moving up to the top of your head.
- **Mindful Walking**: Take a short walk, focusing on the sensation of your feet touching the ground, the movement of your legs, and the rhythm of your breath.

6.2. Self-Care: Prioritizing Your Well-being

Self-care is essential for managing stress and preventing anger. By prioritizing your well-being, you'll be better equipped to handle challenging situations. Here are some self-care strategies to incorporate into your daily routine:

- **Schedule Downtime**: Make time for activities that bring you joy and relaxation, such as reading, listening to music, or taking a warm bath.
- **Exercise Regularly**: Engage in physical activities that help reduce stress and anxiety, such as yoga, walking, or jogging.
- **Healthy Eating**: Focus on consuming a balanced diet rich in whole foods, fruits, and vegetables.

6.3. Time Management: Mastering Your Schedule

Effective time management is crucial for reducing stress and preventing anger. By prioritizing tasks, setting realistic goals, and avoiding procrastination, you can maintain a sense of control and accomplishment. Here are some time management tips to try:

- **Prioritize Tasks**: Identify critical tasks and tackle them first, breaking them down into manageable chunks.
- **Set Realistic Goals**: Establish achievable goals, avoiding overcommitting and setting yourself up for failure.
- **Take Regular Breaks**: Take short breaks throughout the day to recharge and refocus.

By incorporating these stress management techniques into your daily routine, you'll be better equipped to handle the demands of the library workplace, reducing the likelihood of anger and promoting a safe and productive environment for yourself and your colleagues.

7. Cultural Competence and Anger: The Role of Cultural Competence in Understanding and Addressing Anger in the Library Workplace

Cultural competence is the ability to understand, appreciate, and interact with individuals from diverse backgrounds, cultures, and identities. In the library workplace, cultural competence plays a crucial role in understanding and addressing anger, as it helps librarians recognize and respond to the unique needs and experiences of their colleagues and patrons. In this section, we'll explore the importance of cultural competence in managing anger and promoting a safe and inclusive work environment.

Understanding Cultural Differences in Anger Expression

Anger is a universal emotion, but its expression and perception vary across cultures. What may be considered an acceptable way to express anger in one culture may be viewed as inappropriate or even aggressive in another. Cultural competence helps librarians understand these differences and respond appropriately. For example:

- **Collectivist vs. Individualist Cultures**: In collectivist cultures, such as many Asian and Latin American communities, anger is often suppressed to maintain group harmony. In individualist cultures, such as many Western societies, anger is more likely to be expressed openly.
- **High-Context vs. Low-Context Cultures**: In high-context cultures, such as many African and Indigenous communities, anger is often communicated through nonverbal cues and indirect messages. In low-context cultures, such as many Western societies, anger is more likely to be expressed directly and explicitly.

7.1. The Impact of Microaggressions on Anger

Microaggressions are subtle, often unintentional expressions of bias or prejudice that can contribute to feelings of anger, frustration, and marginalization. Cultural competence helps librarians recognize and address microaggressions, creating a safer and more inclusive work environment. Examples of microaggressions include:

- **Assuming a person's identity or experience**: Asking someone where they're "really from" or assuming their gender identity or sexual orientation.
- **Making stereotypical comments or jokes**: Making comments about a person's race, ethnicity, or cultural background, or telling jokes that rely on stereotypes.

7.2. Strategies for Culturally Competent Anger Management

Cultural competence is essential for effective anger management in the library workplace. Here are some strategies to promote culturally competent anger management:

- **Active Listening**: Listen carefully to colleagues and patrons, seeking to understand their perspectives and experiences.
- **Empathy and Validation**: Acknowledge and validate the emotions and experiences of others, avoiding judgment or dismissal.
- **Cultural Humility**: Recognize the limitations of your own cultural knowledge and be willing to learn from others.
- **Inclusive Communication**: Use inclusive language and communication styles, avoiding cultural or linguistic barriers.

By incorporating cultural competence into anger management strategies, librarians can create a safer, more inclusive, and more productive work environment. By recognizing and responding to the unique needs and experiences of colleagues and patrons, librarians can reduce anger and promote a culture of respect, empathy, and understanding.

8. Case Studies: Real-Life Examples of Anger Management in Libraries, Including Successes and Challenges

Anger management is a crucial aspect of maintaining a peaceful and productive environment in libraries. As discussed in previous chapters, anger can manifest in various forms, and its consequences can be severe if left unaddressed. In this chapter, we will explore real-life examples of anger management in libraries, highlighting successes and challenges. These case studies will provide valuable insights into the complexities of anger management and offer practical solutions for librarians and library administrators.

8.1. Case Study 1: The Angry Patron

Background

A middle-aged man, who was a regular visitor to a public library, became agitated when he was unable to find a specific book on the shelves. He approached the circulation desk, visibly upset, and demanded that the librarian find the book immediately. The librarian, who had been trained in conflict resolution, listened attentively to the patron's concerns and empathized with his frustration.

Intervention

The librarian offered to help the patron search for the book and explained the library's cataloging system. She also suggested alternative sources, such as interlibrary loan services, to obtain the book. The patron, still agitated, began to calm down as the librarian continued to listen and offer solutions.

Outcome

The patron eventually found the book, and his anger dissipated. He apologized for his behavior and thanked the librarian for her assistance. The librarian's calm and empathetic demeanor had de-escalated the situation, preventing further conflict.

Takeaway

This case study highlights the importance of active listening and empathy in anger management. By acknowledging the patron's frustration and offering solutions, the librarian was able to diffuse the situation and provide a positive outcome.

8.2. Case Study 2: The Staff Conflict

Background

Two library staff members, who had been working together for several years, began to experience tension in their working relationship. The tension escalated into heated arguments, which affected the overall work environment. The library administrator, recognizing the impact on staff morale, intervened to address the conflict.

Intervention

The administrator facilitated a mediation session between the two staff members, focusing on open communication and conflict resolution strategies. The staff members were encouraged to express their concerns and feelings, and the administrator helped them identify common goals and interests.

Outcome

The mediation session led to a better understanding between the staff members, and they were able to resolve their differences. The work environment improved, and staff morale increased.

Takeaway

This case study demonstrates the effectiveness of mediation in resolving staff conflicts. By providing a safe and neutral environment for open communication, the administrator was able to help the staff members resolve their differences and improve their working relationship.

8.3 Case Study 3: The Library as a Safe Space

Background

A community library, located in a high-crime area, experienced frequent incidents of violence and aggression within the library. The library staff, concerned for their safety and the safety of patrons, sought to create a safer environment.

Intervention

The library administration collaborated with local law enforcement and community organizations to develop a comprehensive safety plan. The plan included increased security measures, such as cameras and panic buttons, as well as training for staff on conflict resolution and de-escalation techniques.

Outcome

The safety plan led to a significant reduction in violent incidents within the library. Patrons and staff reported feeling safer, and the library became a welcoming and inclusive space for the community.

Takeaway

This case study highlights the importance of collaboration and community engagement in creating a safe and peaceful environment. By working together with local organizations and law enforcement, the library was able to address the root causes of violence and create a safer space for everyone.

9. Conclusion

These case studies demonstrate the complexities of anger management in libraries and the importance of effective conflict resolution strategies. By providing a safe and supportive environment, librarians and library administrators can help individuals manage their anger and improve overall well-being. The successes and challenges presented in these case studies offer valuable lessons for librarians and library administrators seeking to create a peaceful and productive environment in their libraries.

References

American Psychological Association. (2020). Stress in America: Coping with change. Retrieved from https://www.apa.org/news/press/releases/stress/2020/stress-in-america-2020.pdf

Covey, S. R. (2004). The 7 habits of highly effective people: Restoring the character ethic. Simon & Schuster.

Deutsch, M. (1985). Distributive justice: A social-psychological perspective. Yale University Press.

Drucker, P. F. (2006). The effective executive: The definitive guide to getting the right things done. HarperCollins.

Ekman, P. (2003). Emotions revealed: Recognizing faces and feelings to improve communication and emotional life. Henry Holt and Company.

Fisher, R., & Ury, W. (1981). Getting to yes: Negotiating agreement without giving in. Penguin Books.

Goleman, D. (1995). Emotional intelligence: Why it can matter more than IQ. Bantam Books.

Hall, E. T. (1966). The hidden dimension. Doubleday.

Hogan, R., & Roberts, B. W. (2004). A socioanalytic perspective on anger. In M. Potegal, G. Stemmler, & C. Spielberger (Eds.), International handbook of anger: Constituent and concomitant biological, psychological, and social processes (pp. 297-314). Springer.

Kabat-Zinn, J. (2003). Mindfulness-based interventions in context: Past, present, and future. Clinical Psychology: Science and Practice, 10(2), 144-156.

Kabat-Zinn, J. (2003). Mindfulness for beginners: Reclaiming the present moment- and your life. Sounds True.

Lazarus, R. S., & Folkman, S. (1984). Stress, appraisal, and coping. Springer.

Lerner, R. M. (1985). On the nature of human plasticity. Cambridge University Press.

Linehan, M. M. (2015). DBT skills training manual (2nd ed.). Guilford Press.

Maslach, C., & Leiter, M. P. (2017). Burnout: A multidimensional perspective. In C. L. Cooper (Ed.), The Oxford handbook of work and organizational psychology (pp. 355-376). Oxford University Press.

Moore, C. W. (2014). The mediation process: Practical strategies for resolving conflict. Jossey-Bass.

Myers, J. E., & Sweeney, T. J. (2005). Counseling for wellness: Theory, research, and practice. American Counseling Association.

Rogers, C. R. (1951). Client-centered therapy: Its current practice, implications, and theory. Houghton Mifflin.

Seligman, M. E. P. (1975). Helplessness: On depression, development, and death. Freeman.

Spielberger, C. D. (1988). State-trait anger expression inventory. Psychological Assessment Resources.

Sue, D. W. (2010). Microaggressions in everyday life: Race, gender, and sexual orientation. John Wiley & Sons.

Tervalon, M., & Murray-García, J. (1998). Cultural humility versus cultural competence: A critical distinction in defining physician training outcomes in multicultural education. Journal of Health Care for the Poor and Underserved, 9(2), 117-125.

Wehbi, S. (2009). Cultural competence: A conceptual framework for library and information science professionals. Library Quarterly, 79(2), 169-193.

12

Green Libraries: Encouraging Sustainability and Eco-Friendly Practices in Modern Library Services

Rajshree Dutta

MLIS, PGDCA, LIS Intern, Central Reference Library, Kolkata - 700027 West Bengal, India, E-mail: rajsheedutta35@gmail.com

Abstract

One of the most prominent developments in modern library services has been known as the "Green Library" concept, characterized by sustainability and eco-friendliness. This research paper takes into consideration the comprehensive and theoretical aspects of green libraries based on how they incorporate issues of sustainability into their operation and infrastructure. This descriptive research approach is used to analyze the practices and innovations of green libraries all over the world related to being energy-efficient, using eco-friendly materials, and landscaping. It shall also discuss how these libraries influence their communities in terms of programs and projects organized, which help raise environmental awareness and community engagement. Green libraries provide the communities they serve with important places for education, which includes workshops and activities, besides resources that enable them to inculcate a culture of sustainability. Reflected in this paper are also the problems of green libraries, such as higher initial costs and the need for continuous community participation, with the possible actions that can be taken to surmount such obstacles. The findings underline the great role that green libraries play in reminding society of their environmental duty and their potential to make even broader societal change. Not only do they reduce the ecological footprint, but they also set an example, by integrating the green principle for other institutions to follow in order to make a difference. This paper, therefore, had underestimated the place of continued innovation and strategic planning in the green library development in reality as the means towards a more sustainable and environmentally conscious outcome in the future.

Keywords: *Green libraries, Sustainable library practices, eco-friendly library services, Resource conservation, Green infrastructure, Environmental awareness.*

1. Introduction

Libraries have an outstanding role in the advocacy of sustainability and green practices, now that the place of environmental awareness is on the rise. Green libraries provide a commitment, through sustainable architectural practices, efficient systems in energy saving, measures taken with environmental consciousness to preserve resources, and ways to conserve resources in the work they do. The libraries focus on ensuring they provide healthy indoor spaces and educate their communities on sustainability. Admittedly, it is through the adoption of these practices that libraries minimize their ecological footprint and provide a beacon of environmental responsibility. Green libraries are facilities that take ecologically friendly approaches in carrying out operations and programming in activities targeting environmental responsibility. Such libraries try their best to use methods that diminish energy consumption and lower resource use to move closer to sustainability at its core, which may have less of an impact on the environment. Green libraries have focused on recycling, reduction of trash, renewable sources of energy like solar panels, energy efficiency in lighting and space heating, sustainable building materials, and architectural style. They also, as an effort, frequently participate in green issues sensitization and promoting green good practices among self and others within the library. An overall green library would set itself as an example of sustainability within the community and exhibit leadership in the proposition set forth to manage the environment. So, this philosophy of a "green library," which was now most recently adapted as the "sustainable library," is rapidly spreading around the globe within the library profession.

It encompasses building friendly environmental library structures, greening already-existing facilities, provision for sustainable library services, as well as adopting eco-friendly library practices. The universe unites to make sure the climate is preserved after finding out of global warming and its negative effects. Publishers, therefore, have been very keen on issues of sustainability and green libraries. Libraries have been centers of learning and information. Nevertheless, in recent years, the realization of their full potential in enabling sustainability and sustainable behavior has seen a lot of progress. It is important for libraries and information centers to make their operations sustainable due to their potential to cut down on carbon printouts, support community greening and environmental protection, and a significant contribution in the quest for environmental conservation. Libraries are greatly characterized by physical infrastructures and resource use, resulting in their environmental performance. Libraries can reduce wastage, conserve energy, and water and lessen their professional impact on the environment by practicing sustainability.

2. Research Objectives

This paper aims to:

1. Examine the concept and principles related to green libraries.
2. Present research on the sustainable practices and innovative designs for green libraries.
3. Analyze what sort of impact the green libraries have on the community.
4. Provide information about the role that libraries can play in increasing environmental awareness.
5. Outline challenges and strategies which might be put into action to execute a green library practice.

3. Research Methodology

The research method will be based on both comprehensive and conceptual approaches. The comprehensive approach shall be majorly on the intensive study of literature reviews comprising articles, case studies, and best practices revolving around sustainability and eco-friendliness in contemporary library service. This provides a detailed understanding of the various concepts of a green library through the synthesis of sources. The conceptual approach applies relevant environmental and organizational theories to analyze and interpret the principles behind the adoption of sustainable practices in libraries. These approaches thus provide a strong theoretical framework for the investigation of challenges and strategies associated with the promotion of sustainability and eco-friendliness in modern library services, without the need for new data collection.

4. Literature Review

Mishra, Sudhakar (2023), in this paper "Greening Libraries for a Sustainable Future: A Comparative Analysis of Green and Traditional Library Practices" discusses the future of libraries by contrasting established green library practices with those that are only getting started. Libraries are changing more sustainable procedures to lessen their ecological footprint. The article focuses on the distinctions between conventional libraries and green libraries, uses of smart building technologies, incorporation of natural elements, adoption of renewable energy sources, workshops and educational programs, partnerships with neighborhood organizations, and promotion of sustainable policies and initiatives. This report offers a comparative comparison of standard libraries and green libraries by examining these distinctions.

Bangar, Machhindra S. (2018), is this study "Knowledge Librarian" An International Peer Reviewed Bilingual E-Journal of Library and Information

Science Special Issue" focuses on the 21st century is being defined by the current trends of green libraries. Movement of green libraries is to reduce environmental pollution and green libraries by joining librarians, libraries, cities, and universities etc. This study highlighted the conditions of green libraries, role of librarians and the characteristics of green libraries. The study finds that librarians need to keep up with the current advances in sustainability.

Hasan & Panda (2023), in his paper " Charting a Sustainable Path: Empowering Green Libraries for a Greener Future in India" exploring about the concept of green or sustainable libraries has growing prominence in response to the global challenge of changes climate and increasing focus on conservation of environment. Libraries recognize their features in encouraging environmentally conscious behavior and sustainability as centers of knowledge and studies. This study analyzes case studies, literature review, and contacts experts to look into the state of environmentally friendly library maneuvers in India.

Kulkarni, Prafulla Datta P. (2018), in his paper "Green Library: concept, features and elements" focuses on about the idea of a "green library" is relatively current developing trends.This article defines the term "green library," includes its advantages, and consequently talks about its components. Due to its dynamic nature, this concept is becoming more and more well-liked among academics, researchers, and library professionals etc. Although it is currently small, the amount of knowledge on "Green Library" and "Sustainability" is expanding. The main focus of this paper of sustainability and green libraries have been extensively discussed about the aimed of international librarians in recent years.

Hauke & Werner (2013), in this paper "Going green as a marketing tool for libraries: environmentally sustainable management practices" study examines libraries' understanding of ecological sustainability as a component of a high-impact marketing plan for stakeholders and customers. Libraries in particular have a special responsibility to set an example for others to follow as well as to spread knowledge about environmental sustainability. A library's reputation can be greatly impacted by even small ecological initiatives. Collaboration with nonprofit organizations, Friends of the Library groups, and other groups can facilitate the development of activities in this field. In order to illustrate how to establish a green identity that projects a desirable market image, the presentation will include examples from libraries across the globe.

Lokhande, Rajesh S. (2022), in this paper "Green Library: Sustainable Development" study defines the term "green library," enumerates its attributes, and inevitably talks about its components. A "green" or "sustainable library" is one that is created, constructed, remodeled, run, or repurposed in a way that is resource- and eco-conscious. The goal of the green library is to reduce global warming via the combined efforts of all people. Other green library components include site selection, construction material selection, energy conservation, indoor air quality, and water conservation. As they have historically played a significant part in community education initiatives, librarians should naturally assume a leading position in the sustainability movement. A modern library that maximizes the use of renewable resources like air, sunshine, and trees while minimizing the usage of electricity is called a "green" or "sustainable" library.

5. Green Library: A View

The green library is transformational in the library sector with respect to sustainability and environmental responsibility through operations and community engagement. Such types of libraries are, therefore, characterized by proactive adoption of sustainable practices, including energy-efficient technologies, designs for buildings that are sustainable, and concerted strategies of waste management for reduced impacts on the environment. Beyond this infrastructure, however, green libraries act as educational platforms for running programs and workshops aimed at raising environmental literacy and thus empowering communities to adopt sustainable behaviors. The complicated role that green libraries play here not only establishes them as preservers of knowledge but also as drivers of societal change, motivating peoples' positive selves to take action toward active environmental stewardship.

Historically, the development of green libraries is like a more general turn toward the greening of institutional practices. Operational efficiencies are where most modern green libraries focus their efforts on holistic sustainability through global initiatives and partnerships. In efforts geared toward advancing environmental goals through fostering collaboration, sharing best practices, and advocating for policies that support sustainable development, green libraries greatly shape a more sustainable future. The future can only hold further green technology innovation, community outreach, and additional education to continue putting them at the forefront of sustainable practices disputes in the library sector and beyond.

6. Green Library: Principles

Green libraries are committed to sustainability through energy efficiency, waste reduction, and community involvement. It is committed to these principles of stewardship through the leadership of environmental stewardship within the library sector and beyond.

- **Sustainable Practices:** Green libraries are those that have inclusively laid-down strategies on sustainability. This entails the use of energy-efficient technologies, such as LED lighting and efficient HVAC(Heating, Ventilation, and Air Conditioning), tapping of renewable energy sources like solar panels, and eco-friendly building materials like recycled insulation and sustainable wood. This further extends to the adoption of green building certifications that ensure that their environmental standards are observed to reduce their ecological footprint, enhance operational efficiency, and foster good environmental stewardship within their communities.
- **Environmental Stewardship:** Green libraries reduce their impacts on the environment through effective management. This involves having in place efficient programs of waste reduction, water-saving technologies and practices, and landscaping that establishes sustainability. Each of these efforts, therefore, shows their commitment to sustainability in a manner that has further enhanced eco-friendly practices toward a healthy environment and sustainable community development.
- **Community Engagement:** Green libraries are dedicated to community engagement for programs aimed at raising education and consciousness towards the environment. Workshops, seminars, and interactive exhibits mirror sustainable life goals, creating an atmosphere where environmental responsibility is deeply felt by all patrons. In these regards, green libraries become hubs of learning and advocacy; they empower people to engage in and promote sustainability within their various contexts at local and global levels.
- **Innovation and Leadership:** Green libraries pioneer advanced solutions in environmental stewardship through the use of smart building technologies and library greening. Within this, they uphold high standards in green architecture and resource management for libraries; this fulfills higher goals through reach to professionalism in striving to achieve global sustainability efforts in the library sector..
- **Advocacy and Collaboration:** Green libraries join international stakeholders' networks to collaborate in advocacy for pressing sustainable policies and practices around the globe. They protect the environment through strategic alliances, moving toward promoting sustainable development objectives within the library sector and beyond.
- **Educational Role:** Green libraries educate on environmental issues by making available versatile resources and interactive learning encounters. They provide means for community participatory involvement in

sustainability through workshops, digital resources, and joint initiatives that raise environmental awareness and promote activity

- **Continuous Improvement:** Green libraries shall continue improving in their sustainability practices through the adoption of new technologies and best management practices that work toward sustainable environmental performance and community impact.
- **Transparency and Accountability:** Green libraries maintain transparency in environment-related activities and are answerable to the stakeholders through periodic reporting on their progress toward the set sustainability goals.

7. Sustainable Practices in Green Libraries

Energy efficiency is one of the primary principles in green library sustainable practices, and it reduces energy use to a minimum while having the least effect on the environment. Some of the key strategies put in place by green libraries to ensure energy efficiency are:

7.1. Renewable Energy Source Utilization

- **Solar Panels:** Most green libraries are equipped with solar panels, either on the rooftop or anywhere else on their property, to help tap solar energy. This can provide a very clean form of energy and nearly decrease non-renewable energy consumption, reducing electricity costs.
- **Wind Turbines:** Wind turbines can be installed in appropriate places in libraries where the speed of the wind is fit enough to run the machines, thus providing them with electricity. This will add to a diversified renewable energy portfolio.

7.2. Implement Energy-Efficient Lighting

- **LED Lighting:** Light Emitting Diodes are favored because they consume far less energy and also have a longer lifetime than the conventional lighting procedures that include incandescent and fluorescent lights. The LEDs also give off less heat, thus reducing the expenses of cooling.
- **Smart Lighting Systems:** Those introducing sensors for self-adjustment, depending on occupancy and the availability of daylight, in order to meet the lighting requirement by using lights only where necessary.

7.3. Maximizing Natural Lighting and Ventilation

- **Day lighting Strategies:** Green libraries espouse a design that allows maximum inflow of natural light through large window fronts, skylights, and light wells. This would allow for lesser usage of artificial lighting during daylight hours.

- **Natural Ventilation:** Provision for operable windows and ventilation shafts allows natural airflow, reducing the requirement of mechanical ventilation systems. The energy use thus decreases, while air quality is also enhanced.

7.4. Upgrades to Insulation and Building Envelope

- **High-Performance Insulation:** The same should be achieved for this purpose through new insulation materials and methods to maintain a more uniform temperature inside houses, hence eliminating part of the heating and cooling.
- **Thermal Mass:** Incorporating materials with high thermal mass—such as concrete and brick—into the building shell helps stabilize indoor temperature through the absorption of heat and then a slow release of it.

7.5. Energy Monitoring and Management Systems

- **Building Management Systems:** Such systems monitor and control the utilization of energy in real-time, providing data that helps optimize energy consumption. They can auto-modulate lighting, HVAC, and other systems for peak efficiency.
- **Energy Audits and Benchmarking:** The periodic energy audit identifies areas for improvement, while benchmarking against similar buildings helps set performance goals and the tracking of progress.

8. Innovative Designs for Green Libraries

Innovative designs for green libraries provide the triad of sustainability, flexibility, and engagement; they reduce the library's negative influence on the environment while maximizing its utility for users. It is a design that puts forward those elements that would enhance efficiency in resources use, their conservation, and community engagement.

9. Green Building Design

9.1. Biophilic Design Principles

- **Green Walls and Roofs:** Green walls and roofs could control the temperature, enhance air quality, and have an added aesthetic value to the atmosphere of the library. These also provide habitat spaces, increasing the share of local wildlife and, hence, offering urban biodiversity.
- **Indoor Plants:** Indoor plants throughout give not only better air quality but also a sense of tranquility and a feel of welcome to visitors inside the library.

9.2. Flexible, Multi-Purpose Spaces

- **Adaptable Interiors:** Design interiors that are easily reconfigurable to support a variety of functions and activities to provide the library with the ability to further evolve as needs change. Movable walls, modular furniture and multi-functional areas accomplish this.
- **Community Spaces:** Community events, workshops and meeting spaces make a library a centre of community life and help tie the building to the community.

9.3. Passive Solar Design

- **Orientation and Layout:** The orientation of the building, to receive maximum winter sun and heat and at the same time, avoid it in summertime ensures that artificial lighting and climate controls are not needed.
- **Thermal Mass Materials:** A material that absorbs heat gradually and gives it off at a comparable rate helps in the stability of inside temperatures. This limits the requirement for mechanical heating and cooling.

10. Smart Technologies

10.1. Smart Building Management Systems

- **Energy Monitoring:** Real-time monitoring of energy use provides insight into what changes should be made to optimize consumption and where in general to make improvements.
- **Automated Controls:** Sensors and automation in the control of lighting, heating, cooling, and ventilation ensure that the mentioned systems are only running when they need to be. This complements better energy efficiency.

10.2. Internet of Things Devices

- **Smart Lighting:** IoT Lighting Systems modulate light intensity and color based on occupancy, available natural light, and personal preference.
- **Climate Control:** IoT devices track temperature and humidity levels in real time to provide comfort with the least possible energy consumption.

10.3. Interactive and Automated Systems

- **Self-Service Kiosks:** Automated check-in and check-out systems are quicker and use less physical resources. Digital Wayfinding: Interactive, library-wide maps/guides further enhance the user experience while reducing a myriad of printed matter. Community Engagement and Education in Sustainability Programs/Workshops

- **Informational Programs:** The community can organize seminars on sustainable living, such as energy conservation, recycling, and other relevant programs on green living to instill in the community an awareness of sustainability.
- **Outdoor Learning Spaces:** Outdoor learning and community space programming area bridges users with nature and further anchors the library's commitment to sustainability.

10.4. Partnership with Local Organizations

- **Partnerships:** Collaborate with other local environmental organizations, schools, and community groups in hosting activities and events to foster a culture of sustainability and tight links with the community.
- **Community Gardens:** Community gardens are quite practical for teaching sustainable agriculture production and growing food.

11. Sustainable Collections and Services

11.1. Digital Collections

- **E-books and Online Resources:** Expanding digital collections reduces the need for physical materials the library consumes, thus reducing its carbon footprint, and also readily provides access to information for users.
- **Virtual Programs:** Online workshops, webinars, and virtual tours increase participation while decreasing associated travel impacts on the environment.

11.2. Green Information Literacy

- **Sustainability Resources:** The development of collections related to sustainability and the environment helps inform users and furthers research in that area.
- **Green Practices:** Green research practices—promoting digital resources and minimizing paper use—keep library operations in line with sustainable goals.

11.3. Innovative Services

- **Tool Libraries:** Provides access to tools and other equipment, allowing sharing, thus reducing the need to own, and hence promotes sustainable ways of life.

- **Seed Libraries:** Lending out seeds to plant and return helps local agriculture, biodiversity, and a connection with nature.

12. The Impact of Green Libraries on Communities

Green libraries influence the community at various levels. The influence stretches not only to social and economic areas but also to educational aspects, way beyond the goal of environmental sustainability. Some of the major impacts are analyzed here:

12.1. Environmental Impact

- **Energy Efficiency and Resource Conservation:** Green libraries are designed by using efficient lighting, heating, and cooling systems and also by reducing water consumption through applying water-saving fittings. This helps in reducing the carbon footprint and conserving resources for the library.
- **Sustainable Building Practices:** Green libraries are being designed and built using materials and practices that reduce their impact on the environment and create healthier interior environments.
- **Waste Reduction:** Green libraries reduce waste through recycling programs and digital resources, trying to develop a culture of reusing and recycling.

12.2. Social Impact

- **Community Engagement:** Green libraries serve as community engagement hubs through the delivery of programs and workshops regarding sustainability, gardening, and recycling—all "green" practices binding people together and instilling in them a sense of shared responsibility toward the environment.
- **Health and Well-being:** The emphasis on natural lighting, air quality, and non-toxic materials used in the libraries contributes to the healthiness of the spaces for patrons and staff alike, thereby impacting well-being.
- **Inclusive Spaces:** Green libraries often make use of inclusive design so that spaces are equally available and open to everybody in the community, including those who have any kind of disability.

12.3. Economic Impact

- **Cost Savings:** In the long term, green libraries offer a lot of cost savings in utilities and maintenance by using energy efficiency measures and having sustainable practices. These can be redirected toward library services and programs.

- **Local Economic Development:** By working with local businesses and organizations on green initiatives, libraries sustain local economic development and job creation.

12.4. Educational Impact

- **Environmental Education:** A green library is one that includes educational resources and programming on issues around environmental sustainability and green practices awareness, education, and promotion of their benefits to the community.
- **Lifelong Learning:** Green libraries, through a variety of programs and resources, help foster lifelong learning and environmental literacy for informed decisions on sustainability.
- **Research and Innovation:** Green libraries can become living laboratories of sustainability research and innovation while creating partnerships across higher education institutions and promoting the development of new green technologies and practices.

12.5. Cultural Impact

- **Environmental Stewardship Promotion:** Green libraries give a feel and support for sustainable practices that usher in a culture of environmental stewardship in the community.
- **Cultural Preservation:** Green libraries often support cultural preservation through the conservation of local culture and heritage by following sustainable practices and initiatives that would ensure the preservation of cultural resources for future generations.

13. Libraries' Role in Enhancing Environmental Awareness

Through a lot of activities and other initiatives, libraries have huge contributions toward the raising of environmental awareness. Here is how they do it:

13.1. Programs and Workshops

- **Sustainability Education:** Libraries offer a lot of programming to raise awareness of the environment, global warming, recycling, and conversation. Such activities may be in the form of lectures, workshops, or even entertainment activities that ensure the understanding and involvement of the community members.
- **Green Living Tips:** The workshops teaching energy reduction, waste reduction, and growing local food provide realistic advice on living green. These sessions will help in developing habits that are friendly to the environment.

13.2. Access to Resources

- **Books and Media:** Library access to books, documentaries, and online resources concerning the environment. These materials increase awareness and in-depth knowledge about many of the environmental issues.
- **Digital Resources:** Electronic databases and e-books on environment and green technologies make information and research of great help to the patron.

13.3. Community Events

- **Environmental Fairs and Exhibitions:** Arrangements for fairs and exhibitions about local organizations involved in the environment, green businesses, and sustainable products would thus link several members of the community with sources and avenues for environmental activity.
- **Clean-Up Drives:** Arranging for clean-up activities in the community or recycling drives would foster active participation in efforts toward the preservation of the environment.

13.4. Partnerships and Collaboration

- **Local Organizations:** Many libraries are working with local environmental groups, schools, and government agencies to advance sustainability and environmental awareness through shared programs.
- **Educational Institutions:** Partnerships with schools and colleges could also help to strengthen environmental education programming by contributing more resources and talent.

13.5. Green Practices

- **Innovative Building Design:** Of the green building practices, some that the libraries can embrace are those of energy-efficient systems, use of eco-friendly materials, and those of waste minimization techniques. Each of these shows leadership related to environmental stewardship.
- **Sustainable Operations:** Events such as recycling, paper-use reduction, and energy conservation all combine to allow the library to act as role models in society to show they really do "care for the environment."

13.6. Advocacy and Awareness Campaigns

- **Environmental Advocacy:** The libraries raise their voice and support policies and plans made for the protection and preservation of the environment.

- **Awareness Programs:** Arrangements like exhibitions, displays, and information desks are organized in libraries to make people aware of and understand the problems and their solutions regarding the environment.

13.7. Youth Engagement

- **Children's Programs:** This includes story hours, crafts, activities with a nature and environment theme that impart values in children.
- **Teen Initiatives:** Engaging the youth in environmental clubs, volunteer opportunities, and other activities makes them active about the environment and leaders in sustainability.

13.8. Innovation and Research

- **Sustainability Research:** Conducting research on subjects related to the environment and green technologies through access to journals, reports, and case studies, across formats such as digital and print.
- **Innovation Hubs:** These would be some libraries that are hubs of innovation in the development and testing of new sustainable technologies and practices in improving environmental solutions.

14. Challenges in Implementing Green Library Practices and Strategies

14.1. Challenges

- **Financing:** Many green practices involve large up-front investments which too many libraries cannot afford to make, be it energy-efficient systems or sustainable materials.
- **Lack of Awareness and Training:** In case library staff do not have proper understanding and training in the green practices being implemented, they may apply them inconsistently or ineffectively.
- **Resistance to Change:** There may be resistance to new practices among the staff and stakeholders, either due to comfort with the existing procedures or skepticism regarding the actual benefits of green measures.
- **Inadequate Infrastructure:** New sustainable practices are not supported by the existing infrastructure of libraries; for example, changes would require expensive renovations or upgrades within the current structure of libraries.
- **Inadequate Measurement and Evaluation:** Due to the lack of standardized metrics, there are measurement and evaluation problems for green initiatives.

14.2. Strategies

- **Formulate the Green Library Plan:** It shall involve the framing of the detailed plan with clear sustainability objectives, actions, timelines for accomplishment, etc. Include stakeholders in the planning process and update the same time-to-time basis.
- **Seek Funding and Partnerships:** Seek grants and other funding opportunities, and form partnerships with environmental organizations for projects related to going green. Partner with other libraries to help accomplish projects that share common resources and best practices.
- **Improve Staff Training and Engagement:** Educate the library staff on issues concerning sustainability and embrace a green culture through workshops and information sessions on the same.
- **Infrastructure and Facility Improvement:** Green infrastructure and facilities up-gradation that would be needed, with the most essential and viable upgrades in the priority funding queue.
- **Set up monitoring and reporting systems:** Monitoring and reporting systems are institutionalized to track and report performance on green practices. Development of metrics that will quantify or measure progress and point out those areas that need improvement.

15. Conclusion

In Conclusion, green library practice research tends to underline that, although the beginning investments are huge for the implementation of improvements that are sustainable, often long-term fiscal and ecological advantages compensate for these costs. Proper staff training will hold the key to successful integration of green practices, coupled with overt communication for overcoming resistance to change. Sensitive planning and modern adaptations can overcome difficulties related to infrastructure adaptation. Moreover, monitoring and evaluation are integral parts of successful efforts toward sustainability through standardized metrics. Lastly, green libraries are important in mitigating the impact on the environment and steering communities in terms of environmental stewardship and sustainability education. This chapter has inscribed the basic ingredients for financial planning, staff involvement, infrastructure adaptation, and constant evaluation needed to ensure the practical implementation of green library practices.

References

Ahn, I. J., Kwak, C. W., Noh, Y., & Park, M. (2012). How do they make libraries green: A case based study on building green libraries. Journal of Information Management, 43(1), 135-158.

De MoraisNeto, P. L., & da Silva, A. J. (2019). Towards a sustainable library: Evaluation of the environmental performance of libraries. Journal of Cleaner Production, 235, 35-44.

Divya, P. I. &Vijayakumar, K. P. (2013) Green libraries for sustainable development. New information technology interfaces in libraries and information centres. KLA, National Seminar Souvenir, 67.

Fedorowicz-Kruszewska, M. (2023). Green libraries: barriers to concept development. Library Management, 44(1). https://doi.org/10.1108/lm-04-2022-0041

Gaffar, S., N, S., & Kumar, S. (2021). The green library initiative in Indian perspective: A study. Library Philosophy and Practice (E-Journal), 1–11. https://digitalcommons.unl.edu/libphilprac/6189

http://www.mugeakbulut.com/bby721/wpcontent/uploads/2017/03/21-09-2013- Prohit.pdf 23.

https://en.wikipedia.org/wiki/Green_librar y/ Retrieved on 26.12.2017, 17.22pm

LeRue, J. & LeRue, S. (1991). The green librarian. Wilson Library Bulletin, 65, 27-33.

Sharifian, S., & Chandrasekaran, S. (2019). Sustainable library buildings and infrastructure: A review of the literature. Library Management, 40(1/2), 101-114.

13

From Dial-Up to Data Streams: Navigating the Evolution from Web 1.0 to Web 4.0

Sunil Madhukar Gauns

Librarian, GVM's Dr. Dada Vaidya College of Education, Farmagudi Ponda-403 401, Goa, India, E-mail: 1479smg@gmail.com

Abstract

Beginning with its inception and continuing up to its current stage, the evolution of the World Wide Web symbolizes a fascinating journey through technological innovation and the enhancement of the user experience. Exploring the transformative phases of the World Wide Web, this article, which is titled "From Dial-Up to Data Streams: Navigating the Evolution from Web 1.0 to Web 4.0," investigates how each iteration of the World Wide Web has transformed digital interactions and the distribution of content. The underlying architecture for the internet was created by Web 1.0, which was characterized by static sites that could only be viewed. The introduction of Web 2.0 made it possible for social media platforms and dynamic websites to be created, which in turn helped to develop worldwide connectedness. Interactivity and user-generated content were key to the experience. A new era of semantic web technologies was ushered in with the move to Web 3.0. These technologies place an emphasis on data interoperability and tailored experiences through the utilization of powerful artificial intelligence and machine learning. Web 4.0 is currently in the process of developing, and it is characterized by the incorporation of the Internet of Things (IoT), augmented reality, and immersive settings. These elements make it possible to achieve data interactivity and automation levels that have never been seen before. The purpose of this study is to present a detailed review of these evolutionary stages, illustrate them with relevant examples, and analyze the consequences for the growth of the web in the future. The purpose of this article is to provide insights into the continuous improvements that continue to impact our digital landscape by tracing these changes and providing background information.

***Keywords**: Web based services, Web 1.0, Web 2.0, Web 3.0, Web 4.0, Web Archiving, World Wide Web Consortium, Ambient Intelligence Systems.*

1. Web Based Services

People can reach the web, which is short for "World Wide Web," which is a huge network of papers and tools that are all linked together. The web, which was created by Tim Berners-Lee in the late 1980s, changed the way people around the world share and receives information.

The growth of technology has had an impact not only on the formats and sources of the information, but also on the manner in which library services are provided and the locations where they are provided. A portion of libraries and the resources they provide have been relocated to the digital realm created by the Internet. The web works on a client-server model: web computers (clients) ask web servers for web pages and other data and then show them. There are connections between the text, pictures, movies, and apps that make up these tools.

Web development is based on key tools like HTML (Hypertext Markup Language), CSS (Cascading Style Sheets), and JavaScript, which make it possible to make web pages that look good and let you interact with them. Over time, improvements like flexible web design (which makes sure that websites work on all devices) and web APIs (which let different software systems share data) have made the web experience even better.

The web has become an important part of daily life because it makes it easier to communicate, do business, have fun, learn, and do a lot more. With the rise of web apps, cloud computing, and the Internet of Things (IoT), it will continue to change and grow, bringing us even more connection and usefulness in the digital age. The web continues to shape how we meet and talk to each other in the modern world because it is a living, growing thing.

2. Evolution of Web Pages

Websites have changed dramatically since the World Wide Web's introduction. Web sites were mostly text-based in the early 1990s, with little formatting and no pictures. HTML, which emphasizes hyperlinks, was the main content language. Web growth in the mid-1990s led to HTML developments that allowed pictures and formatting using tables and early CSS. Despite being simplistic, web sites became increasingly visually attractive throughout this time. Dynamic web pages that changed content without reloading (AJAX) improved user experience and enabled web apps in this period.

From the mid-2000s, Web 2.0 emphasized user-generated content and social media integration. Interactive and adaptable web pages adapted to numerous devices and screen sizes. This time also witnessed the widespread use of CMS like WordPress and Drupal, simplifying complicated website building and maintenance. Responsive web design has improved viewing experiences on computers, cellphones, and other devices in recent years.

3. Web Archiving and Evolution of Web Archiving

Web archiving refers to the process of capturing and preserving web pages, websites, and other online content for future access and historical record. It addresses the ephemeral nature of the web, where content can change or disappear over time due to updates, deletions, or server failures. Early efforts in web archiving emerged in the late 1990s and early 2000s, driven by organizations like the Internet Archive, which introduced the Wayback Machine in 2001. This tool allowed users to browse archived versions of web pages across different points in time, offering a valuable resource for research and historical preservation.

As the web grew in complexity and scale, so it was very much challenging to do the web archiving. Institutions such as national libraries, universities, and cultural heritage organizations started developing comprehensive archiving initiatives. These efforts aimed not only to capture static web pages but also to preserve dynamic content, multimedia elements, and interactive features. Technological advancements have been pivotal in the evolution of web archiving. Automated crawlers and web harvesting tools enable systematic and continuous capture of web content. Standards like WARC (Web ARChive) format have been established to facilitate the storage, management, and exchange of archived data in a structured manner.

4. Web 1.0

The term "Web 1.0" was first presented by Tim Berners-Lee in the year 1989. In the history of the World Wide Web, it is considered to be the very first version. In terms of social interaction, individuals were only able to view the material that was provided by websites; hence, this time period is also known as the "Read-Only Web." Pages on the web that were developed during the Web 1.0 era were static and did not go through any frequent content updates. For the purpose of marketing their products or services, manufacturers and service providers have begun publishing catalogs on the internet. Providing information to a large number of people at any given time and establishing a prominent presence on the internet was the key goals of the websites. As a result of the development of shopping carts, customers have been encouraged to make purchases online rather than purchasing items from traditional retailers situated in physical locations. Due to the fact that customers were unable to engage or assist in the production of content by providing their valuable feedback, the information distribution to the consumers was carried out using the "Push Model." The Hypertext Transfer Protocol (HTTP) was the principal communication mechanism on Web 1.0 sites, which were written using HTML.

Examples of Web 1.0

Early Yahoo!

Yahoo! began as a science, arts, and technology web directory in 1994. Before search engines, it was an organized index to help people discover relevant sites. Lists and links replaced interactive features in the UI. Yahoo! was a Web 1.0 app before it became a comprehensive web gateway with more services.

GeoCities

GeoCities, launched in 1994, let users design HTML templates for web pages. People without technological expertise might host personal websites with the service. Topics or interests were used to classify user pages as "neighborhoods". A popular Web 1.0 site for personal expression and creativity, GeoCities offered basic customization and content storage.

The World Wide Web Consortium (W3C)

The 1994-founded W3C website provided web standards and protocol information. Through standardization, it sought web expansion and interoperability. Static material included web developer specs, research papers, and guidelines. It was essential for early web developers seeking to preserve the web's technical basis.

Companies' Basic Websites

First-generation IBM and Microsoft websites provided company information. Sites have static pages containing company history, product descriptions, and contact info. They presented information simply and lacked interactive elements. Early corporate sites were crucial for web firms.

Directory Mozilla (DMOZ)

DMOZ, a hierarchical open-content directory, launched in 1998. A team of volunteer editors examined and listed sites to help visitors find relevant material. Before search engines like Google, DMOZ was one of the most comprehensive web directories, categorizing website finding.

5. Web 2.0

In 2004, web pioneer and O'Reilly VP Dale Dougherty invented Web 2.0 at a conference brainstorming session with Media Live International. Wisdom, people-centric, participatory, and read/write Web 2.0 are some names for it.

Web 2.0 changed the approach from publishers providing content to customers to a dynamic participatory one where most web page authors could update their media-rich sites as frequently as they wanted. Content producers and viewers shared information. Web site popularity is measured by hit counts, while user comments show involvement. Blogging, video sharing, chatting, hosted services, web applications, voice over Internet Protocol (VoIP), emails, instant messages, social bookmarking, podcasting, picture sharing, weblogs, mashups, and folksonomies are just some of the forms of online interaction that have been made possible and successful by user-generated content and massive social media interaction. Applications and technology related to Web 2.0 are displayed in the graphic below. The most significant contribution that Web 2.0 has made to the human race is the creation of social media websites like as Facebook, Twitter, Linked-In, Zynga, Google+, Flickr, Foursquare, and YouTube. Social media has linked people worldwide in ways imagined decades ago. Web 2.0 is driven by new Web technology and standards. Users had a rich and exciting interactive experience utilizing Ajax, JavaScript, CSS, DOM, XHTML, XSL Transformations (XSLT)/ XML, and Adobe Flash. These solutions render Web services like desktop applications, hiding dispersed processing issues.

The emergence of Web 2.0 in the early 2000s marked a notable departure from the static characteristics of Web 1.0, towards a more dynamic, interactive, and user-centric virtual environment. This phase incorporated functionalities such as user-generated content, social networking, and heightened interactivity. Web 2.0 technology facilitated the provision of immersive and interactive experiences through the utilization of dynamic content and real-time updates. In recent years, websites have increasingly embraced collaboration platforms, multimedia integration, and user interaction, thereby cultivating a more interconnected and interactive online environment.

Examples of Web 2.0

Facebook

Introduced in 2004, Facebook brought about a significant transformation in the realm of social networking by enabling users to establish personal profiles, disseminate updates, upload photographs, and engage with others via comments and likes. The primary focus was on user-generated content and the facilitation of real-time communication. The incorporation of features such as News Feed, groups, and events has established it as a prominent platform for facilitating social interaction, enabling worldwide connectivity and nurturing virtual communities. The dynamic interface and user-driven content of Facebook serve as prime examples of the fundamental characteristics of Web 2.0.

YouTube

Established in 2005, YouTube offered a digital platform that facilitated the uploading, sharing, and viewing of videos by its users. This development facilitated a novel mode of media consumption and production, wherein individuals were empowered to assume the role of content creators. The website incorporated user-generated content, comments, and video ratings, so serving as a central platform for a wide range of multimedia content. The incorporation of interactive elements, such as playlists and recommendations, exemplified the transition towards user engagement and tailored experiences within the context of Web 2.0.

Wikipedia

The website Wikipedia Established in 2001, Wikipedia is an interactive digital encyclopedia that facilitates user participation, editing, and revision of articles. Ensuring correctness and comprehensiveness is achieved through the utilization of community-driven content production and peer review processes. The open editing methodology employed by Wikipedia facilitates the ongoing enhancement and augmentation of information knowledge. The success of this phenomenon serves as evidence of Web 2.0's focus on collective intelligence and user interaction in the creation and management of information.

Twitter

Twitter was founded in 2006 and emerged as a pioneering platform for microblogging, enabling users to share concise and immediate updates referred to as tweets. The platform prioritized swift dissemination of content and facilitated social engagement through replies, retweets, and hashtag applications. The platform provided by Twitter facilitated instantaneous interactions and networking, exerting an impact on prevalent trends and public conversation. The sense of immediacy and engagement inherent in Twitter constitute fundamental elements of the Web 2.0 phenomenon.

Instagram

Instagram, which was introduced in 2010, strategically prioritized the sharing of photos and videos, placing significant emphasis on visual content and fostering social engagement. The platform allows users to upload photographs, employ filters, and interact with other users by means of likes, comments, and direct chat. The comprehensive incorporation of multimedia, user-generated content, and social networking functionalities inside the platform exemplifies the dynamic and participatory characteristics inherent in Web 2.0. Instagram's emphasis on visual content has emerged as a vital component in the realm of digital communication and online culture.

6. WEB 3.0

Web 1.0 and Web 2.0 did not have the machine readability functionality that was introduced to the web pages by Web 3.0. This feature was available in Web 3.0. The term "semantic web" is another name for Web 3.0. Tim Berners-Lee, the person who came up with the idea for the World Wide Web, came up with it. Considering that the word semantic is derived from the phrase "meaning of," the term "semantic web" may be taken to imply "to add meaning to the web." The World Wide Web Consortium (W3C) states that "The Semantic Web provides a common framework that allows data to be shared and reused across application, enterprise, and community boundaries."

An effective method of representing data on the World Wide Web, also known as a database that is globally connected, in a form that can be understood by computers, to the content of documents that are found on the World Wide Web, is known as the Semantic Web. Meaning is represented by semantic technologies via the use of ontologies, and reasoning is provided by means of the connections, rules, logic, and conditions that are represented in those ontologies. In spite of the fact that the semantic Web is not a brand-new Web, it is an extension of the Web that already exists.

For instance, if someone is looking for the flight times of a particular country, a semantic search will return flight times to that country in addition to the specifics of the weather conditions that will be present during the time of visit, maps, city guides, and other helpful information such as reservations for hotels, restaurants, and automobiles. This applies to a wide range of information. RDF, which stands for Resource Description Framework, is a graphical representation of the data that is stored on the Semantic Web. This representation is based on the syntax of XML. RDFS, which stands for "RDF Schema," is a representation of taxonomies that include classes and properties. Words that are considered standard are provided by Web Ontology Language (OWL). AsidefromRDFSandOWL,weareabletocreateruleswiththehelpofRIF/SWRL. For the purpose of querying RDF data, the Simple Protocol and RDF Query Language (SPARQL) is utilized. Digital signatures are used to verify the provenance of the sources from which the input data was obtained. The applications that are used by users are developed at the highest level of all layers.

The architecture that the Semantic Web offers allows for the development of a large number of web apps, which will minimize the amount of human work that is necessary to search for goods or services or websites. Some examples include Google Squared, Hakia, Watson, and Wolfram Alpha. The significance of semantic web search engines and other applications, such as Zemanta (a browser plugin), Facebook (a feature that allows users to "like" posts), and TripIt (an e-commerce travel service website), has been demonstrated by the fact that they have provided people with conveniences in their day-to-day activities on the internet.

Illustrations of Web 3.0

Ethereum

Ethereum Since its inception in 2015, Ethereum has emerged as a decentralized blockchain platform that facilitates the construction and implementation of smart contracts and decentralized apps (dApps) by developers. In contrast to conventional block chains, Ethereum leverages its smart contract capabilities to enable programmable transactions, hence facilitating the automation and enforcement of agreements without the need for middlemen. Decentralization and programmability play a pivotal role in Web 3.0's overarching vision of a more autonomous and transparent internet.

The Inter Planetary File System (IPFS)

The Internet File System (IPFS) is a decentralized and robust peer-to-peer network that has been specifically created for the public web. Initiated in 2015, this platform facilitates the storage and dissemination of files in a decentralized fashion, employing a content-addressed framework instead of depending on centralized repository systems. By diminishing dependence on conventional web servers and augmenting the durability of data, IPFS conforms to the ideas of decentralization and enhanced data accessibility associated with Web 3.0.

Brave Browser

The Brave Browser, which was launched in 2019, incorporates Web 3.0 principles by providing a decentralized surfing interface with the inclusion of privacy-enhancing functionalities. The application operates by obstructing unwanted advertisements and trackers, while also enabling users to accumulate Basic Attention Tokens (BAT) through voluntary ad viewing. Brave's emphasis on user sovereignty and privacy, in conjunction with blockchain-based ad payments, exemplifies the key objectives of Web 3.0, which prioritize the augmentation of user autonomy and data protection.

Chainlink

Chainlink, established in 2017, is a decentralized oracle network that facilitates the integration of smart contracts with real-world data, including price feeds and APIs. This technology facilitates the secure and dependable interaction between blockchain-based applications and external data sources. The technology developed by Chainlink plays a pivotal role in the development of decentralized financial (DeFi) applications and other services based on smart contracts. This underscores the integration of external data with blockchain systems, serving as a key aspect of Web 3.0.

Decentraland

Decentraland, a virtual reality platform hosted on the Ethereum block chain, was introduced in 2020. This platform enables users to engage in the purchase, development, and trading of virtual land and assets. The platform facilitates a decentralized virtual environment wherein people possess ownership and authority over their digital places and interactions. Decentraland serves as a prime illustration of Web 3.0's conceptualization of a digital context characterized by enhanced immersion and user autonomy, wherein ownership and creativity are distributed across multiple entities.

7. WEB 4.0,

Version V However, Web 4.0 is still in the process of becoming a revolutionary notion. As seen in Figure 3, the year 2020 will mark the year when Web 4.0 will become a reality. Web 4.0 is described as a "always-on" society in which people are able to "self-upgrade" via the introduction of technological extensions, as stated by Nils Muller, the Chief Executive Officer of Trend One, a German Microtrend analysis organization. By the time we reach Web 4.0, the operating system will be hosted in the cloud, and participation on the web will be an absolute must. In order to get the data, we will have a number of options available to us, including desktop computers, laptops, netbooks, mobile phones, tablets, and even iTV. Web 4.0 is often referred to as the "Symbiotic Web" in certain circles. The symbiotic web is based on the concept that once the metadata are structured by web 3.0, interfaces that are controlled by the mind may be used by both humans and robots to connect with one another. In order to load websites quickly while maintaining a high level of quality and performance, the machines would be able to scan the contents of the web with a high level of intelligence and then respond by executing and determining what to execute first. This would allow them to construct interfaces that are more commanding.

Examples of Web 4.0

Ambient Intelligence Systems

Ambient Intelligence (AmI) systems are specifically engineered to facilitate smooth and flawless interaction between users and their surroundings by utilizing intelligent technology. For instance, smart home systems such as Google Nest incorporate artificial intelligence (AI) and the Internet of Things (IoT) to acquire knowledge about user preferences, automate household configurations, and offer predictive analytics. In response to user actions and environmental changes, these systems possess the capability to adapt, thereby providing a live experience that is more customized and intuitive. The objective is to establish settings that exhibit intelligent responsiveness to user requirements, so exemplifying the emphasis of Web 4.0 on technology that is both context-aware and proactive.

Autonomous Vehicles

Autonomous vehicles, exemplified by the advancements made by firms like Tesla and Waymo, employ artificial intelligence (AI), the Internet of Things (IoT), and real-time data processing to conduct navigation and make driving decisions without the need for human interaction. These cars employ a variety of technologies to analyze their environment, utilize intricate decision-making processes, and guarantee safety. The use of sophisticated sensors and machine learning techniques empowers cars to acquire knowledge from their surroundings and enhance their performance progressively, aligning with the principles of Web 4.0's concept of highly intelligent and adaptable robotic systems.

AI-Powered Personal Assistants

AI-powered personal assistants, such as Apple's Siri, Amazon's Alexa, and Google Assistant, serve as prime examples of Web 4.0 principles through their ability to offer intelligent and contextually aware interactions. These virtual assistants employ natural language processing (NLP) and machine learning techniques to comprehend and predict customer requirements, providing proactive suggestions and automating various administrative duties. The intelligent and adaptive character of Web 4.0 technology is exemplified by its capacity to seamlessly interface with diverse Internet of Things (IoT) devices and acquire knowledge through user interactions.

Smart Cities

Smart city efforts, exemplified by the implementation in cities such as Barcelona and Singapore, employ a network of interconnected sensors, Internet of Things (IoT) devices, and artificial intelligence (AI) platforms to augment the quality of urban life. These systems enable the management of traffic, optimization of energy use, and enhancement of public services through the analysis of real-time data and the prediction of future requirements. The primary objective is to establish urban settings that are characterized by enhanced efficiency, responsiveness, and sustainability, aligning with the principles of Web 4.0 which emphasize the integration of technology to enhance both the quality of life and operational efficiency.

Advanced Robotics

The field of advanced robotics is witnessing a growing integration of autonomous and intelligent systems in various domains such as industry, healthcare, and household settings. Robots, such as those developed by Boston Dynamics, employ artificial intelligence (AI) and machine learning techniques to execute intricate jobs, adjust to novel circumstances, and engage with humans in a manner that closely resembles natural interactions. These robots serve as a prime example of Web 4.0's conceptualization of a mutually beneficial association between humans and machines, wherein technology not only facilitates but also amplifies human abilities and insights.

8. Developmental stages of web 1.0 to 4.0

Web Generation	Time Period	Focus	Technology	User Role	Examples	Characteristics
Web 1.0	1989-2005	Read-only, static information	HTML, HTTP, URI	Passive consumers	Early websites, Craig's List	Static pages, limited interactivity
Web 2.0	2005-present	Interactive, user-generated content	HTML5, CSS3, JavaScript	Active participants	Social media platforms, blogs, YouTube	Dynamic content, user engagement
Web 3.0	Present	Semantic integration, AI, machine learning	Linked Data, Ontologies, AI technologies	Enhanced personalization	Smart assistants, recommendation systems	Context-aware, intelligent services
Web 4.0	Future	Decentralized, IoT, advanced virtual reality	Advanced AI, NLP, Quantum Computing	Integrated immersive experiences	Virtual Reality, Augmented Reality, Digital Twins	Seamless interaction, symbiotic web

9. Conclusion

The World Wide Web is always undergoing change. Over the course of the last several years, new technologies have arisen that have made it possible for people to enjoy experiences that are more interactive and immersive. In this aspect, the Internet has become more pervasive as a result of the expanding knowledge of wireless networks as well as the growing number of smart phones and tablets. The term "Internet of Things" was coined to describe the phenomenon in which people are not the only ones who can connect to the internet; other physical items, gadgets, and vehicles may also connect to the internet. These new realities resulted in the development of a new paradigm for the web, which is referred to as Web 4.0.

Web 4.0 is a notion that cannot be pinned down and is subject to change. Since the year 2009, a number of writers have proposed a number of different definitions for the Web 4.0. On the other hand, in contrast to the previously widely acknowledged definitions of Web 1.0, Web 2.0, and Web 3.0, there is not a complete agreement over a good definition of Web 4.0. Taking everything into consideration, it is possible to assert that Web 4.0 encompasses a number of different aspects. Every one of these aspects provides a unique perspective on the Web 4.0 paradigm, while at the same time providing a thorough overview of it. It should be noted that applications like social networks, as well as technologies like the Internet of Things, Big Data, artificial intelligence, and machine-to-machine (M2M) and other similar technologies, play a significant part in the acceptance and implementation of Web 4.0.

References

Academia.edu-Share research (n.d.). https://www.academia.edu/

Chamria, R. (2022, September 29). Evolution of the Internet - from web1.0 to web3. https://www.linkedin.com/pulse/evolution-internet-from-web10-web3-ravi-chamria/

Ebrahim, N. A. (2023, November 22). Integration of AI in academic research. https://www.linkedin.com/pulse/integration-ai-academic-research-nader-ale-ebrahim-sgt3f/

Evans, M. (2007). The evolution of the Web-from Web 1.0 to Web 4.0. Reading: University of Reading.

Kumar, A., & Saroj, K. (2022). Role of social media in Marketing 4.0. Establishing Sustainability Context: First Step for Businesses on Their Sustainability Journey, 7(1), 213.

Levy, M. (2009). WEB 2.0 implications on knowledge management. Journal of knowledge management, 13(1), 120-134.

Noh, Y. (2015). Imagining library 4.0: Creating a model for future libraries. The Journal of Academic Librarianship, 41(6), 786-797.

Patel, K. (2013). Incremental journey for World Wide Web: introduced with Web 1.0 to recent Web 5.0–a survey paper. International Journal of Advanced Research in Computer Science and Software Engineering, 3(10).

Solsjö, C., & Aronsson, S. (2022). Transmedia storytelling & Web 4.0— an upcoming love story: Investigating transmedia storytelling across Web 2.0 & 3.0 to assess its relationship with Web 4.0.

Tucker, S., & Kimbrell, J. (2013). The impact of web 2.0, web 3.0, and web 4.0 technologies used in distance education. In ICERI2013 Proceedings (pp. 6803-6809). IATED.

Wijaya, H. (n.d.). Digital evolution: From IT to ICET and From Web 1.0 to Web 4.0.

Zapater, J.J.S. (2014, April). From web 1.0 to web 4.0: the evolution of the web. In Proceedings of the 7th Euro American Conference on Telematics and Information Systems (pp. 1-1).

14

Disaster Management in the Library Context: A Theoretical Framework

Hatneihoi Changsan[1] and Sagarika Medhi[2]

[1]*Teaching Associate, DLIS, Cotton University, Guwahati - 781 001, Assam, India E-mail: hatneihoichangsan@gmail.com*

[2]*PhD Research Scholar, DLIS, Cotton University, Guwahati-781 001, Assam India E-mail: samudrika1230@gmail.com*

Abstract

Libraries play a crucial role in preserving cultural heritage and serving as hubs of community engagement. However, these institutions face a range of potential disasters that can threaten their operations and collections. Comprehensive disaster prevention strategies are essential to safeguarding library resources and ensuring the continuity of services. This paper explores the key components of effective disaster prevention strategies for libraries, addressing both natural and human-caused threats. It examines the importance of regular building inspections, security equipment maintenance, and emergency preparedness drills to enhance facility resilience. The paper also highlights the significance of communication technologies and early warning systems in disseminating critical information and facilitating proactive disaster responses. Recognizing the growing importance of digital preservation, the paper delves into the role of digitization and robust backup solutions to safeguard electronic records and cultural materials. The paper underscores the diverse range of disaster scenarios libraries may encounter, from natural disasters to cyber-attacks, and emphasizes the need for comprehensive prevention strategies to strengthen the long-term sustainability and preservation of libraries worldwide. The findings provide valuable insights for library administrators, disaster management professionals, and policymakers to develop and implement effective disaster prevention frameworks.

Keywords: *Disaster, Disaster management, Libraries, Prevention, Toxins.*

1. Introduction

Disasters are events that occur over a period and cause widespread destruction, suffering, or distress to everything they affect (Oxford English Dictionary, n.d.). The impact of a disaster may be confined to specific areas or widespread and could last for an extended duration, often requiring assistance from external sources to address the aftermath (International Federation of Red Cross and Red Crescent Societies [IFRC], n.d.). These events can arise from naturally occurring phenomena, such as earthquakes, tornadoes, or floods, as well as man-made occurrences, including industrial accidents (e.g., accidental toxic spills or nuclear power plant events) or acts of terrorism (e.g., bombing or poisoning) (United Nations Office for Disaster Risk Reduction [UNDRR], n.d.). Disasters typically result in significant harm to infrastructure, loss of life, displacement of people, and severe economic consequences.

Disasters can range from minor disruptions, such as termite infestations, to catastrophic events, such as tsunamis, presenting in various forms and scales (UNDRR, n.d.). They are not uncommon occurrences and can arise unpredictably, often wreaking havoc on human populations and the natural environment. Disasters disrupt the normal order of life, leaving behind extensive devastation and causing profound suffering, poverty, and despair (IFRC, n.d.). Understanding the nature, causes, and impacts of disasters is crucial for developing effective strategies to mitigate their effects and build resilient communities.

The study of disasters encompasses a multidisciplinary approach, drawing from fields such as emergency management, environmental science, public health, urban planning, and social sciences (Coppola, 2015). Researchers in these domains have investigated various aspects of disasters, including risk assessment, preparedness and response planning, recovery and reconstruction efforts, and the long-term socioeconomic consequences. By examining past disaster events and their impacts, scholars can contribute to the development of more effective policies, procedures, and technologies to enhance disaster risk reduction and management (Cutter, 2016).

Categorization of some of the disasters are:

Natural Disasters	**Man-Made Disasters**
Earthquakes	Deforestation
Floods	Air/ Water Pollution
Accidents	Chemical Disaster
Landslides	Nuclear and Radiological
Wildfires	Infrastructure Failures

Cyclones/ Typhoons/ Tsunami	Biological Disaster
Droughts	Terrorist Attack/ Bomb Blast
Volcanic Eruptions	War and Armed Conflicts
Disease Epidemics and Insect	Hijacking/ Hostage Situation

2. The Disaster Management Act of 2005: Enhancing India's Resilience

The government of India introduced the Disaster Management Act on December 23, 2005 (Government of India, 2005) following the devastating and catastrophic tsunami of 2004 which claimed thousands of lives in India. This legislative measure aimed to strengthen the nation's capacity to manage disasters efficiently, encompassing the formulation of alleviating strategies, capacity-building efforts, and associated concerns.

The Act recommended the formation of the National Disaster Management Authority (NDMA) under the Prime Minister's leadership and the State Disaster Management Authorities (SDMAs) under the guidance of their respective Chief Ministers (Government of India, 2005). These entities were tasked with leading and executing a comprehensive and coordinated approach to disaster management across India. The Act's vision was "to build a safer and disaster resilient India by a total, pro-active, technology-driven and sustainable development strategy that involves all stakeholders and fosters a culture of prevention, preparedness and mitigation" (NDMA, n.d.).

The Disaster Management Act of 2005 defines Disaster Management as a continuous and integrated approach that encompasses planning, organisation, coordination, and execution of actions essential for:

- Preventing the occurrence of any disaster or potential threat
- Alleviating or minimising the risk, severity, or aftermath of any disaster
- Enhancing capacity through research and knowledge management
- Readiness to address any potential disaster
- Swift response to any imminent disaster scenario
- Evaluating the impact and scale of any disaster
- Conducting evacuation and rescue operations and providing relief
- Facilitating rehabilitation and reconstruction efforts (Government of India, 2005)

Disaster risk management involves the application of policies and strategies to reduce disaster risk, reduction to prevent the creation of existing risks, diminish current risks, and handle any remaining risks (UNDRR, n.d.). This contributes to enhancing resilience and minimising disaster-related losses.

Disaster management does not guarantee the total avoidance or eradication of threats; rather, it focuses on crafting and executing preparedness and other strategies to minimise the impact of disasters and "build back better" (UNDRR, n.d.). Failing to devise and implement such plans could result in harm to lives, assets, and financial losses (UNDRR, n.d.).

3. Library and Disaster Management

In the context of a library, a disaster refers to an event or situation that causes significant damage or disruption to library resources, facilities, or services (IFLA, 2006). This encompasses natural calamities such as floods, fires, earthquakes, and human-caused incidents like vandalism or theft (Kahn, 2012). Libraries are inherently vulnerable to various disasters, as no library is entirely free from risk and the potential for disruption (Feather, 2004). Libraries commonly have disaster management plans in place to reduce the impact of such events and support recovery endeavours (IFLA, 2006).

The renowned Alexandria Library, founded around 300 B.C. which reportedly housed between 400,000 and 700,000 documents at its zenith, served as a prominent historical example of library disaster (El-Abbadi, 1990). Sadly, the library was burned down and destroyed during conflicts in the era of Caesar (100 BC-44 BC) (El-Abbadi, 1990).

In the modern era, disasters in libraries can include natural events like floods fires, earthquakes, and storms, as well as human-made incidents such as vandalism, theft, arson, or structural failures (Kahn, 2012). These events can result in substantial harm to library resources, facilities, and services, requiring extensive efforts for recovery and restoration (IFLA, 2006).

The disaster control plan serves as a comprehensive guide for action, and readiness should entail understanding all its components (IFLA, 2006). Staff members should undergo training and fully comprehend the contents of the disaster plan, understanding their roles and responsibilities within the team (Kahn, 2012). Adequately stocked supplies of equipment and materials should be maintained to address any unforeseen calamitous events effectively (IFLA, 2006). The disaster control plan document should be designed for easy reference and undergo regular review and updates (Kahn, 2012). Identifying potential hazards, risks, and threats both within and outside the library, and prioritising actions related to them, is vital to minimising their impact (IFLA 2006). Incorporating a disaster management plan is essential during the initial planning stages of establishing a library, as it must cover the entire library including staff, patrons, information sources, equipment, and infrastructure requiring a collaborative effort from everyone involved (Feather, 2004).

4. Causes of Disaster in Libraries

The causes of disasters in libraries can be varied, but some of the common factors include:

4.1. Natural Disasters

- **Earthquakes:** Seismic activity can cause structural damage to library buildings, collapse of shelving, and loss or damage of library collections.
- **Floods:** Libraries located in flood-prone areas are at risk of damage or destruction from heavy rains, storm surges, or rising water levels.
- **Hurricanes/Typhoons**: High winds, heavy precipitation, and storm surges associated with these events can severely impact library facilities and collections.
- **Wildfires:** Libraries situated in areas prone to wildfires face the threat of damage or destruction from the flames and smoke.

4.2. Human-Made Disasters

- **Chemical spills**: Accidental spills of hazardous materials, either within the library or in the surrounding area, can contaminate library collections and the environment.
- **Fires:** Electrical malfunctions, arson, or accidents can lead to fires that devastate library collections and facilities.
- **Terrorism and vandalism**: Acts of terrorism, such as bombings, or intentional vandalism can result in physical damage, destruction, or theft of library resources.
- **Water leaks**: Plumbing failures, broken pipes, or roof leaks can cause water damage to library materials and equipment.

4.3. Technological Failures

- **Cybersecurity breaches**: Cyberattacks, such as hacking or ransomware, can compromise library systems, networks, and digital resources.
- **Hardware/software failures:** Malfunctions or crashes of computers, servers, or other digital infrastructure can lead to the loss of digital collections and records.
- **Power outages**: Prolonged power failures can disrupt library operations, damage electronic equipment, and compromise environmental controls for library collections.

4.4. Organizational Factors

- **Insufficient building maintenance:** Neglected or aging infrastructure, such as leaky roofs or outdated electrical systems, can increase the risk of disaster-related incidents.
- **Lack of disaster preparedness**: Libraries with inadequate emergency planning, staff training, or resource allocation may be ill-equipped to respond effectively to disasters.
- **Poor collection management**: Improper storage, handling, or preservation of library materials can make them more vulnerable to damage or loss during a disaster.

5. Disaster Management of Libraries

Libraries play a crucial role in preserving knowledge and cultural heritage, making it imperative that they implement effective disaster management strategies. This comprehensive approach ensures the protection of library resources, safeguards staff and patron safety, and enables a swift recovery in the event of a disaster.

5.1. Disaster Management Plan

Libraries should develop and maintain a comprehensive disaster management plan that is regularly monitored and reviewed. This plan should include the following key elements:

5.2. Directory of External Services and Products

Compile a comprehensive list of local service providers and backup facilities, including their contact information and secondary contacts.

5.3. Emergency Funding Procedures

Develop a plan to secure immediate access to contingency funds for the purchase of additional disaster supplies and other emergency requirements.

5.4. Equipment and Supplies

- Maintain an inventory of essential equipment and emergency disaster supplies, including wet and dry vacuum cleaners, water pumps, fire extinguishers, fans, dehumidifiers, and water hoses.
- Regularly inspect and maintain the functionality of all equipment.
- Implement a termite treatment program to protect the library's physical infrastructure.

5.5. Facility Infrastructure

In addition to the disaster management plan, libraries should prioritize the structural resilience of their buildings. This includes ensuring adequate drainage and flood-proof systems, as well as constructing library structures to withstand seismic activity. Regular inspections for water leakage should also be conducted to mitigate the risk of water damage.

5.6. Human Resource Management

- This clearly defines the roles and responsibilities of the disaster management team, ensuring that all members are aware of their duties during various phases of a disaster.
- To provide regular training and mock drills to ensure the team is prepared to respond effectively.

5.7. Insurance Coverage

Ensure that the library's insurance policies are up-to-date and compliant with all necessary conditions to maintain the validity of potential claims.

5.8. Priority Materials

Identify and prioritize the evacuation of important and/or rare materials in the event of a disaster.

5.9. Recovery Protocols

- Establish timely decision-making processes for salvage procedures, tailored to the specific types and degrees of damage.
- Incorporate techniques such as vacuum-freeze drying and air drying to facilitate the recovery of damaged materials.

5.10. Response Protocols

- Conduct periodic full-scale mock drills to test the functionality of equipment and the preparedness of the library staff.
- Clearly specify the location of disaster supplies and ensure their accessibility in the event of an emergency.
- Prioritize human safety as the primary objective in any disaster scenario.

6. Disaster Prevention in Libraries

Safeguarding library resources and ensuring the continuity of services in the face of emergencies or disasters is the primary goal of comprehensive disaster prevention strategies for libraries. These strategies encompass range of bold measures that aim to lessen the impact of potential disasters and maintain the library's operations.

7. Building Inspections and Maintenance

Regular building inspections are a crucial component of disaster prevention. Libraries should specify the inspection methods and frequency intervals, ensuring that these assessments are conducted consistently. If external contractors are responsible for the inspections, their contact information should be readily available.

8. Communication and Early Alert Systems

Communication technologies, including the internet, mobile phones, email, radio, and television, can be utilized to disseminate and exchange relevant information before and during disaster situations. Additionally, a robust forecasting and warning system can play a pivotal role in shaping potential disaster responses, facilitating the full-on evacuation of at-risk areas and the relocation of individuals to safer places before a disaster occurs.

9. Digital Preservation and Backup

Digitization is a crucial strategy for conserving priceless heritage and cultural materials. Digital archiving ensures that materials are preserved in formats that resist decay, necessitating ongoing migration efforts to maintain their integrity. Storing information on servers outside the library building and using cloud servers to back up essential information, such as collection lists, digital copies of materials, and contact information for staff members, is crucial. Digitization can help to preserve electronic records, images, catalog records, and other critical library assets.

10. Security and Emergency Preparedness

Routine emergency and evacuation drills are essential to verify the functionality of emergency measures, exits, and equipment, such as fire alarms. Regular inspections of facilities, like fire extinguishers and security alarms, are necessary to ensure their proper functioning.

11. Conclusion

Effective management of disasters requires a comprehensive approach that covers both prevention and response. Disasters, whether originating from natural or human causes, can lead to significant disruption and damage, making it crucial for libraries and other institutions to have well-developed disaster management strategies. Libraries, as important centers of knowledge and culture, are particularly vulnerable to diverse disaster scenarios and must implement specific and targeted measures to protect the resources, ensure the safety of the staffs and patrons, and facilitate immediate recovery. Disasters typically arise as unforeseen occurrences, carrying unpredictable and potentially catastrophic consequences. Disaster preparedness demands dynamic processes, necessitating effective cooperation and coordination among various professionals. Providing education and raising awareness about disaster management is necessary and holds immense importance. Collaboration with libraries and disaster management can take various forms, such as organizing joint training programs and acquiring and storing emergency equipment. With the integration of communication systems, regular building maintenance, staff training, and digital preservation, libraries can enhance their resilience to disasters. Thus, libraries must consistently anticipate and adeptly address potential crises. The primary aim of disaster management is to diminish the deterioration of library resources and facilitate the library's recovery even in the aftermath of a catastrophic event. To sum up, a bold and integrated approach to disaster management is essential for safeguarding the valuable resources and services that libraries provide and to maintain the integrity of library services amidst unforeseen challenges.

References

Abbadi, M.E. (2024, March 7). Library of Alexandria. Encyclopedia Britannica. https://www.britannica.com/topic/Library-of-Alexandria

Bansal, J. (2015). Disaster management in libraries: An overview. Gyankosh- The Journal of Library and Information Management. 6(1).9.DOI:10.5958/2249-3182.2015.00002.7

Coppola, D. P. (2015). Introduction to international disaster management (3rd ed.). Butterworth-Heinemann.

Cutter, S. L. (2016). The landscape of disaster resilience indicators in the USA. Natural Hazards, 80(2), 741-758.

El-Abbadi, M. (1990). The life and fate of the ancient library of Alexandria. UNESCO.

Feather, J. (2004). Managing preservation for libraries and archives: Current practice and future developments. Routledge.

Garlinghouse, T. (2022). The Rise and Fall of the Great Library of Alexandria. Live Science. https://www.livescience.com/rise-and-fall-of-the-great-alexandria-library#:~:text=%22The%20library%20was%20probably%20created,University%20of%20California%2C%20Los%20Angeles.

IFLA. (2006). Disaster preparedness and planning: A brief manual. International Federation of Library Associations and Institutions.

IFRC. (2024, May 5). What is a disaster?https://www.ifrc.org/our-work/disasters-climate-and-crises/what-disaster

Insights IAS. (2024, May 5). Disaster Management. https://www.insightsonindia.com/disaster-management/

International Federation of Red Cross and Red Crescent Societies. (n.d.). What is a disaster? https://www.ifrc.org/en/what-we-do/disaster-management/about-disasters/what-is-a-disaster/

Kahn, M. B. (2012). Disaster response and planning for libraries. American Library Association.

Mane, S. (2017). Disaster management in college libraries. Aayushi International Interdisciplinary Research Journal. 4(12). 233-237.https://www.academia.edu/43499288/Disaster_Management_in_College_Libraries?sm=b

National Institute of Disaster Management (2024, May 5). Understanding Disasters. https://nidm.gov.in/PDF/Disaster_about.pdf

Oxford English Dictionary. (n.d.). Disaster. In OED Online. Oxford University Press. https://www.oed.com/view/Entry/54626

Oxford English Dictionary. (2024, May 5). Disaster.https://www.oed.com/search/dictionary/?scope=Entries&q=disaster

Oyeniran, K.G. (2023). A survey of disaster management practices in selected public university libraries in Nigeria. Iiorin Varsity International Journal of Library & Information Science, 6(1&2).https://www.researchgate.net/publication/379178600_A_survey_ of_Disaster_Management_Practices_in_Selected_Public_University_Libraries_in_Nigeria

Physiopedia (2024, May 4). Disaster Management. https://www.physio-pedia.com/Disaster_Management

Puducherry District (2024, May 5). Disaster Basics.https://puducherry-dt.gov.in/disaster-basics/

Rode, G. (2023). Disaster management in college libraries: A study. International Journal for Multidisciplinary Research. 1-4. https://www.ijfmr.com/special-issues/2/131.pdf

Syracuse University (2024, May 7). How Libraries Can Prepare for Natural Disasters and Preserve Community History. https://onlinegrad.syracuse.edu/blog/emergency-preparedness/

Tulane University School of Public Health and Tropical Medicine (2021, September 3). What is Disaster Management? Understanding Emergencies from Prevention to Mitigation. https://publichealth.tulane.edu/blog/what-is-disaster-management/

United Nations Office for Disaster Risk Reduction (2024, May 5). Disaster. https://www.undrr.org/terminology/disaster

United Nations Office for Disaster Risk Reduction (2024, May 5). Disaster risk management. https://www.undrr.org/terminology/disaster-risk-management

United Nations Office for Disaster Risk Reduction (n.d.). Disaster. https://www.undrr.org/terminology/disaster

15

Navigating the Future of Libraries: Embracing Digital, Virtual, and Hybrid Modes with the Focus on Digital Literacy for Library Users

Basudeb Jana

Assistant Librarian, Kalinga University, Raipur - 492 101, Chhattisgarh, India
E-mail: basujana5@gmail.com

Abstract

The key component for the global information explosion has been built by the invention and quick development of information and communication technologies (ICT). As a result, it becomes challenging to choose or discover the necessary information in a timely manner. The traditional print medium of information sources is giving way to electronic or digital mediums. As a result, both libraries and their collections now have a different personality. Libraries now provide services and functions beyond those of typical libraries in order to address this problem. Libraries, such as traditional libraries, automated libraries, hybrid libraries, electronic libraries, digital libraries, virtual libraries, and paperless libraries, have evolved in response to the evolving nature of collections. Virtual libraries provided their services to Users virtually via networks at any time and from any location in the globe. This study aims to draw attention to the function of digital, virtual, and hybrid libraries in Indian higher education.Digital literacy necessitates the easy acquisition, exploration, manage, present and use of digital content. To function well in the digital age, one must be able to use and operate digitally enabled tools and software, which calls for cognitive abilities, technological soundness, and web proficiency.

This study aims to aware about the notion of Digital, Virtual, and Hybrid libraries, their benefits and functions, as well as the issues surrounding them. It also focuses on the digital literacy skills for library users to accessing the Digital, Virtual and Hybrid library.

Keywords: *Digital library, Virtual library, Hybrid library, Digital literacy, ICT, Computer literacy.*

1. Introduction

Traditionally, libraries were considered book storage facilities for kings and other prominent members of society. However, the idea of conventional libraries was altered with the advent of ICT.The development of the Internet, the World Wide Web, and information technology (IT) are assisting in increasing financing for studies on creating, accessing, and administeringelectronic information resources. As a result, a new era of digital and virtual libraries has emerged. These technological developments have made it possible for a new generation of information professionals to successfully select, organize, retrieve, and communicate digital content to their target audience (Nazim & Saraf, n.d.). The concept emerged again when computers were connected to tremendous networks to create education and research cantered on creating digital libraries accessible to anybody all over the World. Terms like "virtual library", "electronic library", "library without walls",and "digital library" have been used to characterize this broad concept (Cleveland, n.d.). Digital Libraries are enormous collections of organized information resources. Elegant "Digital Library" software can enable everyone to perceive, organize, develop, and disseminate new information collection.Libraries have long been seen as safeguards of information, having developed over centuries from private repositories of manuscripts to community centres teeming with activity and education. This evolution has accelerated in the digital age, forcing libraries to negotiate the challenging flows of digital evolution. The way that library's function has to alter as a result of the significant changes in information-consuming patterns brought about by the growth of the internet and digital media. The physical libraries of the past have given way to digital ones, which signifies a significant change in the way knowledge and information are gathered, accessed, and shared. This transition is a slow, methodical process rather than happening suddenly. The digitalization of library catalogue systems was a significant first step away from the card catalogue drawers that needed to be physically browsed. The introduction of eBooks, online databases, scholarly articles, and multimedia resources—all accessible through library websites and digital platforms—was made possible by this first stage of the digital revolution in libraries. (Zilla, 2024). The classic library, a sanctuary of printed books and quiet conversations in an age of digital innovations faces revolutionary opportunities and challenges. Libraries are changing as we move through the digital era, changing their functions, offerings, and basic nature. Join us as we explore what libraries might look like in this rapidly evolving digital landscape (Prepare, 2023).

2. Objective of the Study

The primary objective of this paper is to discuss the importance of virtual, digital and library in higher education. Beside this the other important objectives of this paper are as follows:

i. To know the concept of digital, virtual, and hybrid library
ii. To know the functions and services of digital, virtual, and hybrid library
iii. To find out the initiatives in India
iv. To identify the advantages of digital, virtual, and hybrid library
v. To know about digital literacy of accessing digital, virtual, and hybrid library

3. Methodology

Several primary documentary sources, including research journals and conference proceedings, as well as secondary sources, such as test books, have been explored in order to identify and examine the relevant information. In the absence of this, other websites have been checked out for this information. Following a thorough and in-depth analysis of several sources, pertinent material has been gathered. The subject's evolution has led to the organization and discussion of the content that has been gathered. Following a thorough consideration of the subject, a decision has been formed.

4. What is Digital Library?

A digital library is a library that has collections kept in an electronic or digital format that computers and other electronic devices can access. A "Digital Library"is sometimes called an online library or internet library for digital collections such as e-textbooks, images, audio, video, and other digital documents (Library Science, 2022). The digital library also provides a framework for maintaining the database management of its services. Additionally, a digital library is an electronic device, system, or format containing a collection of regulated and accessible material using universal patterns and modes. Similarly, a digital library is a paperless space that relies entirely on computer technology. It offers all the services and activities of a library without requiring any specialized understanding of librarianship. Although digital libraries still have a ways to go before becoming worldwide digital libraries, ICT has dramatically advanced the field of library services, and it would be very beneficial to realize the promise of digital library services. Most importantly, a digital library should provide ways to arrange, save, and retrieve its contents. The contents of the e-Library are centrally stored to enable access across many devices (TechSolvePrac, 2020). Some important examples of digital libraries in India are:

Fig. 1: Digital Library: What & Why

Source: https://www.sydle.com/blog/digital-library-63e521409e3c274f1050e336

1. "National Digital Library of India IIT Kharagpur"
2. "Traditional Knowledge Digital LibraryMinistry of Health and Family Welfare, Government of India"
3. "Indira Gandhi National Center for the Arts, Digital LibraryNew Delhi"
4. "Vidyanidhi Library and Digital Library of the Department of Informatics University of Mysore, Mysore"

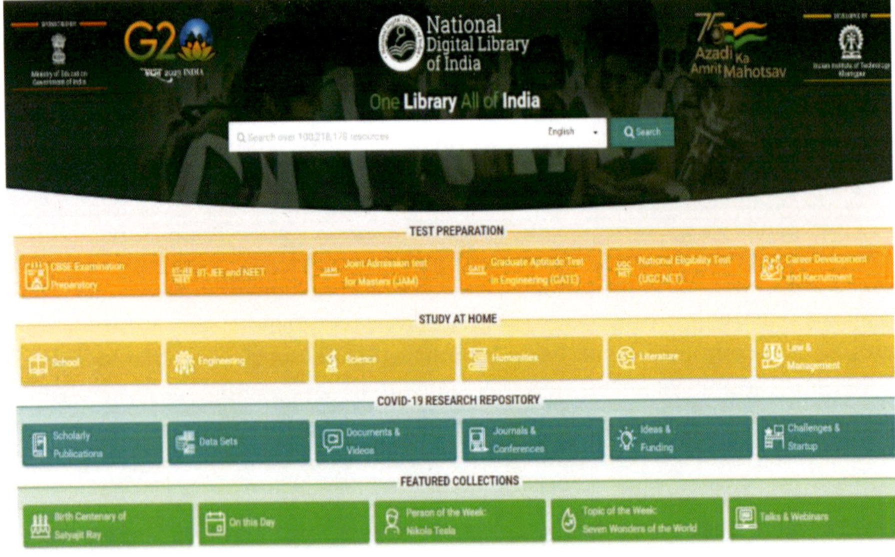

Fig. 2: NDLI

Source: https://ndl.iitkgp.ac.in/

Fig. 3: TKDL

Source: https://www.tkdl.res.in/tkdl/langdefault/common/Home.asp?GL=Eng

Components of Digital Library

- Hardware and Software
- Good Bandwidth WIFI
- Storage devices – Optical Storage devices, CD-ROM
- Computer Server – PC with Multimedia
- Network – LAN, WAN, MAN, etc.
- Telecommunication facility
- Human Resources

Objectives of Digital Library

- To gather, preserve, process, and disseminate data.
- To offer the user need-based and retroactive search services.
- To provide a networking and resource-sharing facility.
- To access national and international documentsthat are published in machine-readable format.
- To digitize the documents for the purpose of preserving and saving space.
- To increase the productivity of library operations.

5. Role of Digital Library in Education and Research

Libraries have traditionally played a vital role in education. The library can be seen as a centrally located collection of resources that comprise human services and a wide range of media formats, including text, video, and hypermedia. Learning via digital libraries has become more commonplace in education and research as most components have been replaced by digitalization. Meanwhile, students may prefer obtaining information and reading content without physically visiting libraries (Sarkar, 2022).

Presently, most libraries are digitizing their materials for access by the user community 24/7. Let us take a look at the main benefits of digital libraries for education and research:

- "Digital Libraries"basically offer infinite resources and opportunities for accessing a wide range of content.
- Online libraries provide the scientific community withcrucial research data, information, and findings.
- With an essential technical device like a PC, tablet, or smartphone students can access digital libraries from anywhere at any time as long as they can access the Internet.
- Digital libraries have developed various search capabilities, including Boolean, proximity operators, and truncation. For example, students may conduct complex searches for various queries and quickly access information and data collections.
- Students and researchers can also access current material from digital libraries that is beneficial to their studies.

6. Salient Features of Digital Library

- There are two distinct categories of users: one for the end-users of the digital library and the other for the system administrators.
- Several people and organizations can use the same resource simultaneously.
- Digital libraries reduce regional and national boundaries. For example, assume a person in one country wants to access documents in another. In this case, he or she can quickly obtain the book from a digital library.
- Digital Libraries possess a significantly more comprehensive range of content and are not limited by physical constraints. Additionally, they require minimal upkeep expenses in comparison to traditional libraries.
- Digital libraries are another solution to return and renewal problems. Pay a small monthly fee and gain unrestricted access to all their content till

the subscription period ends. There is no need to worry about returning the book punctually.

- In addition to possessing a diverse range of texts and books, a digital library also offers texts in multiple languages.
- An unlimited storage capacity is available in a digital library. Thousands of resources can be stored without difficulty on DLs due to the low cost of media storage.
- Digital libraries offerassistance to their users through software and desktop configurations that are universally applicable.
- A variety of indexes and catalogs are available for the purpose of information retrieval. These indexes support and administer varioussearch strategies.

7. What is Virtual Library?

"Virtual Library" is a platform that allows library professionals, teachers, students, and other users to share and access a wide range of items, including books, papers, journals, films, and other media types. These resources are accessible from any location, at any time, and on any device (Zwetsloot, 2023). The term "Virtual Library" can also refer toa collection of digital documents, that are readily available to access on the Internet.It can also refer to a location/ place where the documents can be accessed. Computers, mobile devices, and the Internet are examples of such places (Gonzales, 2023). A virtual library is an online compilation of digital resources that offers access to various digital content, including databases, e-books, audio and video files, and other digital media. These digital materials are arranged logically, so that users may easily and rapidly search for specific content. This offers a basic search form where users can type any keywords or search phrases quickly. Additionally, they might employ more sophisticated search strategies that let users narrow down results based on criteria such as, author, publishing date, or topic matter. Virtual libraries also have some tools and services that facilitate study, research, and offerusers a vast array of resources and information. It is a valuable resource for education and research, since they offer access to a wide range of materials that would not be found in traditional library collections. They are also an essential resource for people who, for whatever reason, cannot visit physical libraries (LIS edunetwork & LIS edunetwork, 2023). The Web-based interfaces are commonly used to organize and access these libraries. It offers users search and retrieval features to facilitate the quick and effortless discovery of desired information. Essentially, virtual libraries give users instantaccess to a wide range of digital resources, which improves the flexibility, interactivity, and accessibility of learning and research.

Figure 4: Virtual Libraries

Source: https://home.edweb.net/virtual-libraries/

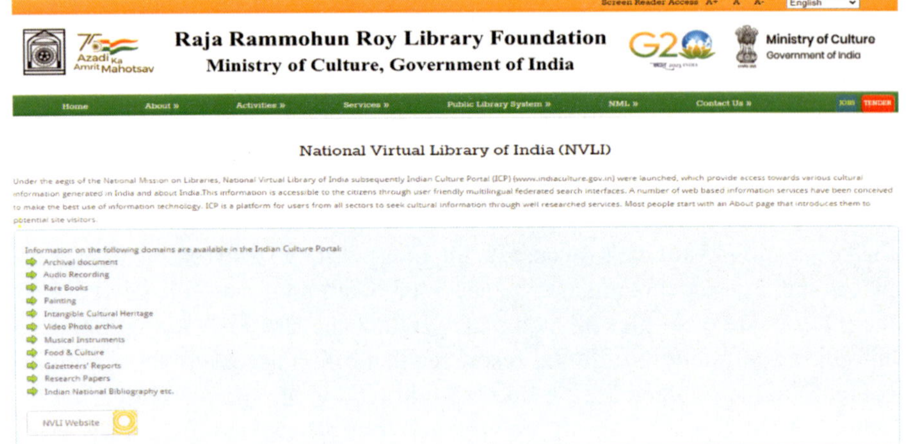

Fig. 5: National Virtual Library of India

Source: http://rrrlf.nic.in/NML/NVLI.aspx

8. Role of Virtual Library in User Services

Virtual libraries have become essential to education and research, enabling easy access to a wealth of knowledge and resources needed for study, research, and personal development. Some of the importance of virtual libraries for education and research are given below:

- Virtual libraries provide research activities with supplementary tools and services, such as software for managing references, citations, and collaboration, which facilitates researcher collaboration.
- To make learning more flexible, interactive, and accessible, virtual libraries provide e-learning resources including tutorials, lectures, and other instructional materials.

- Students' learning can also be improved via virtual libraries. Sometimes it is difficult for students to understand a topic; in that case, a learner may benefit bywatching a short video or listening audio in the virtual library that breaks down the difficulties of that particular topic.
- Students can acquire the critical independent learning abilities needed for future employment andhandle critical projects and assignments on complex subjects.
- Virtual libraries are more than just repositories of information. Scholars, researchers, and students can also connect them. The virtual library enables users to exchange ideas and opinions with the community through collaborative efforts on particular topics and links to forums.

9. Salient Features of Virtual Library

The Virtual Library has so many essential features that make it unique. Some of the important features are described below.

- Virtual libraries are always open, in contrast to traditional libraries.The virtual library can assist you anytime, regardless of your schedule.
- The virtual library may be accessed from any location, which is particularly helpful for those residing in rural areas.
- Documents are available in electronic format, so information is updated immediately.
- An effective search and browse facility is provided by the virtual library, making it easy for users to search and retrieve their requirements.
- Accessible for disabled persons who have physical difficulty. They also can access the library regularly.
- Provide global access to up-to-date information in a timely and thorough manner.
- Well-designed and organized collections increased the productivity and efficiency of the users.
- It improves the quality of life by utilizing digitally available materials.
- Virtual libraries enable quick and easy access to national and international knowledge.
- The virtual library revolutionizedconventional library systems, which were limited to categorizing solely, printed resources.

10. Hybrid Library

Hybrid libraries combine electronic resources like e-books, electronic journals, and downloadable audiobooks with traditional print resources like print books and periodicals. These days, the majority of academic and public libraries are hybrid.It appears that Chris Rusbridge originally used the term "hybrid library" in a 1998. When electronic materials became more widely available for libraries to purchase for public use in the 1990s, hybrid libraries began to take shape. These electronic resources were usually searches of specialized databases or access to content provided on media like CD-ROM. By offering cooperating libraries a single technology resource, OCLC encouraged libraries to acquire digital resources.

The "Hybrid Library" idea is a novel method to information access and learning that combines state-of-the-art digital technologies with traditional physical resources in a seamless manner. Amidst a time when data is becoming more digital and easily accessed via the internet, the Hybrid Library presents itself as a flexible way to connect the real and virtual worlds of knowledge. By combining digital databases, multimedia resources, interactive platforms, and conventional print materials like books and manuscripts, the Hybrid Library aims to create an enhanced learning environment that accommodates a range of learning styles and preferences. This innovation redefines how people engage with and explore information while simultaneously preserving the value of traditional library experiences and utilizing the possibilities of digital breakthroughs. Examining the complexities of the Hybrid Library reveals a paradigm shift that preserves the timeless value of academic inquiry while adjusting to the changing requirements of contemporary learners (Ashikuzzaman & Ashikuzzaman, 2023).

A hybrid library is an adaptable and responsive information center that meets the many demands and preferences of a modern audience.The fundamental idea behind a hybrid library is the peaceful coexistence of digital and physical resources. Books, manuscripts, and magazines are examples of physical holdings that are still essential to the modern library paradigm. However, they are carefully incorporated with a range of digital resources, including interactive platforms, online databases, multimedia content, and e-books. The combination of the dynamic potential of the digital world and the depth of traditional scholarship generates an ecosystem.

Moreover, a personalized and interactive aspect defines the Hybrid Library experience. Users can personalize how they interact with resources, bookmark online pages, annotate materials, and even take part in online chats with other students. Through the adoption of digital technologies, the library transforms from a mere information repository into a center for interactive knowledge exchange and interaction.

Fig 6: Hybrid Library

Source: https://www.bdcnetwork.com/blog/welcome-hybrid-library

11. Significance Hybrid Library

The image of a library has long been associated with endless rows of bookshelves and the peaceful silence of academic study. But just as the digital revolution is still changing every aspect of our lives, libraries have experienced a significant metamorphosis of their own. Let me introduce you to the hybrid library, a dynamic fusion of innovation and tradition that redefines the fundamentals of learning and information access.

11.1. Smooth Combination of Digital and Physical Resources

A key component of the Hybrid Library is the skillful blending of digital and physical materials. This creative method combines the endless possibilities of digital information with the physical joy of turning pages in a book. Readers can dip into the depths of vintage print collections and move fluidly between interactive databases, multimedia presentations, and immersive e-books.

11.2. Seamless Accessibility

A hybrid library's ability to exist outside of physical boundaries is one of its most appealing qualities. Learners and researchers are no longer restricted to a particular area because of digital resources that are accessible remotely. The Hybrid Library enables a global exchange of information, whether users are working with peers locally or utilizing resources from the comfort of their homes.

11.3. Interactive Data Analyses

By utilizing data analytics, the Hybrid Library can shed light on user preferences and activity. Libraries can improve their services, curate collections, and foresee changing requirements by gathering insights into how users engage with their resources.

11.4. The Convergence of Multimedia Content

A hybrid library incorporates a variety of multimedia content in addition to printed materials. This convergence of media typesfrom interactive simulations and instructional films to podcasts and visual presentations, accommodates different learning preferences and engages users.

The emergence of the hybrid library embraces the limitless possibilities of the digital age while highlighting the continuing importance of traditional libraries. Its seamless fusion of digital and physical resources, accessibility from anywhere in the world, individualized interactions, data-driven insights, multimedia convergence, and cultural preservation.

12. Essence of Digital, Virtual, and Hybrid Library in the Current Era

An electronic library, or e-library, is the same as a "Digital Library". It is a platform or website that offers 24/7 internet access to digitized text, audio-video content. It gives users access to books, journals, and other materials. Conversely, a virtual library is merely an online platform housed on a website; it is not real but virtual. Users can access it instantly from anywhere and anytime with an internet connection.

A digital library is generally defined as an electronic book collection, online journals, multimedia files, and digital copies of printed books or other documents. Conversely, a virtual library usually refers to a particular application or implementation of a digital library inside an organization. Frequently, it is customized to fulfil the requirements of the user group, such as educators, researchers, learners, etc.In contrast to a virtual library, which is a library that only exists online and does not exist in the real world and is made up of elements arranged using computers and computer networks in a virtual environment, a digital library contains a location-based collection of electronic resources and may even have print resources.

A virtual library offers access to various remote sites through a local online catalogue or any other gateway, such as the Internet. In contrast, a digital library serves its Users through direct or indirect access to systematically organized collections of digital objects. However, a "Digital Library" is a collection of digital and paper documents.

The Virtual Library depends entirely on the Internet and cannot exist anywhere without Internet access. However, the Digital Library is a digital version of printed information and the internet base. It can be compatible with CDs and DVDs.This is an assortment of materials, books, or other kinds of content that may be used rather than just seen. It is in charge of keeping up-to-date data so that it can consistently satisfy clients' needs. A library provides print or digital access of materials and can be either a physical building or an online space, or both. Written textbooks, a variety of physical resources including CDs, DVDs, and Disc, as well as access to content, music, and other materials stored in databases, can all be found in a library's collection.

In addition to offering quiet, comfortable study spaces and common areas for group study sessions and collaboration, library buildings frequently allow the public to use its technology assets, such as computers and Internet access. The type of library determines its patrons and, consequently, the services it offers to users, for example, a general library and a specialized librarian have different expectations. Libraries can act as community hubs where people can participate in programs and courses for lifetime education. Advanced libraries provide a range of services. books accessible electronically, allowing them to provide their services outside of a facility's physical wallseven from home through over the Internet. Consequently, data preservation in the electronic age has become very simple, given that users can access or submit material electronically from distant locations, even if they are far from the library (B. Singh *et al.*, 2022).

13. Digital Literacy

Digital literacy encompasses the proficiency to effectively utilize technology, along with the requisite knowledge and skills to do so responsibly and securely. The term "digital" encompasses a broad spectrum of technology products and services, such as computers, the internet, smart home systems, cell phones, airport check-in kiosks, and other such entities. Literacy is the ability to utilize this technology.The competencies needed to use technology in a responsible, efficient, and safe manner are included in digital literacy. Acquiring digital literacy skills is becoming more and more crucial as technology continues to permeate every aspect of daily life (Team, 2023). Libraries have been serving as educational and research hubs, and in the twenty-first century, they are crucial for advancing digital literacy. The term "digital literacy" describes the capacity to locate, assess, utilize, and produce digital information ethically and successfully. Digital resources that can aid in the development of users. Digital literacy include e-books, online databases, and digital archives, all of which are accessible through libraries. Libraries can provide workshops and technology training programs to assist people in gaining more comfort with computers, the internet, and other digital tools. Libraries can also serve as a conduit for persons with more sophisticated abilities to interact with those who have limited access to technology. By giving underprivileged populations access to technology and training, they can contribute to closing the digital divide. Libraries can also encourage digital literacy by way of their holdings.

They can offer books and online tutorials as well as activities and events that emphasize digital literacy. Library can also offer materials that educate digital literacy skills to the users.In general, libraries are crucial for advancing 21st-century digital literacy. They give people access to tools, resources, and instruction that can help them acquire the abilities required to function well and morally in the digital world (Inamdar, S.H., 2021).

14. Digital Literacy of Library Users

In the twenty-first century, the majority of libraries are creating digital and virtual collections. The majority of libraries continue to operate in hybrid mode. However, users are facing difficulties accessing those. The cause of this issue is digital obsolence. Users and other library personnel must become digitally literate in order to overcome Digital Obsolence. In libraries, digital literacy is an innovative idea. Professionals and librarians coordinate training and orientation sessions to give library users the practical skills they need to access all digital material, e-resources, databases, etc. Digital literacy is especially important for library users since, in addition to being open-access surroundings these days.Libraries are more likely to subscribe to online resources than print resources. Therefore, most of the time useraccess the resources on their own.Technology is becoming a vital instrument in the rapidly changing educational scene, helping to shape the development of literacy, particularly in rural areas. We can see how technology is changing reading abilities. The dynamic interplay between literacy and technology emphasizes the part nonprofits play in utilizing these tools to further education (Evans, 2024). Digital Literacy introduces and integrates technology into rural schools, colleges, universities, and other academic institutions. By doing this, we hope to close the gap in access to technology and allow library users more chances for better library experiences. Particularly in areas with limited resources, nonprofit groups are essential in ensuring that technology is a tool for empowerment rather than luxury (Evans, 2024b).

Digital resources are information sources that can only be accessed via computers and other ICT devices. Examples of these resources include CD-ROMs, the Internet, OPAC, e-books, e-journals, and e-databases. These days, libraries are creating digital libraries and institutional repositories that are available through their websites. These resources might call for the usage of an external device that is directly attached to a computer, like a CD-ROM drive or a network or Internet connection (Mulat,T. & Natarajan, M., 2020).

One important way that libraries support digital literacy skills is by giving people access to technology. Libraries provide computers, internet connection, and other technology tools for public use, allowing people to experiment with and learn digital skills. By giving people who might not have access to their own devices or internet connectivity at home, libraries play a critical role in closing the digital divide (Josie *et al.*, 2018). This serves to lessen access gaps and guarantee that everyone has the chance to acquire digital literacy abilities (Gorham & Bertot, 2018). Libraries assist level the playing field and provide equitable opportunities for people to participate in digital activities, access online resources, and gain necessary digital skills by providing access to technology (Jaeger *et al.*, 2015). Public computer workstations with internet connection are available at libraries, enabling users to surf the web, use productivity tools, access online databases, and take part in online educational programs. People can learn the fundamentals of computer literacy and become more comfortable using digital technologies because to this access (Small, 2019).

15. Conclusion

Library services have developed throughout time to meet the demands of a diverse society and new technological advances. From the perspective of academics, researchers, and educational institutions, digital and virtual libraries have brought many advantages and made substantial contributions. Virtual and digital libraries are digitizing information resources and services worldwide. However, as user expectations and technological capabilities allow in this fast-paced world, the initial digital and virtual libraries resulting in digitization projects or even virtual collections are being improved.

Thus, the primary goal of the librarianship profession in the current scenario should be to create a cohesive force that fosters understanding and communication among the various components of digitization. That makes up the professional development as a whole to prevent and understand the recent trend of user requirements.Virtual libraries support self-directed and casual learning for individuals. Individuals are utilizing virtual libraries directly, with little help from librarians or other middlemen.Informationliteracy,end-user training, ongoing professional education and training, appropriate infrastructure, and adequate infrastructure are all necessary to guarantee the democratization of knowledge and information.

References

Ashikuzzaman, M., & Ashikuzzaman, M. (2023b, December 17). What is Hybrid Library? Library & Information Science Education Network.https://www.lisedunetwork.com/what-is-hybrid-library/#google_vignette

Cleveland, G. (n.d.). Digital libraries: Definitions, issues and challenges.

Evans. (2024, January 29). How technology shapes literacy development in the digital Age - Angaza Technology Literacy Center. Angaza Technology Literacy Center - Empower students in rural African secondary with technology literacy crucial for attaining economic and social prosperity. https://angazacenter.org/how-technology-shapes-literacy-development-in-the-digital-age/?gad_source=1&gclid=CjwKCAjwiaa2BhAiEiwAQBgyHiGOpxKJdGVJJjjrOWV2y-L9h7pzWwfxIYNQQ0UWTD_vcr1wTG6mAhoC-v4QAvD_BwE

Gonzales, S. (2023, November 4). What is a Virtual Library? Easy Tech Junkie. https://www.easytechjunkie.com/what-is-a-virtual-library.htm#google_vignette

Gorham, U., & Bertot, J. C. (2018). Social innovation in public libraries: Solving communitychallenges. The Library Quarterly, 88(3), 203–207. doi:10.1086/697701

Inamdar, S.H., (2021). The role of libraries in promoting digital literacy in the 21st century. Journal of Emerging Technologies And Innovative Research (JETIR), 8(8), 2349–5162. https://elsevier-ssrn-document-store-prod.s3.amazonaws.com/23/05/04/ssrn_id4437672_code5529446.pdf?

Jaeger, P. T., Gorham, U., Taylor, N. G., & Bertot, J. C. (2015). teaching informationpolicy in the digital age: Issues, strategies, and innovation. Journal of Education for Library and Information Science, 56(3), 175–189. doi:10.3138/jelis.56.3.175

Josie, J., Fang, C., Chetty, K., Qigui, L., Gcora, N., & Wenwei, L. (2018). Bridging the digital divide: measuring digital literacy. Economics: The Open-Access, Open-Assessment EJournalhttps://doi.org/10.5018/economics-ejournal.ja.2018-23

Library Science. (2022, March 24). What is Digital Library? | What is meant by Digital Library? LibraryScience.in - Platform for Library and Information Science. https://www.libraryscience.in/2022/03/digital-library.html

LISedunetwork & LISedunetwork. (2023, October 25). Virtual Library. Library & Information Science Education Network. https://www.lisedunetwork.com/virtual-library/

Main benefits of digital libraries for Higher Education. (2021, September 29). https://in.pearson.com/blog/2021/09/5-main-benefits-of-digital-libraries-for-higher-education.html.

Mulat, T & Natarajan, M., (2020). Digital literacy skills among library professionals in Jimma university libraries. Library Philosophy and Practice (e-journal). 4629. https://digitalcommons.unl.edu/libphilprac/4629

National Digital Library of India. (n.d.). National Digital Library of India. https://ndl.iitkgp.ac.in/

Nazim, M., & Saraf, S. (n.d.). Digital libraries: contents & services.

Prepare, S. (2023, October 23). The future of libraries in the digital age. Skill Prepare. https://skillprepare.com/the-future-of-libraries-in-the-digital-age/

Sarkar, L. (2022). Importance of virtual library in higher education in present era: A challenge for Indian educational. ResearchGate. https://www.researchgate.net/publication/361207969_Importance_of_Virtual_Library_in_Higher_Education_in_present_era_a_challenge_for_Indian_educational_library_system_Importance_of_Virtual_Library_in_Higher_Education_in_present_era_a_challenge_for_India.

Singh, B., Gangwar, S., Sharma, M., & Devi, M. (2022). An overview of hybrid, digital and virtual library. World Journal of English Language, 12(3), 32. https://doi.org/10.5430/wjel.v12n3p32

Team, L. (2023, October 5). What is Digital Literacy: Definition and Uses in Daily Life. Learning. https://www.learning.com/blog/what-is-digital-literacy-definition-and-uses-in-daily-life/

TechSolvePrac. (2020, July 4). What is a Digital Library or e-Library?TechSolvePrac. https://www.techsolveprac.com/digital-library-e-library/

TKDLTraditional Knowledge Digital Library. (n.d.). https://www.tkdl.res.in/tkdl/langdefault/common/Home.asp?GL=Eng

What is a hybrid library? How does it differ from a typical library? - LibAnswers. (n.d.). https://libanswers.gateway.kctcs.edu/faq/54050

Welcome to the hybrid library (2022, June 8). Building Design + Construction. https://www.bdcnetwork.com/blog/welcome-hybrid-library

Zilla. (2024b, April 23). The Future of Libraries: Navigating Digital Transformations. Princh. https://princh.com/blog-the-future-of-libraries-navigating-digital-transformations/

Zwetsloot, J. (2023, April 25). What is a Virtual Library? How can it benefit schools and educators? https://blog.symbaloo.com/blogs/113/what-is-a-virtual-library-how-can-it-benefit-schools-and-educators

16

Supporting AI Technologies for Libraries to Achieve Access to Information in the Context of SDGs

Rashmi Shewale

Assistant Librarian, National Institute of Technology, Goa - 403 703, India
E-mail: shewalerashmi@gmail.com

Abstract

Libraries play a crucial role in providing access to information for economic, social, and cultural development. They curate a wide range of resources, including books, journals, databases, and multimedia materials, to meet users' varied needs. Libraries also offer public access to computers and high-speed internet, enabling people to access online information and research. They provide specialized assistance in locating and accessing relevant resources, and engage in interlibrary loan programs to expand the scope of information.The Sustainable Development Goals (SDGs) are crucial for libraries to achieve their mission of information access. By aligning their efforts with these goals, libraries can contribute to broader development objectives, mobilize partnerships and funding, measure and report progress, foster sustainable practices, and promote holistic development.Quality education is essential for libraries, as it ensures basic reading skills, free and required education, and the availability of resources in regional languages. Libraries can also provide public access to health and welfare information and participate in first aid training. Gender equality is another goal, aiming to empower women and girls and address issues such as sexual abuse and exploitation. By embracing the SDGs as a guiding framework, libraries can strengthen their position as essential institutions contributing to global sustainable development objectives and enhance their ability to provide universal and equitable access to information for all. Artificial Intelligence (AI) is a technology that enables computers and machines to perform tasks that require human intelligence. AI can significantly contribute to libraries' sustainable growth by aligning with the UN Sustainable Development Goals (SDGs). AI can help libraries provide personalized educational resources, identify knowledge gaps, ensure diverse representation, and provide job matching systemsto effectively adopt AI solutions, libraries should prioritize data protection, ensure transparency, conduct digital literacy training, collaborate with other libraries and technology organizations, and regularly evaluate their impact on the SDGs.

Keywords: *Libraries, Artificial Intelligence, Sustainable development goals (SDGs), Access to information.*

1. Introduction

Information is necessary for the economic, social, and cultural development of individuals and communities, and libraries play a significant role in providing access to information, which is essential for this growth. These institutions function as centers of knowledge, offering a wide range of resources to their users, including books, journals, databases, and multimedia materials, in order to meet the varied requirements of their users.

In this day and age, the function of libraries has evolved beyond the confines of physical locations to encompass activities such as online platforms and digital collections. Because of the large amount of information that is accessible on the internet, people rely on libraries to help them navigate through this sea of data and locate sources that are credible and reliable.

2. Role of Library to Access the Information Through Several Means

- Collections: Libraries curate a wide range of books, periodicals, newspapers, digital resources, and other materials to meet the informational requirements of its users. These collections offer a broad array of knowledge and opinions.
- Libraries provide public access to computers and high-speed internet, enabling individuals without these resources at home to access online information, do research, and utilize digital services.
- Librarians, who are skilled information specialists, offer specialized assistance to patrons in locating and accessing the most pertinent resources for their information requirements. The library provides reference services, research advice, and teaching on how to properly utilize library resources.
- Interlibrary Loan: Libraries engage in interlibrary loan programs, enabling clients to obtain materials from other libraries that may not be accessible locally. This broadens the scope of information that is available to the community.
- Outreach: Libraries frequently engage in cooperative efforts with schools, community organizations, and other institutions to expand their influence

and offer informational resources to marginalized demographics, including children, seniors, and individuals with disabilities.

- Libraries provides access to digital resources, like e-books, online databases, streaming media, and digital archives, in addition to physical materials. These digital resources improve the ability to obtain information from a distance or while on the move.
- Libraries offer educational programs, conferences, and training sessions to assist patrons in developing information literacy abilities. These skills enable individuals to efficiently search, assess, and utilize information from many sources.
- Libraries aim to establish inclusive environments that are open and accessible to people from various backgrounds, promoting a feeling of community and ensuring fair access to information.

3. Challenges Faced by Libraries in Providing Access to Information

- Libraries play an important role in making sure people can get information, but they have to deal with a number of problems to do this well. Libraries often run into problems like not having enough tools, money, or trained staff. It is also hard for libraries to keep up with the changing ways that information is shared because the amount of information is growing at an exponential rate and technology is improving quickly.
- Additionally, libraries still face a big problem: making sure that everyone has equal access to knowledge, even underserved communities and marginalized groups. The goal of giving everyone access to knowledge is slowed down by things like the digital divide, language barriers, and people who don't know about library services.

4. What are SDGs

SDGs, which stand for "Sustainable Development Goals," are a group of 17 global goals that were set by the UN in 2015 to solve the world's most important problems by 2030. In easy terms, the SDGs are a plan for making everyone's future better and more stable.

These goals are about a lot of different things, like ending hunger and poverty, promoting sustainable development, protecting the environment, and supporting peace and justice. The SDGs are a way for governments, businesses, civil society, and people to work together to solve these problems on a global scale.

The main goal of the SDGs is to make the world more fair, wealthy, and long-lasting for everyone by 2030. They give everyone a way to work together to get things done and a way to measure progress.

Fig. 1: SDGs

5. What is Importance of SDGs to Attain the Information Access in Context to Libraries

Here are some key reasons why the SDGs are vital for libraries to achieve their mission of information access:

1. **Aligning with Global Priorities**: By aligning their efforts with the SDGs, libraries can ensure that their information access initiatives are contributing to broader development objectives that have been collectively endorsed by the international community.
2. **Mobilizing Partnerships and Funding**: The SDGs have become a rallying point for governments, international organizations, and private sector actors to collaborate and invest in sustainable development initiatives. By demonstrating their contributions to the SDGs, libraries can more effectively secure funding, form strategic partnerships, and amplify their impact on information access.
3. **Measuring and Reporting Progress**: The SDGs provide a comprehensive set of measurable targets and indicators that libraries can use to track their progress in improving information access. This enables libraries to better evaluate the effectiveness of their initiatives, optimize their strategies, and communicate their contributions to stakeholders and policymakers.

4. **Fostering Sustainable Practices**: Certain SDGs, such as SDG 11 (Sustainable Cities and Communities) and SDG 12 (Responsible Consumption and Production), highlight the importance of environmental sustainability. By aligning their operations and services with these goals, libraries can model sustainable practices and inspire their communities to adopt more environmentally-friendly behaviors related to information access and consumption.
5. **Promoting Holistic Development**: The interconnected nature of the SDGs underscores the need for a holistic, multidimensional approach to development. Libraries, by addressing information access in the context of the SDGs, can recognize and leverage the synergies between their work and other critical areas, such as health, gender equality, and climate action.

By embracing the SDGs as a guiding framework, libraries can strengthen their position as essential institutions that contribute to the achievement of global sustainable development objectives, ultimately enhancing their ability to provide universal and equitable access to information for all.

6. How Libraries can Contribute to the Goals of SDGs

1. **No Poverty:** Poverty is primarily caused by factors such as unemployment, social exclusion, disasters, and diseases. This goal aims to ensure that a nation is free from all the problems caused by poverty. Libraries can contribute to this goal by providing open access to all users, allowing anyone to benefit from this access and improve their own situation. Libraries can also facilitate training in new skills for education and employment, provide information to support decision-making in poverty alleviation, and establish mobile libraries to reach underserved communities.
2. **Zero Hunger:** Conflicts, climate change, escalating living costs, civil unrest, and decreasing food consumption are a few of the factors contributing to the issue of zero hunger. In order to achieve this objective, it is essential that countries eradicate hunger. The library can achieve this objective by offering access to research and data on crops market information, and farming techniques. Facilitating the availability of information in regional languages. The libraries of the Agriculture Research Institute can provide farmers with training on information and communication technology (ICT) and new farming technologies enabling them to engage in farming more efficiently.
3. **Good Health and Well-being:** It ensures the maintenance of good health and the promotion of overall well-being for individuals of all

ages. Some of the initiatives include health coverage, access to safe and affordable medicines, vaccination rates, and improved immunization. In order to assist medical practices, libraries can get both print and online resources that are relevant to medical practice. Additionally, libraries can provide public access to health and welfare information and also take part in arranging and engaging in first aid training.

4. **Quality Education:** Basic reading skills, free and required education, hiring more teachers, making basic school facilities better, and accepting the digital change are all parts of it. There should be trained and devoted staff in the library who teach people how to use information and how to become information consumers. Everyone should be able to use the library for free.
5. **Gender Equality:** Instances of sexual abuse and exploitation, unequal distribution of unpaid care and domestic chores, and workplace discrimination are widespread. Therefore, this objective aims to prioritize the attainment of gender equality and the empowerment of all women and girls. The library can enhance its role as a secure and inviting environment for reading and meetings, while ensuring equitable access and services for all user groups.
6. **Clean Water and Sanitization** & 7. **Affordable and Clean Energy**: The entire global community is currently struggling with environmental degradation and a shortage of water resources. Additionally, it encompasses concerns regarding the contamination of water, the deterioration of ecosystems associated to water, and the need for collaboration in managing water basins that cross national boundaries. To address these concerns, libraries can utilize the components of a green library, such as sustainable locations, water efficiency, energy and atmosphere, materials and resources, and indoor environmental quality.
8. **Decent Work and Economic Growth:** There are many challenges during work like lack of decent work opportunities, insufficient investments and Under-consumption contribute to the erosion of the basic social contract, expanding infrastructure and upgrading technology etc., which affects the economic growth.Library can offer career services reserves multiple rooms early each spring for mock job interviews in the library and students have used library study rooms for virtual interviews for internships and jobs.
9. **Reduce Inequalities:** Library can aid to offer programs, services, or collections to promote awareness and education around equity, diversity, and inclusion. It can provide employee training around equity, diversity, and inclusion.

7. What is AI and AI Technologies

Artificial intelligence (AI) is a field of technology that enables computers and machines to carry out activities that often necessitate human intelligence. These tasks include comprehending language, identifying patterns, making judgments, and acquiring knowledge via experience.

The examples of AI technologies are:

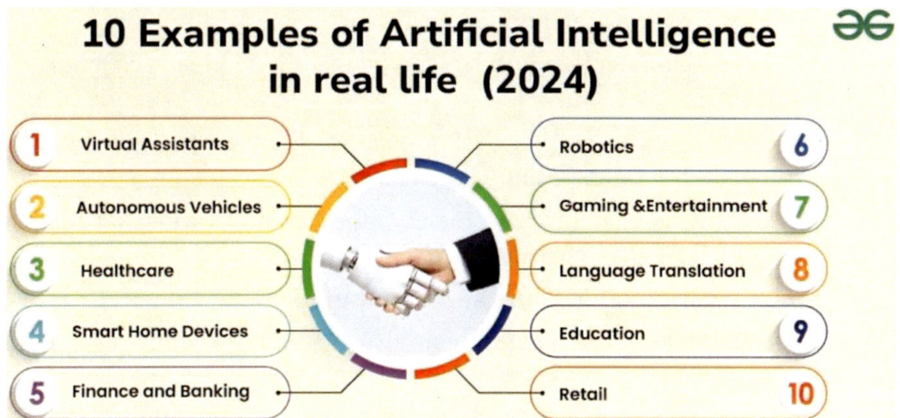

Fig no.2 Examples of AI Technologies

8. How AI Can Help Libraries to Grow Sustainably

AI can significantly contribute to libraries achieving access to information in the context of the UN Sustainable Development Goals (SDGs). Here's how AI can help libraries align with and support various SDGs:

1. **SDG 4: Quality Education**
   - AI-powered adaptive learning platforms can provide personalized educational resources.
   - Virtual tutors can offer 24/7 learning support.
   - AI can help identify knowledge gaps in collections and suggest resources to fill them.
2. **SDG 5: Gender Equality**
   - AI can analyze collection data to ensure diverse representation in materials.
   - Automated translation tools can make women's empowerment resources available in multiple languages.
   - AI can help identify and flag gender biases in existing materials.

3. **SDG 8: Decent Work and Economic Growth**
    - AI-driven job matching systems can connect patrons with employment opportunities.
    - Personalized skill development recommendations based on job market trends.
    - Automated resume builders and interview preparation tools.
4. **SDG 10: Reduced Inequalities**
    - AI can help identify underserved communities and suggest outreach strategies.
    - Text-to-speech and speech-to-text technologies can improve accessibility for differently-abled patrons.
    - Multilingual chatbots can provide information in various languages, reducing language barriers.
5. **SDG 11: Sustainable Cities and Communities**
    - AI can analyze community data to help libraries tailor services to local needs.
    - Smart building technologies can make library spaces more energy-efficient and sustainable.
    - AI-powered community forums can facilitate local engagement and information sharing.
6. **SDG 16: Peace, Justice, and Strong Institutions**
    - AI can help organize and make accessible government documents and public records.
    - Fact-checking algorithms can help combat misinformation.
    - AI-driven content moderation for online library forums to promote respectful dialogue.
7. **SDG 17: Partnerships for the Goals**
    - AI can facilitate knowledge sharing between libraries globally.
    - Automated translation of research papers can promote international collaboration.
    - AI-powered platforms can match libraries with potential partners for SDG-related initiatives.

In order to implement for the purpose of libraries to effectively adopt these artificial intelligence solutions, they should:

i. Give data protection and ethical usage of AI the highest priority.

ii. Ensure that AI systems are transparent and explainable.
iii. Conduct training sessions on digital literacy for both employees and customers.
iv. Collaborate with other libraries and technology organizations in order to share resources and best practices through collaboration.
v. Evaluate the impact of AI projects on the Sustainable Development Goals (SDGs) on a regular basis.

9. Conclusion

AI has a huge amount of potential to change libraries and make them better at giving people access to knowledge, which is in line with the Sustainable Development Goals. Libraries can solve problems, make the user experience better, and help reach the goal of everyone having access to knowledge by using AI technologies. As libraries continue to change in the digital age, incorporating AI into their processes will be very important to make sure they stay useful and meet the needs of their communities and users.By utilizing AI in these ways, libraries have the potential to strengthen their function as information hubs and make a significant contribution to the achievement of the Sustainable Development Goals (SDGs) in their communities and beyond.

References

10 Examples of Artificial Intelligence in Real-Life (2024, June 24). GeeksforGeeks. https://www.geeksforgeeks.org/10-examples-of-artificial-intelligence-in-real-life-2024/

Applying Carbon Standards to Sustainable Development Goals. (n.d.). Ecosystem Marketplace. https://www.ecosystemmarketplace.com/articles/applying-carbon-standards-sustainable-development-goals/

Dei, D.-G. J., & Asante, F. Y. (2022). Role of academic libraries in the achievement of quality education as a sustainable development goal. Library Management, 43(6/7), 439–459. https://doi.org/10.1108/LM-02-2022-0013

Development and Access to Information (DA2I). (n.d.). Development and access to information (DA2I). https://da2i.ifla.org

Mathiasson, M. H., & Jochumsen, H. (2022). Libraries, sustainability and sustainable development: A review of the research literature. Journal of Documentation, 78(6), 1278–1304. https://doi.org/10.1108/JD-11-2021-0226

Poluru, L. (2020). Sustainable Development Goals (SDGs): The Role of Libraries and Access of Information for Transformation.

Yap, J. M., & Kamilova, Y. (2020). Toward becoming an inclusive library: Integrating Sustainable Development Goal 5 in the library agenda. Library Management, 41(2/3), 53–66. https://doi.org/10.1108/LM-08-2019-0057

17

Revolutionizing Libraries: The Role of Robotics in Modern Library Services

***Sudhir Baburav Halvegar*[1], *Srikanth H.G.*[2] *and Padmavati Tubachi*[3]**

[1]*Librarian, MES Vasant Joshi College Zuari Nagar, Vasco - 403 726, Goa, India*
E-mail: halvegarsb@rediffmail.com

[2]*Librarian, Soundarya Institute of Management and Science, Soundarya Nagar Sidedahalli, Bengaluru - 560 073, Karnataka, India*
E-mail: srikanth.hg@soundarya institutions.in

[3]*Librarian, DPM's Shree Mallikarjun & Shri Chetan Manju Desai College, Delem Canacona - 403 702, Goa, India*
E-mail: padmavati.tubachi@shreemallikarjuncollege.ac.in

Abstract

Robots have been used in libraries to facilitate a variety of library functions, including STEM education initiatives. This chapter explores how robotics is revolutionizing libraries, focusing on key applications, benefits, challenges, and future directions. This chapter also explores the significant advancements in robotics within the context of modern library services. Historically rooted in industrial applications, robotics has now expanded its influence into various sectors, including libraries, which have begun to harness these technologies to enhance both operational efficiency and user experience. The paper delves into the evolution of robotics in libraries, from early automation systems to the contemporary use of humanoid robots like NAO and Pepper, which engage with patrons and assist in a variety of tasks. These AI-driven systems have revolutionized traditional library operations by automating routine tasks, improving inventory management, and offering personalized user interactions. As libraries continue to evolve in the 21st century, the future of robotics promises even more sophisticated AI integration, enhanced user interaction, and expanded accessibility, positioning libraries as dynamic hubs of knowledge and learning.

Keywords: *Artificial Intelligence, Automation, Library services, Robotics, STEM education, User interaction.*

1. Introduction

Modern technologies like artificial intelligence, machine learning, and big data have made robotics, a concept that dates back to ancient times, much more advanced. The area of robotics, originally used in industrial settings, has now expanded its scope to include a diverse range of industries, such as industrial robots, medical robots, military robots, space robots, and humanoid robots that adopt human-like characteristics.

In accordance with this existing structure, libraries have additionally adopted the use of robots, which have shown enhanced efficacy and capabilities in executing diverse library functions. A significant number of libraries around the world have used humanoid robots to facilitate a variety of library functions, including STEM education initiatives. Academic institutions such as the Chicago Public Library and the University of Rhode Island Library organize programs centered on robotics.

Libraries may become obsolete in the 21st century if they do not begin to harness new technology and improve their information and service delivery. Robots come in a variety of shapes and sizes, and they serve a variety of functions. It is difficult to define the purpose of a specific robot and its categories because each robot has its distinct feature and varies in size, shape, color, and components.

The rapid increase in book consumption has led to libraries assuming a significant role as primary intermediaries for knowledge acquisition. The advent of humanoid robots equipped with artificial intelligence has instigated a substantial transformation in the organization and functioning of information and knowledge infrastructure. The advancements in technologies such as big data, enhanced processing power, and deep learning have significantly augmented the capabilities of AI-enabled robots.

2. AI and Robotics in Libraries

AI is used in a wide range of industries, including medicine, the military, commerce, education, and gaming; AI systems in libraries were proposed for the first time in 1990. AI in humanoid robots is already available in libraries in both developed and developing countries, which is no longer news. The robots engage in activities such as scrambling, rolling, soaring, and climbing. These robots have an impact on libraries and the larger information (and social) environment in which we all live. Artificial Intelligence (AI) refers to the simulation of human intelligence in machines programmed to think and learn. Robotics involves the design and use of robots to perform tasks. When combined, AI and robotics create systems capable of sophisticated operations and interactions, leading to substantial advancements in library services.

Artificial Intelligence (AI) and robotics are revolutionizing libraries by introducing advanced capabilities that enhance both operational efficiency and user experiences. AI-driven systems and robots streamline tasks such as book sorting and shelf management through automation, significantly reducing manual labor and improving accuracy. Artificial intelligence (AI) is being used in libraries to improve services and enhance user experience. AI-powered systems are being used to provide knowledge-based services to both library employees and visitors. AI is also being used in descriptive cataloging, subject indexing, reference services, technical services, shelf reading, collection development, and information retrieval systems. Service robots equipped with AI can assist patrons with real-time information, guide them through the library, and even offer personalized book recommendations based on individual preferences. Additionally, libraries use robotics to host educational workshops and competitions, fostering engagement with technology and STEM education. Despite the benefits, challenges such as high costs, staff training needs, and data privacy concerns must be addressed. Overall, the integration of AI and robotics in libraries is transforming these institutions into modern, efficient hubs of knowledge and learning, poised to meet the evolving needs of their communities.

3. Evolution of Robotics in Library

The evolution of robotics in libraries has been a significant development in recent years. Robotics has had an immediate impact on libraries, influencing the larger information and social environment in which libraries and librarians operate. The adoption of robotic technologies in academic libraries, for instance, has led to the development of new roles such as designing library apps and data analysis. Advances in reading list software have created a whole new aspect of library work.

- **Early Beginnings:** Automated Storage and Retrieval Systems (ASRS): One of the earliest uses of robotics in libraries was the implementation of ASRS, which automated the storage and retrieval of books. This system allowed libraries to store a large number of books in a compact space and retrieve them quickly when needed.
- **Advancements in the 21st Century:** Libraries began experimenting with humanoid robots like NAO and Pepper. These robots interact with patrons, provide information, and even conduct educational workshops. Robots were introduced to sort returned books and place them back on shelves, significantly reducing the workload of library staff and minimizing errors.

- **Recent Innovations:** Modern libraries now use robots that can guide patrons to specific sections, answer queries, and even engage in conversations. For example, Dewey at Palo Alto City Library and Libby at University of Pretoria Libraries.Robots like Finch and Ozobot are used in libraries to teach coding and robotics to children and adults, making libraries a hub for STEM education.

4. The Impact of Robotics on Library Services

The presence of robots in libraries is giving librarians more time to focus on other important information service delivery to today's dynamic needs. Robotics has an immediate impact on libraries and the larger information (and social) environment in which libraries and librarians of all types work. Some of the areas where robotic technologies have been applied to library operations include:

- Shelving and locating library materials
- Security
- Inquiries and answering of repetitive reference and directional queries
- Outreach and public relation (PR) via library tour
- Information illiteracy instruction

Moreover, robots are useful for automated storage and retrieval systems, which have aided in the management of library space.

5. Types of Robots Used in Libraries

i. **Accessibility Robots:** Robots that assist patrons with disabilities, such as robotic wheelchairs or robots that provide real-time transcription services. Example: The Toronto Public Library uses a robotic wheelchair called "Whill" to assist patrons with mobility impairments.

ii. **Book-Bot:** A robotic book retrieval system that can store and retrieve books from a compact storage system, freeing up floor space and reducing labor costs. Example: The University of California, Berkeley's Moffitt Library uses a Book-Bot to retrieve books from a compact storage system, freeing up floor space and reducing labor costs.

iii. **Delivery Robots:** Robots that deliver books and other materials to patrons within the library or even to their homes. Example: The Singapore National Library Board uses a delivery robot called "NLB Bot" to deliver books and other materials to patrons within the library.

iv. **Entertainment Robots:** Robots that offer entertainment options, such as robotic pets or interactive games, to create a welcoming atmosphere in the library. Example: The San Francisco Public Library uses a robot called "Baxter" to offer interactive games and activities for children.

v. **Information Robots:** Robots that provide information and answer frequently asked questions, freeing up librarians to focus on more complex queries. Example: The University of Tokyo's Komaba Library uses a robot called "Robi" to provide information and answer frequently asked questions, freeing up librarians to focus on more complex queries.

vi. **Programming Robots:** Robots that assist with children's programming, such as reading stories, teaching coding, and leading STEM activities. Example: The Kansas City Public Library uses a robot called "Dash" to assist with children's programming, such as reading stories and teaching coding.

vii. **Research Robots:** Robots that assist with research tasks, such as data collection, data analysis, and content creation. Example: The University of California, Los Angeles's (UCLA) Library uses a robot called "ROSIE" to assist with research tasks, such as data collection and data analysis.

viii. **Security Robots:** Robots that patrol the library, monitoring for suspicious activity and providing an added layer of security. Example: The Huntington Beach Public Library in California uses a security robot called "Knightscope K5" to patrol the library and monitor for suspicious activity.

ix. **Self-Service Robots:** Robots that allow patrons to check out books, pay fines, and access other library services without human intervention. Example: The Hong Kong Public Libraries use a self-service robot called "Libby" to allow patrons to check out books, pay fines, and access other library services without human intervention.

x. **Shelf-Reading Robots:** Robots that scan bookshelves to identify misshelved books, saving librarians time and ensuring that books are easily accessible to patrons. Example: The University of Nevada, Las Vegas's Lied Library uses a shelf-reading robot to scan bookshelves and identify misshelved books, saving librarians time and ensuring that books are easily accessible to patrons.

6. The Emergence of Robotics in Libraries

The integration of robotics into libraries began with simple automation systems but has rapidly expanded into sophisticated solutions that handle various tasks. Robotics in libraries can be categorized into several areas:

6.1. Automated Book Handling

One of the most significant impacts of robotics in libraries is the automation of book handling processes. Traditional library tasks such as sorting, shelving, and retrieving books are labor-intensive and time-consuming. Robotics has introduced several innovations in this area:

a) **Automated Sorting Systems:** Automated sorting systems, like the ones used in some modern libraries, utilize conveyor belts and robotic arms to sort returned books into their respective categories. These systems are capable of handling large volumes of books with high efficiency and accuracy. For example, the book sorters at the Seattle Public Library can process thousands of books per hour, significantly reducing the time and effort required for manual sorting.

b) **Robotic Shelf Management:** Robotic systems for shelf management are designed to assist with the physical task of shelving books. These robots navigate the library's aisles, retrieving books from return bins and placing them on the appropriate shelves. This not only speeds up the process but also helps maintain the organization of the collection. The integration of such robots allows library staff to focus on more complex tasks, such as customer service and programming.

6.2. Service Robots

Service robots in libraries enhance the user experience by providing assistance and information. These robots are designed to interact with patrons and perform various functions:

a) **Information and Guidance:** Robots like "Libratus," used in some libraries, are equipped with AI to assist patrons with locating books, answering questions, and providing information about library services. They can guide visitors to specific sections of the library and help them navigate complex library systems. This not only improves user experience but also makes libraries more accessible, particularly for individuals with disabilities or those unfamiliar with library layouts.

b) **Interactive Learning:** Some libraries have introduced educational robots that engage children and adults in interactive learning experiences. These robots can host storytelling sessions, lead educational workshops, and provide hands-on learning opportunities. Programs like these promote STEM (Science, Technology, Engineering, and Mathematics) education and foster a love for learning in a fun, interactive way.

7. Robotics for Education and Engagement

Beyond operational efficiency and service, robotics in libraries play a crucial role in educational outreach and community engagement:

a) **Robotics Workshops and Clubs:** Libraries often host robotics workshops and clubs where patrons, especially children and teens, can learn about robotics and programming. These programs may include building and programming robots, participating in competitions, and exploring various robotics technologies. Such initiatives support educational development and encourage interest in technology careers.

b) **Robotics Competitions:** Libraries may also organize or participate in robotics competitions, providing a platform for individuals and teams to showcase their skills and innovations. These events can inspire creativity, problem-solving, and collaboration among participants, while also promoting the library as a center for technological learning and innovation.

8. Benefits of Robots in Libraries

- **Automation of Routine Tasks:** Robots can handle repetitive tasks such as sorting and shelving books, freeing up librarians to focus on more complex and engaging activities.
- **Enhanced User Assistance:** Robots can serve as information desks, helping patrons locate books, answer simple queries, and provide directions within the library.
- **Improved Accessibility:** Telepresence robots can assist users who are unable to visit the library in person, allowing them to participate in events and access resources remotely.
- **Educational Opportunities:** Robots can be used in educational programs to teach coding and robotics, making the library a hub for STEM learning.
- **Efficient Space Management:** Automated storage and retrieval systems (ASRS) can optimize the use of space in libraries, making it easier to manage large collections.
- **Enhanced Security:** Robots can help monitor library premises, ensuring the safety and security of both patrons and resources.

9. Challenges and Considerations Robotics in Modern Library Services

- **Accessibility and Inclusivity:** Libraries must ensure that robotic services are accessible to all patrons, including those with disabilities. This may involve incorporating accessible design principles into the development of robotic interfaces and ensuring that alternative services are available for patrons who cannot use robots.

- **Assessment and Evaluation:** Libraries must develop methods to assess the effectiveness of robotic services and evaluate their impact on library operations and patron satisfaction. This may involve collecting data on usage, conducting user surveys, and monitoring feedback.
- **Cost and Funding:** Implementing robotics in libraries can be expensive. Libraries must consider the initial cost of purchasing robots, as well as ongoing maintenance and upgrade expenses. Securing funding for these initiatives can be challenging, especially for smaller libraries with limited budgets.
- **Ethical Considerations:** The use of robots in libraries raises ethical considerations, such as the potential for job displacement or the impact on the human experience of library services. Libraries must consider these issues and develop strategies to mitigate any negative consequences.
- **Integration with Existing Systems:** Robots may need to be integrated with existing library systems, such as the integrated library system (ILS) or online public access catalog (OPAC). Ensuring seamless integration can be a technical challenge.
- **Patron Acceptance and Engagement:** Some patrons may be hesitant to interact with robots, preferring human assistance. Libraries must consider how to encourage patron acceptance and engagement with robotic services, potentially through user-friendly interfaces and clear instructions.
- **Privacy and Security:** Robots may collect data about patron interactions, raising concerns about privacy and data security. Libraries must develop policies and procedures to protect patron data and ensure that robots are used in a way that respects patron privacy.
- **Scalability and Flexibility:** Libraries must consider the scalability and flexibility of robotic solutions, ensuring that they can adapt to changing library needs and priorities over time.
- **Staff Training and Support:** Library staff may need training to effectively integrate robots into their workflows and provide support to patrons using robotic services. This requires an investment in staff development and ongoing support.
- **Technical Issues and Maintenance:** Robots require regular maintenance to function properly. Technical issues can arise, requiring libraries to have a plan in place for troubleshooting and repair. This may involve training staff or contracting with external vendors for support.

10. Application of Robots in Library

- **"TuBao"- Robot in Nanjing University Library Lobby:** TuBao is an automated receiving device located in Nanjing University Library's main lobby. The main purpose of the robot is to support library visitors by providing them with information and delivering assistance within the library. The service "TuBao" has the ability to provide responses to inquiries on the library's services, book locations, and other essential general information. The "TuBao" system, which incorporates laser sensors, has the ability to independently navigate the library, assisting library users in locating various areas or individual volumes.
- **"Pepper"-Community Robot in Roanoke County library:** The Roanoke County Public Library in Virginia has just included "Pepper," a humanoid robot, among its permanent staff. This initiative establishes the library as the first institution in the United States to enlist a robot of this nature. Specifically designed to engage with patrons, Pepper facilitates resource allocation towards less frequented sections of the library and provides a diverse range of educational initiatives. The previously mentioned initiatives include early literacy, genealogy research, consumer health information, and computer programming and robotics instruction. The Pepper robot has the ability to recognize human features, interpret human emotions, and participate in conversations, making it a distinctive instrument for fostering children's reading and educational development.
- **Automated Retrieval Center (ARC) in J.Willard Marriott Library:** The Automated Retrieval Center (ARC), located in the J. Willard Marriott Library, University of Utah, is an innovative storage facility specifically designed to effectively accommodate library resources with a high use rate. Library staff uses four self-navigating robots to identify and deliver bins, after which they distribute the goods to clients.The efficiency of this procedure is evident as it takes less than five minutes to retrieve the requested objects. The library's digital catalog allows members of the public to make requests for materials housed in the ARC.
- **Smart Talking Robot "Xiao Tu" in Tsinghua University:** Tsinghua University Library has developed Xiao Tu, a virtual intelligent and talking robot, to assist scholars in their search activities. The referred-to intelligent robot has the capacity to acquire knowledge and enhance its responses by engaging in interactions with others. This system employs artificial intelligence to provide responses to library inquiries

using natural, spoken language. The main goal of Xiao Tu is to provide users with a multimedia experience similar to engaging with a human librarian, rather than just an automated system. However, the interruption of services was necessary due to the large number of visitors seeking to engage in conversations lacking library-related topics. Hence, Xiao Tu's vocabulary expanded beyond the intended audience. The Xiao Tu system detected cases of "unsuitable language" among website users and then underwent a data cleansing process to rectify the issue.

11. The Future of Robotics in Libraries

The future of robotics in libraries promises to be as dynamic and transformative as the technology itself. As robotics and artificial intelligence (AI) evolve, libraries are poised to benefit from increasingly sophisticated systems that can enhance operational efficiency, improve user experiences, and redefine traditional library services.

i. **Advanced Artificial Intelligence Integration:** Future robots in libraries will likely leverage advanced AI to offer highly personalized services. AI-driven robots could analyze patrons' borrowing histories, reading preferences, and even real-time interactions to provide tailored book recommendations. This level of personalization could extend to educational materials, event suggestions, and more. Robots with advanced NLP capabilities will be able to engage in more nuanced conversations with patrons. They could answer complex queries, provide detailed explanations, and interact with users in a more human-like manner, making library assistance more accessible and user-friendly.

ii. **Enhanced User Interaction:** Robots could become central to interactive learning environments, particularly for children and students. Advanced educational robots might facilitate interactive story sessions, coding workshops, or even virtual field trips. They could use augmented reality (AR) and virtual reality (VR) to create immersive learning experiences within the library setting. Future robots might also be designed to provide emotional and social support. They could assist patrons with disabilities or mental health needs by offering companionship, engagement, and customized resources, helping to create a more inclusive and supportive library environment.

iii. **Autonomous Library Management:** Robotic systems equipped with AI could manage inventory with even greater precision. Future robots might autonomously update catalog records, identify and address discrepancies, and predict trends in book usage, allowing libraries to optimize their collections and space utilization effectively. Robots could be used to dynamically adjust library spaces based on usage patterns. For instance, robotic systems could rearrange furniture or reconfigure spaces for different events and activities, optimizing the library environment in real time to meet changing needs.

iv. **Robotic Collaboration and Integration:** Libraries could employ collaborative robots (cobots) that work alongside human staff to perform complex tasks. These robots could handle repetitive or strenuous tasks while humans focus on more creative or interpersonal activities, creating a synergistic work environment. Robots in libraries will increasingly integrate with other emerging technologies, such as Internet of Things (IoT) devices and smart systems. This integration could lead to more seamless operations, where robots communicate with sensors and other devices to manage lighting, temperature, and security, enhancing overall library functionality.

v. **Expanded Services and Accessibility:** Robots might play a role in expanding library services beyond physical spaces. For instance, telepresence robots could allow users to interact with librarians remotely, attend virtual events, or access resources from anywhere in the world. This could significantly increase the reach and impact of library services. Robots could further enhance accessibility in libraries by providing physical support to patrons with mobility issues, offering language translation services, or assisting with other special needs. This would help make libraries more inclusive and accommodating to all users.

vi. **Ethical and Practical Considerations:** As robots become more integrated into library systems, privacy and security concerns will need to be addressed. Libraries will need to ensure that data collected by robots is handled responsibly, with robust measures in place to protect user information and maintain trust. The implementation of advanced robotics can be costly, and libraries will need to balance these expenses with budget constraints. Future developments will need to focus on cost-effective solutions and sustainable models to ensure that robotics can be a viable option for libraries of all sizes. Staff training will continue to be essential as new robotic technologies are introduced. Libraries will need to invest in ongoing professional development to help staff adapt to new tools and workflows, ensuring that the integration of robotics enhances rather than disrupts library operations.

12. Conclusion

The integration of artificial intelligence and robotics into libraries represents a significant leap forward in modernizing library services and enhancing user experiences. By automating routine tasks, providing interactive assistance, and supporting educational initiatives, these technologies offer numerous benefits that align with the evolving demands of contemporary library environments. As libraries continue to embrace AI and robotics, they will play an increasingly vital role in fostering knowledge, innovation, and community engagement in the digital age. Robotics is fundamentally transforming libraries, making them more efficient, interactive, and accessible. From automating book handling to enhancing patron services and maintaining cleanliness, robots are revolutionizing how libraries operate and serve their communities. While there are challenges to overcome, the benefits of robotics in libraries are clear. As technology continues to advance, the role of robotics in libraries will likely expand, offering new possibilities and further enhancing the library experience for users worldwide.

References

DilipYadav, S., & Bundele, C. (2021). Importance of robotic technology in different fields. 06, 986–992. www.irjmets.com

Dresvyannikova, V. (2020). IFLA Statement on Libraries and Artificial Intelligence.

El Kadiri, S., Grabot, B., Thoben, K. D., Hribernik, K., Emmanouilidis, C., Von Cieminski,G., & Kiritsis, D. (2016). Current trends on ICT technologies for enterprise information systems. Computers in Industry, 79, 14–33. https://doi.org/10.1016/j.compind.2015.06.008

Faife, I. (2020). IFLA statement on libraries and artificial intelligence.

Gasparetto, A., & Scalera, L. (2019). A brief history of industrial robotics in the 20th century. Advances in Historical Studies, 08(01), 24–35. https://doi.org/10.4236/ahs.2019.81002

Liau, Y. C. (2017). Transforming library operation with robotics.

https://archive.unews.utah.edu/news_releases/u-of-u039s-automated-library-system-proves-technology-and-books-go-well-together/

https://lib.utah.edu/services/automated-retrieval-center.php

https://wset.com/news/local/roanoke-libraries-recruits-humanoid-robot

Husbands, P. (2016). Evolutionary robotics. In: Sammut, C., Webb, G. (eds) Encyclopedia of Machine Learning and Data Mining. Springer, Boston, MA. https://doi.org/10.1007/978-1-4899-7502-7_94-1

Tella, A. (2020). Robots are coming to the libraries: are librarians ready to accommodate them? Library Hi Tech News, 37(8), 13–17. https://doi.org/10.1108/LHTN-05-2020-0047

Tella, A., & Ajani, Y.A. (2022). Robots and public libraries. Library Hi Tech News, 39(7), 15-18. https://doi.org/10.1108/LHTN-05-2022-0072

Vysakh, C., & Rajendra, B. (2019, August). Application of artificially intelligent robot in libraries. In E-Proceedings of the International Conference on Reshaping Librarianship: Innovations and Transformations (ICRLIT–2019) 16th & (pp. 59-63).

Wang, Z. (2017). How do library staff view librarian robotics? Librarian staff's ignored humanistic views on the impact and threat of robotics adoption.

18

Exploring the Impact of Digital Technologies on Education: A Comprehensive Analysis

Dr. P. Krishnamoorthy

Assistant Librarian, R.M.K. College of Engineering and Technology, Puduvoyal, Thiruvallur, Chennai - 601 206, Tamil Nadu, India, E-mail: krishlib@rmkcet.ac.in

Abstract

The growing incorporation of digital technologies into education has altered the teaching and learning landscape. This in-depth examination dives into the multidimensional impact of digital technology on education, looking at both the benefits and challenges they provide. This research paper tries to shed light on the dynamic link between digital technologies and education through an examination of historical backdrop, literature review, case studies, and emerging trends.

We analyze the positive effects of digital technology in this study, such as increased accessibility, personalization of learning experiences, and increased involvement in the educational process. In addition, we highlight and address the issues inherent in this digital revolution, such as privacy, security, and the digital divide.

The educational consequences of this research are also highlighted, illustrating how digital technology have altered teaching methods and spurred a movement toward student-centered, active learning. We give specific case studies and examples of educational institutions that have successfully integrated digital technologies, shedding light on the practical impact on student results.

Furthermore, we hypothesize about future educational technology trends and developments, such as the role of artificial intelligence, blockchain, and the Internet of Things.

Finally, this study gives insights into the tremendous impact of digital technologies on education and makes recommendations for educators, governments, and institutions on how to harness the potential of these technologies to improve the educational experience. The symbiotic relationship between education and technology is clearly expanding, impacting the future of learning.

***Keywords**: Education, Digital technologies, Impact analysis, Teaching and learning.*

1. Introduction

Education is undergoing a digital transformation. The incorporation of digital technologies into the educational scene over the last two decades has radically changed the way we teach and learn. This transition has started a revolution with far-reaching repercussions, both positive and negative, for students, educators, and institutions alike.

The growing adoption of digital technologies in educational settings has created a world of possibilities. Online platforms, learning management systems, educational apps, virtual reality, and augmented reality are just a handful of the innovations that have transformed the educational experience. This comprehensive investigation tries to investigate the intricate and multifaceted impact of various digital technologies on schooling.

This study goes beyond anecdotal findings to present a comprehensive, evidence-based understanding of the digital change in education. It aims to provide answers to fundamental concerns regarding how digital technologies shape the learning process, what opportunities they provide, what obstacles they bring, and what implications they have for pedagogy and educational results.

This study will begin with a historical framework, tracking the evolution of digital technology in education from experimental tools to essential components of modern learning. A detailed literature study will next lay the groundwork, outlining the abundance of existing information and research on the issue. Following that, we will look at practical examples and case studies that demonstrate the real-world impact of digital. In educational settings, technologies are used.

Furthermore, this investigation will critically assess both the good and negative effects of digital transformation. On the one hand, digital technologies have greatly expanded access to educational resources, allowing students from all walks of life to pursue knowledge. They've also enabled tailored learning experiences, shattering traditional one-size-fits-all educational approaches. Furthermore, digital tools have the ability to increase student involvement and collaboration, transforming the learning process into a more participatory and dynamic experience.

This digital transformation, however, is not without problems. Concerns about privacy and data security have developed as educational institutions rely more on digital platforms. The digital divide remains an issue, with discrepancies in access to technology and the internet limiting students' potential. Furthermore, educators are navigating the shifting terrain of education, adapting to new pedagogical approaches, and gaining digital literacy abilities.

As the investigation progresses, it will investigate the pedagogical consequences of digital technology on teaching methods, revealing a dynamic movement toward student-centered and active learning. The study will also speculate on the future of education, analyzing current trends and developments such as the expanding influence of artificial intelligence, blockchain, and the Internet of Things.

Finally, this comprehensive analysis aims to provide a full knowledge of the impact of digital technology on education. By examining the past, present, and future of this digital revolution, we hope to provide educators, policymakers, and institutions with the insights and recommendations they need to navigate this transformative journey and ensure that the benefits of digital technologies are harnessed for the betterment of education. The symbiotic link between education and technology is constantly evolving, and it is our objective to provide light on the way forward.

2. Literature Review

2.1. The Development of Digital Technologies in Education

The incorporation of digital technologies in education constitutes a significant paradigm change, changing traditional classroom settings into dynamic, technology-infused learning spaces. This transition has its roots in the early 1960s, when the introduction of computers in education was greeted with great enthusiasm. Early supporters of computer-assisted instruction (CAI) predicted that it would change teaching and learning by providing tailored, self-paced training.

However, it wasn't until the late 20th and early 21st centuries that the true potential of digital technology became apparent. The internet created new potential for distant learning and collaborative educational platforms. Personal computer proliferation, followed by the rise of smartphones and tablets, substantially enlarged the reach of digital education.

2.2. The Implications for Accessibility and Inclusion

The increased accessibility of learning materials and resources is one of the most notable effects of digital technologies in education. Online platforms, digital libraries, and massive open online courses (MOOCs) have made high-quality education accessible to people all over the world, regardless of geography. This has provided opportunities for lifelong learners, working professionals, and underserved groups who would not otherwise have had access to traditional educational resources.

Furthermore, assistive technology has given students with impairments resources for tailored learning experiences. Screen readers, speech recognition software, and captioning services are among the innovations that are making education more inclusive than ever before.

2.3. Learning Experience Personalization

Personalization has been made possible by digital technologies. Algorithms are used in adaptive learning platforms to evaluate individual students' progress and modify information to their personal needs. This level of customisation appeals to a wide range of learning styles and skills, allowing each learner to advance at their own speed. Personalized learning boosts not only engagement but also learning results dramatically.

2.4. Improved Participation and Collaboration

Increased student involvement has resulted from interactive learning platforms, gamification, and the use of multimedia resources. Learning becomes more interactive and dynamic as a result of digital technologies, breaking away from traditional lecture-based teaching approaches. They enable collaborative projects, discussions, and virtual classes, establishing a sense of community among students even when they are not in the same location.

However, the impact of digital technology on education is not without obstacles. The parts that follow will look at the problems and issues raised by this shift, such as privacy and security, the digital gap, and the changing role of education. We may obtain a more thorough grasp of the digital revolution in education by evaluating both the benefits and the obstacles.

3. Methodology

3.1. Research Approach

We used a mixed-methods study strategy to thoroughly examine the impact of digital technology on schooling. This strategy combines qualitative and quantitative methods to create a more complete picture of the complex interaction between digital technology and education.

3.2. Data Collection

The study began with a thorough examination of the literature. This phase included a comprehensive evaluation of scholarly publications, books, reports and pertinent studies published during the last two decades. This was critical for establishing a solid theoretical foundation and identifying major issues, trends, and debates in the field of digital technology in education.

3.3. Surveys and Questionnaires

Quantitative data was gathered through surveys and questionnaires delivered to instructors, students, and educational administrators. The surveys sought information on the present deployment of digital technology in educational institutions, the perceived benefits, and the problems encountered.

3.4. Case Studies

We conducted case studies in several educational environments, including K-12 schools, universities, and online learning platforms, to give in-depth, qualitative insights. Interviews with teachers, students, and administrators, as well as classroom observations, were conducted as part of these case studies. The purpose was to understand how digital technologies are being integrated into real-world educational situations, as well as the influence they have on teaching and learning.

3.5. Data Analysis

Survey data was examined quantitatively using statistical software. Qualitative data from interviews and observations were transcribed and thematically examined to discover recurring patterns and themes. The mixed-methods technique enabled data triangulation, which increased the reliability and validity of our findings.

3.6. Ethical Considerations

All data collection techniques followed ethical standards, ensuring that participants' privacy and informed permission were respected. When minors were involved, parental consent was obtained.

3.7. Limitations

It is critical to recognize the research's limitations. The breadth of the research was constrained by time and resources, and the fast-evolving nature of technology in education implies that conclusions may become out of date. Furthermore, the study did not investigate every possible sub-domain of digital technology in education, but instead concentrated on a representative sample.

Despite these limits, this extensive analysis offers useful insights into the dynamic link between digital technologies and education, providing a comprehensive assessment of their influence, difficulties, and future promise. The following sections of this article will go into the specific conclusions produced from the data collected and evaluated using the approach described.

4. Digital Technologies and Education

Digital technologies have become inseparable from education, revolutionizing how knowledge is transmitted, gained, and applied. This section delves into the multidimensional function of digital technologies in education, shedding light on how they affect the learning process, instructional approaches, and institutional structures.

4.1. The Electronic Classroom

Traditional classrooms have developed into digital hubs in which students interact with a variety of devices. Learning Management Systems (LMS) like Moodle, Canvas, and Blackboard have streamlined course delivery, assessment, and communication. These systems provide teachers with a centralized platform for uploading resources, creating assignments, and facilitating discussions. Students, in turn, have simple access to instructional resources, grades, and a collaborative virtual space.

4.2. Virtual Learning Environments

Education has become more democratic as a result of the spread of online learning platforms. Institutions and organizations such as Coursera, edX, and Khan Academy make a wide range of courses available to learners globally, frequently for free or at a low cost. These platforms cover a wide range of topics, allowing for lifelong learning and professional development. Learners can study when and where they want, making education more flexible to their schedules.

4.3. Electronic Learning Resources

Digital resources have supplemented and, in some cases, overtaken textbooks as the principal source of educational content. E-books, interactive simulations, open educational resources (OER), and video lectures give multimedia-rich, current, and easily accessible information to students. These materials cater to a variety of learning styles and encourage participation through interactive content.

4.4. Personalized Education

Personalized learning is now a reality because to advances in digital technology. Algorithms in adaptive learning software assess individual students' progress and alter the complexity of information accordingly. This method adapts to pupils' individual strengths and shortcomings, resulting in a more in-depth comprehension of the subject matter. Personalized learning not only increases engagement but also leads to better learning outcomes.

4.5. Ease of Access and Inclusion

Incorporating digital technologies has the ability to reduce educational hurdles. Assistive technology such as screen readers, speech recognition software, and captioning services serve students with disabilities by ensuring that instructional resources are accessible to all. Furthermore, online courses and digital libraries enable those in remote or underdeveloped locations to gain access to education, bridging the geographical difference in learning chances.

5. Interactivity and Participation

The incorporation of multimedia components, gamification, and interactive features into instructional content has changed the game in terms of student engagement. Active engagement and critical thinking are promoted through online quizzes, discussion forums, and virtual labs. Gamified components also make studying more interesting and encourage pupils to master difficult ideas.

As digital technologies advance, their significance in education is certain to grow, influencing not only what is taught but also how it is taught. However, as we will see in later parts, this shift is not without its difficulties. Because of the multidimensional character of this impact, a thorough analysis of both the benefits and challenges that digital technologies provide in education is required.

6. Digital Technologies' Beneficial Effects on Education

The use of digital technologies into education has resulted in a slew of positive changes that have transformed the learning experience for students, educators, and institutions alike. This section focuses on some of the most significant positive effects of digital technologies on education.

6.1. Enhanced Accessibility

By removing geographical and social obstacles, digital technologies have democratized education. Quality education is now available to people all around the world thanks to online courses, open educational resources, and digital libraries. Learners in remote or underserved locations can have access to a plethora of knowledge and educational opportunities, thereby leveling the playing field for everyone.

6.2. Learning Personalization

Personalized learning experiences are possible thanks to digital technologies. Data-driven algorithms are used by adaptive learning platforms to analyze each student's progress and tailor the content and pace of training to their specific needs. This method accommodates a wide range of learning styles and skills, allowing each learner to advance at their own speed and properly absorb subjects.

6.3. Active Participation and Learning

Passive learning has been changed into an active, engaging activity thanks to digital technologies. Learning is made more interesting and engaging through interactive features, multimedia content, and gamification components. Students are encouraged to explore, experiment, and contribute in virtual labs, simulations, and online forums, which improves their grasp of the content.

6.4. Extensive Learning Resources

The digital ecosystem provides a plethora of learning options that go far beyond traditional textbooks. E-books, videos, podcasts, and interactive simulations give students with multimedia-rich content that caters to a variety of learning methods. In comparison to traditional print materials, these resources are frequently more up to date, easily available, and cost-effective.

6.5. Adaptability and Convenience

Digital technology has made teaching more adaptable and convenient. Online learning systems and tools allow students to access educational content at their leisure, supporting a variety of schedules and lifestyles. This is especially useful for working professionals, adult learners, and those juggling schooling with other commitments.

6.6. Networking and Collaboration

Students and teachers that use digital tools are more connected and collaborative. Communication and teamwork are made possible regardless of geographical distance by virtual classes, discussion boards, and collaborative project tools. This promotes communication among students, builds a sense of community, and broadens their perspectives.

6.7. Real-Time Evaluation and Feedback

Real-time evaluation and prompt feedback are made possible by digital tools. Teachers can assess student progress and spot areas that need more attention by using online quizzes, automatic grading, and data analytics. Timely interventions and more precisely targeted training are made possible by this data-driven methodology.

7. Lifelong Learning Possibilities

Lifelong learning is now a part of education thanks to the development of digital technologies. Individuals can seek professional development, explore new hobbies, and keep their skills up-to-date with the help of a wealth of online tools and courses. For both professional and personal development, this has significant ramifications.

These advantages of digital tools for learning highlight how they can revolutionize the educational process. We will discuss these issues and concerns in more detail in the following sections of this research, but it is important to understand that there are drawbacks in addition to these advantages. A thorough grasp of the effects of digital technology on education is made possible by the comprehensive perspective offered here.

8. Problems and Worries

While there is no denying that digital technologies have revolutionized education, they are not without their share of difficulties and worries. Gaining full use of digital technologies in education requires an understanding of and attention to these concerns.

8.1. Data Security and Privacy

The security of student data and privacy is one of the top issues in the digital era of education. There is a higher risk of data breaches and illegal access to sensitive information as more learning occurs in digital contexts. The Family Educational Rights and Privacy Act (FERPA) in the United States is one of the relevant rules that institutions must comply with and implement strong data protection procedures.

8.2. Digital Divide

Because not all pupils have equal access to the required technology and internet connectivity, the digital gap continues to be a serious problem. Accessibility issues can make educational disparities worse by making it more difficult for students to engage in online learning and use digital resources. Equitable education requires attempts to close this gap by bringing technology to marginalized populations.

8.3. Information Saturation

Students may experience information overload as a result of the abundance of digital information available, which makes it difficult for them to adequately analyze, evaluate, and apply the material. To assist kids navigate the sea of online information, it is crucial to foster digital literacy and critical thinking abilities.

8.4. Demise of Traditional Educational Models

Digital technology integration can occasionally upset established teaching paradigms, raising questions about the function of teachers in the educational process. There may be pressure on educators to adopt new pedagogies and develop their digital literacy. To preserve the worth of teaching, it's imperative to strike a balance between technology and interpersonal engagement.

8.5. Tiredness from Online Learning

Online learning fatigue can result from using digital platforms for learning over a lengthy period of time. Due to too much screen time, teachers and students may get disengaged and burn out. It might be difficult to control the structure and speed of online education to avoid burnout.

8.6. Quality Control and Standardization

Concerns concerning uniformity and quality control have been raised due to the rapid growth of digital learning. It is essential to uphold the integrity and worth of digital education that online courses and educational resources adhere to high standards.

8.7. Technical Obsolescence

As a result of the quick evolution of digital technologies, obsolescence issues have been raised. In order to assure compatibility with the most recent software and hardware, educational institutions must constantly update and modify their technology infrastructure, which can be expensive and difficult.

8.8. Copyright and Plagiarism

Due to concerns over plagiarism and copyright infringement, the digital environment has made it simpler for students to obtain and share material. Promoting moral research and appropriate citation techniques is how educators need to solve these problems.

Utilizing digital technology in education to their fullest potential requires navigating these difficulties and worries. It necessitates a comprehensive strategy that integrates technology, policy, education, and constant adjustment to the changing digital landscape. Understanding these difficulties aids in the development of strategies by institutions and educators to maximize the benefits of digital technology while minimizing the drawbacks they may have on education.

9. Implications for Education

The incorporation of digital technologies into education has far-reaching pedagogical consequences, affecting not only how education is given, but also how learners engage with and absorb knowledge. This section investigates the major changes in teaching caused by digital technologies.

9.1. Transition to Student-Centered Learning

Because of digital technologies, traditional teacher-centered instruction has given way to student-centered learning. Students can take greater control of their learning thanks to the availability of online materials, adaptive learning platforms, and collaborative technologies. They can investigate areas of interest, choose their own learning pace, and conduct self-directed research tailoring instruction to their specific needs and interests.

9.2. Experiential and Active Learning

Active and immersive learning is facilitated by digital technologies. Students can engage with course information in a hands-on, practical way by using virtual labs, simulations, and interactive educational games. These tools promote critical thinking, problem solving, and experimenting, which improves subject learning and retention.

9.3. Classrooms that are Flipped

With the assistance of digital technologies, the "flipped classroom" approach has grown in popularity. Before arriving to class, students in this paradigm engage with learning materials independently, frequently through online resources. The remainder of the class period is devoted to discussions, group activities, and problem solving, during which educators can provide targeted assistance. During face-to-face encounters, this method encourages higher-order thinking and active engagement.

9.4. Individualized Feedback and Assessment

Digital technology allows personalized and quick feedback. Automated grading, quizzes, and data analytics technologies enable educators to assess student progress more effectively. Feedback can be personalized to specific needs, assisting students in understanding their strengths and limitations and providing recommendations for progress.

9.5. Collaborative Learning Environments

Online platforms and collaboration tools promote group work and collaborative projects regardless of students' geographical whereabouts. This encourages collaboration, communication, and the exchange of varied viewpoints. It prepares pupils for the collaborative demands of today's workforce.

9.6. Models of Blended Learning

Because of digital technologies, blended learning has grown increasingly viable and popular, combining in-person education with online components. Educators may create courses that blend in-person interactions with online resources and activities, resulting in a flexible and dynamic learning experience.

9.7. Hierarchies are Flattening

Hierarchical hierarchies in education can be broken down by digital technologies. Learners can participate in online debates, forums, and collaborations with experts from all around the world, breaking down traditional barriers. This increases access to knowledge and fosters a culture of lifelong learning.

9.8. Development of Digital Literacy

As digital technologies become more prevalent in education, the development of digital literacy skills has become an essential component of the educational process. Students must learn to critically evaluate material, navigate online environments, and use digital tools efficiently.

It is clear that digital technologies are more than just instruments to improve education; they have radically changed pedagogical practices. Educators must adjust their approaches and curricula to fully utilize the potential of these technologies while ensuring they correspond with learning objectives and provide students with a rich, engaging, and successful educational experience. The educational implications of digital technologies continue to alter as technology improves, making it critical for educators to remain nimble and receptive to innovation.

10. Case Studies and Illustrations

Let's look at specific case studies and examples from various educational settings and institutions to gain a practical grasp of the impact of digital technologies on education:

10.1. Khan Academy (Khan Academy)

Khan Academy is a well-known example of how digital technologies may be used to transform education. This nonprofit platform provides a massive library of instructional videos and practice exercises covering a wide range of subjects, making high-quality education available to anybody with an internet connection. Khan Academy's adaptive learning capabilities enable students to learn at their own pace and receive real-time feedback, emphasizing individualized learning.

10.2. K-12 Education's Flipped Classrooms

The flipped classroom model has been adopted by a large number of K-12 institutions globally. Before attending class in a flipped classroom, students evaluate online educational resources (videos, books, etc.). The remainder of the class period is devoted to discussions, problem solving, and projects. This method not only boosts student participation but also promotes a deeper knowledge of the content.

10.3. Coursera and Massive Open Online Courses (MOOCs)

Massive Open Online Courses (MOOCs) provided by platforms such as Coursera have enabled millions of people to gain access to high-quality courses from prestigious colleges and institutes. Coursera's interactive courses, peer assessments, and discussion boards foster a thriving online learning community. Learners can choose from a variety of subjects such as data science, business, and the humanities.

10.4. Google Classroom in Schools

Google Classroom has become a mainstay of digital education in K-12 classrooms. It gives teachers a place to create, distribute, and grade assignments, as well as engage in discussions with students.

This example shows how educational tools from a tech company have streamlined and improved the learning process in regular classrooms.

10.5. WGU Online Learning

Western Governors University (WGU) is an example of a totally online university that has transformed higher education. WGU offers flexible, self-paced, and affordable education through a competency-based learning paradigm and a variety of digital resources.

10.6. The Use of Virtual Reality in Medical Education

Virtual reality (VR) technology has produced important advances in medical education. VR simulations have been included into the curricula of universities such as Stanford University School of Medicine and Harvard Medical School. Medical students can practice surgical operations and investigate anatomical structures in a risk-free virtual environment, boosting their clinical skills and knowledge.

These case studies and examples highlight the various ways in which digital technologies are revolutionizing education. The impact on teaching and learning is tremendous, whether through easily accessible internet platforms, interactive tools, or immersive technologies like VR. These examples also highlight the necessity of adaptability and creativity in education as institutions and educators continue to explore the possibilities of digital technologies.

11. Innovations and Future Trends

The integration of digital technology in education is a continuous process, and it is critical to anticipate upcoming trends and breakthroughs that will alter the educational landscape.

11.1. AI in Education

AI has the potential to transform education. AI-powered personalized learning systems will grow more advanced, providing adaptive content and evaluations suited to the talents and needs of individual students. Chatbots and virtual assistants driven by AI will improve student support and guidance by offering immediate solutions to questions and concerns.

11.2. Augmented and Virtual Reality

The use of AR and VR technology in education will grow. Immersive technologies provide new levels of engagement by allowing students to investigate historical events, travel to distant locations, and undertake sophisticated experiments in virtual worlds. They will also give essential training experiences for areas such as medicine, engineering, and architecture.

11.3. Blockchain and Credentials

To securely validate and handle educational qualifications and certifications, blockchain technology will be used. This invention protects the legitimacy of academic achievements, simplifies the qualification verification procedure, and avoids credentialing system fraud.

11.4. IoT in Smart Classrooms

Classrooms will be transformed into smart, interconnected settings as a result of IoT. Data on student interactions and learning patterns will be collected by smartboards, IoT-enabled devices, and sensors. This information will allow educators to make data-driven decisions, improve learning environments, and give students with more tailored support.

11.5. Game-Based Learning and Gamification

Gamification as an educational method will continue to expand. Students are motivated and engaged by game-based learning experiences, which also promote problem-solving skills. Educators will have access to progressively more complex instructional games and simulations as the gaming industry continues to innovate.

11.6. Online Exchanges and Global Collaboration

Students and educators will be able to collaborate more globally as a result of digital technologies. Students will be able to learn about different cultures, collaborate on projects, and gain a broader perspective on global issues through online exchanges, video conferencing, and international partnerships.

11.7. Microcredentials and Continuous Learning

Individuals seeking microcredentials and short, focused courses to acquire specific skills or knowledge will seek out the concept of continuous learning. This demand will be met by online platforms, allowing professionals to stay current in rapidly changing industries.

11.8. Ethical Technology Application

The ethical implications of digital technology use will become increasingly important. To navigate the complexities of the digital world, educators will need to teach digital ethics, responsible online behavior, and critical thinking skills.

Education's future is inextricably linked to digital technologies. As these innovations progress, educators, institutions, and policymakers must adapt in order to fully realize their potential. The education sector can stay at the forefront of digital transformation by remaining proactive in adopting emerging technologies and considering their ethical implications.

12. Conclusion

This research article's comprehensive analysis highlights the transformative impact of digital technologies on education. From the introduction of computers in the classroom to the current era of AI, augmented reality, and virtual reality, the incorporation of technology has revolutionized not only what is taught but also how it is taught and learned. Finally, we consider the key takeaways as well as the broader implications of this digital transformation.

Unquestionably, digital technologies have increased access to education, allowing people all over the world to engage in lifelong learning and gain knowledge beyond traditional boundaries. The democratization of education through online platforms and resources has the potential to reduce educational disparities and create opportunities for those who were previously marginalized.

The thorough analysis in this research article emphasizes the transformative impact of digital technologies on education. From the introduction of computers in the classroom to the current era of AI, augmented reality, and virtual reality, technology has revolutionized not only what is taught but also how it is taught and learned. Finally, we consider both the key takeaways and the broader implications of this digital transformation.

Without a doubt, digital technologies have increased access to education, allowing people all over the world to engage in lifelong learning and gain knowledge that transcends traditional boundaries. Education democratization via online platforms and resources has the potential to reduce educational disparities & create opportunities for those who were previously marginalized.

Personalization of learning experiences, enabled by adaptive technologies, is a standout achievement. Students can now chart their own educational path, guided by data-driven insights. Engagement and interactivity have grown, with multimedia-rich content, collaborative tools, and gamification redefining the learning experience.

This digital journey, however, is not without its challenges. Privacy and data security are critical concerns that necessitate stringent safeguards to protect sensitive information. The digital divide persists, necessitating concerted efforts to bridge the gap in technology access. The erosion of traditional teaching models, concerns about online learning fatigue, and the pressure on educators to adapt to the digital landscape are all significant challenges.

In the future, we see a landscape in which artificial intelligence tailors education to individual needs, augmented reality and virtual reality create immersive learning environments, and blockchain ensures the veracity of credentials. The Internet of Things will revolutionize education, gamification will encourage active participation, and global collaboration will become the norm. Continuous learning and microcredentials will keep professionals adaptable in rapidly changing industries, while ethical considerations will guide responsible technology use.

Finally, the symbiotic relationship between education and digital technology is evolving. A richer, more inclusive, and adaptive educational experience is on the horizon. To fully realize this potential, educators, institutions, and policymakers must remain agile, proactive, and committed to responsible innovation. This comprehensive analysis has laid the groundwork for understanding the dynamic relationship between digital technologies and education, guiding the way forward as education embraces the digital age.

As we chart the future of education, it is critical that we remain steadfast in our commitment to providing quality education for all, leveraging the power of technology to make learning a lifelong, accessible, and deeply enriching journey.

13. Recommendations

This research article's comprehensive analysis provides valuable insights into the impact of digital technologies on education. It is critical to translate these insights into actionable recommendations for educators, policymakers, institutions, and education stakeholders. The recommendations that follow provide a road map for effectively leveraging the potential of digital technologies in education:

13.1. Education in Digital Literacy

Promote digital literacy and responsible online behavior as essential components of all levels of education. Give students the ability to critically evaluate information, navigate online environments, and effectively use digital tools.

13.2. Data Security and Privacy

To protect student data, implement strict data privacy and security measures. Compliance with relevant regulations, open data usage policies, and strong security practices should be priorities for educational institutions.

13.3. Closing the Digital Gap

Address the digital divide by investing in initiatives that provide underserved communities with technology and internet access. Ensuring equitable access to digital resources is critical for closing educational gaps.

13.4. Professional Growth

By providing professional development opportunities, educators can be assisted in adapting to the digital landscape. These programs should emphasize digital literacy, pedagogical strategies for online and blended learning, and technology ethics.

13.5. Personalized Education

Encourage the creation and integration of adaptive learning platforms that meet the needs of individual students. These technologies should be simple to use and capable of providing personalized content and assessments.

13.6. Collaboration and Interdisciplinary Learning are also Important

Utilize digital technologies to promote interdisciplinary collaboration. To promote global perspectives and cooperation, virtual collaboration tools and online exchanges should be integrated into the curriculum.

13.7. Integration of Smart Classrooms

Use Internet of Things (IoT) technology to build smart classrooms that collect and analyze data on student interactions and learning patterns. This data-driven approach can help to guide decisions and improve learning environments.

13.8. Microcredentialing and Continuous Learning

By providing short, focused courses and microcredentials, you can embrace the concept of continuous learning. Professionals seeking specific skills and knowledge throughout their careers will require educational institutions to adapt.

13.9. Ethical Application of Technology

Integrate digital ethics education into the curriculum, emphasizing responsible online behavior, the value of data privacy, and the importance of critical thinking when navigating the digital world.

13.10. Research and Development

Support educational technology research and innovation. Encourage educators and institutions to try out new technologies and teaching methods. Collaboration between academia and industry can propel educational technology forward.

13.11. Teacher-Student Collaboration

In the adoption of digital technologies, encourage collaboration and communication between educators and students. Students should have a say in the tools they use, fostering a sense of ownership and engagement.

13.12. Policy Development

Educational policymakers should establish a regulatory framework that promotes innovation while protecting students' rights and privacy. Policies should be flexible enough to accommodate changing technology and pedagogy.

These recommendations take a comprehensive approach to maximizing the benefits of digital technologies in education while addressing the challenges that they entail. By adhering to these guidelines, educators, policymakers, and institutions will be able to confidently navigate the ever-changing digital landscape, ensuring that the future of education is both transformative and inclusive.

References

Anderson, C. A., & Dill, K. E. (2000). Video games and aggressive thoughts, feelings, and behavior in the laboratory and in life. Journal of Personality and Social Psychology, 78(4), 772-790.

Anderson, T., & Dron, J. (2011). Three generations of distance education pedagogy. The International Review of Research in Open and Distributed Learning, 12(3), 80-97.

Baran, E., & Correia, A. P. (2014). A professional development framework for online teaching. Techtrends, 58(5), 95-101.

Dede, C. (2010). Comparing frameworks for 21st century skills. In J. Bellanca & R. Brandt (Eds.), 21st Century Skills: Rethinking How Students Learn (pp. 51-76). Solution Tree Press.

Johnson, L., Adams Becker, S., Estrada, V., & Freeman, A. (2015). NMC Horizon Report: 2015 Higher Education Edition. The New Media Consortium.

Khan Academy (2021). About Khan Academy. https://www.khanacademy.org/about

Oblinger, D. G., & Oblinger, J. L. (2005). Educating the Net Generation. EDUCAUSE.

Picciano, A.G. (2017). The evolution of big data and learning analytics in American higher education. Journal of Asynchronous Learning Networks, 21(5), 3-13.

Siemens, G. (2005). Connectivism: A learning theory for the digital age. International Journal of Instructional Technology and Distance Learning, 2(1), 3-10.

Siemens, G. (2013). Massive open online courses: Innovation in education? In R. McGreal, W. Kinuthia, & S. Marshall (Eds.), Open Educational Resources: Innovation, Research, and Practice (pp. 5-16). Commonwealth of Learning and Athabasca University.

Siemens, G., & Gasevic, D. (2012). Guest editorial - learning and knowledge analytics. Educational Technology & Society, 15(3), 1-2.

West, D.M. (2018). How blockchain technology could change our lives. Brookings. https://www.brookings.edu/research/how-blockchain-technology-could-change-our-lives/

Winkelman, R. (2016). Teacher and student perspectives on a blended learning intensive English program. Journal of Language Teaching and Research, 7(5), 964-977.

World Economic Forum (2019). Schools of the future: Defining new models of education for the fourth industrial revolution. http://www3.weforum.org/docs/WEF_Schools_of_the_Future_Report_2019.pdf

Zhang, D., Zhao, J. L., Zhou, L., & Nunamaker, J. F. (2004). Can e-learning replace classroom learning? Communications of the ACM, 47(5), 75-79.

19

Green Library Initiative in India: Paving the Way for Sustainable Development Goals

***Dr. Sonia Sharma*[1] *and Dr. Piyali Sarkar*[2]**

[1]*Associate Professor, Department of Education, Lovely Professional University Phagwara - 144 411, Punjab, India, E-mail: soniasharma7oct@gmail.com*

[2]*Assistant Head Mistress, (PGT), Barabisha High School (H.S.), Alipurduar - 736 207 West Bengal, India, E-mail: piu12sarkar@gmail.com*

Abstract

One urgent global concern that is caused by human action is climate change. As India strives to meet the Sustainable Development Goals (SDGs), green libraries are becoming more and more significant. These libraries support social responsibility, economic prosperity, and ecologically responsible development. To enhance climate resilience and lower carbon footprints, they make use of renewable energy sources including solar power, sustainable waste management techniques, and environmentally friendly building materials. In addition, green libraries promote lifelong learning, equal access to information, and environmental literacy, which helps to achieve SDGs 11 (Sustainable Cities and Communities) and 4 (Quality Education). Through community-driven programs, seminars, and workshops, green libraries in India function as green public places that enhance urban liability, foster social solidarity, and encourage responsible consumption and production. Via educational events, seminars, and community-driven initiatives, they educate the general public regarding sustainable practices and the importance of reused, recycled, and reducing resources.

Green libraries offer a workable paradigm for public institutions in India, a country with severe resource difficulties, increasing urbanization, and environmental degradation. They serve as role models for sustainability and as stimulants for environmental awareness and action in the community. Green libraries encourage other organizations to adopt a similar approach by providing models of how to integrate sustainability into daily operations and community events. Additionally, they serve as community-wide accelerators for environmental consciousness and action, encouraging a sustainable culture that permeates every aspect of day-to-day existence. Green libraries in India are expanding and developing, and this is a great illustration of how educational organizations can lead the way in sustainable development. Through highlighting the interrelationships among social, environmental, and financial objectives, they inspire cooperation and the development of more sustainable societies. By adopting a comprehensive approach to sustainability, green libraries help to create a more resilient, equitable, and environmentally friendly tomorrow for everyone.

***Keywords**: Green library, Green library in India, Green library and SDGs, Educational technology, SDGs.*

1. Introduction

There is a real climate change, and human activity is its primary driver. Notably, worldwide average temperatures have been rising consistently since the beginning of the Industrial Revolution, posing a danger to life as we know it from climate change. For example, 2019 had record sea levels, melting ice, and unusually high global temperatures as a result of increased human activity. Regrettably, in spite of the fact that climate change is causing the environment to deteriorate more and more, insufficient worldwide pledges have been made to mitigate its effects. Thus, the necessity for green libraries became apparent. In India, green libraries are becoming more and more important as the country works toward achieving the Goals for Sustainable Development (SDGs). These libraries promote a comprehensive approach to environmentally friendly growth by serving as examples of social responsibility, environmental care, and economic success. Green libraries use eco-friendly construction materials, sustainable waste management practices, and renewable energy sources like solar power in their design. They clearly support climate resilience and drastically cut carbon footprints, which is in line with SDG 13 (Climate Action). Additionally, by fostering environmental literacy, lifelong learning for people of all ages, and inclusive and equal access to knowledge and educational resources, green libraries in India actively support SDG 4 (Quality Education). Additionally, by acting as green public spaces that improve urban livability and provide a haven from the bustle of city life while encouraging social cohesion and community participation, these institutions serve SDG 11 (Sustainable Cities and Communities). Green libraries promote SDG 12 (Responsible Consuming and Production) by teaching the public about sustainable practices and the value of recycling, reusing, and reducing materials through seminars, workshops, and community-driven projects. They frequently hold workshops that impart knowledge on sustainable living practices, such water and energy conservation, composting, and energy efficiency, enabling people to make decisions that are less harmful to the environment. In a nation such as India, where severe resource shortages, growing urbanization, and contamination of the environment are problems, green libraries provide a viable model for public institutions.Through their activities, they not only exemplify sustainability, but they also function as catalysts for environmental knowledge and action across the community. The integration of environmentally friendly methods into the facilities and amenities they provide enables green libraries in India to play a crucial role in the national drive to achieve the Sustainable Development Goals (SDGs) and contribute to a larger cultural change towards sustainability. Libraries such as these are more than simply places to store books; they are vibrant hubs of knowledge and creativity that help to create a sustainable culture that permeates the entire community. Green libraries are a shining example of how educational organizations can lead the charge in sustainable development, encouraging people to work cooperatively towards a more environmentally friendly future as India moves closer to its 2030 Framework for Sustainable Development.

In the worldwide push for sustainability, green libraries are becoming more important institutions, especially in nations with severe environmental problems like India. These libraries contribute in many ways to the advancement of the Sustainable Development Goals (SDGs) by demonstrating a dedication to social responsibility, environmental care, and economic sustainability. In line with SDG 13 (Climate Action), the idea of a "green library" incorporates sustainable resource management techniques, energy produced from renewable sources, and environmentally friendly architectural designs. Green libraries drastically lower their carbon footprint by deploying energy-efficient technologies, adding solar panels, and using sustainable construction materials. This promotes environmental sustainability and climate resilience. Additionally, by promoting environmental literacy and offering accessible access to knowledge, green libraries support SDG 4 (Quality Education). To encourage community people to interact with and embrace sustainable lifestyles, they hold educational events, provide a plethora of materials on sustainable practices, and function as centers for lifelong learning. These organizations also contribute to SDG 11 (Sustainable Cities and Communities) by making cities more liveable. Being green public areas, they provide a haven from the hustle and bustle of the city, encouraging community involvement and social cohesiveness. Environmental roofs, rain garden designs, and other environmentally friendly landscaping techniques that manage rainwater and lessen the impacts of urban heat islands are common features of green libraries. Green libraries promote SDG 12 (Responsible Consumption and Production) by teaching the public the value of reducing, reusing, and recycling through workshops, seminars, and cantered around communities' projects. They provide people hands-on instruction in eco-friendly living techniques like composting, reducing water consumption, and energy conservation, thereby enabling them to make decisions that benefit the environment.

Green libraries provide a sustainable paradigm for public institutions in India, where major issues arise from rising urbanization and environmental deterioration. These libraries serve as examples of how to incorporate sustainability into regular operations and community activities, which encourages other organizations to do the same. Additionally, they operate as catalysts for environmental awareness and action across the community, promoting a sustainable culture that penetrates every facet of daily life. Green libraries serve a vital role in teaching and inspiring communities to embrace sustainable practices, in addition to reducing their own environmental effect through sustainable procedures. Green libraries have an effect that goes beyond their local communities. Through the cultivation of a sustainable culture, they facilitate wider social transformations toward more environmentally friendly actions and habits. In order to achieve the SDGs and guarantee long-term social and environmental well-being, a cultural transformation is imperative. India's growing and changing green libraries serve as a prime example of how academic institutions may take the lead in sustainable development. By emphasizing the connections between social, environmental, and economic objectives, they encourage people to collaborate and build more sustainable communities. Green libraries are vibrant hubs of learning, creativity, and community involvement rather than only being places to store books. They serve as an example for other public institutions, exhibiting sustainability in both their operations and infrastructure. Green libraries are essential to the advancement of the Sustainable Development Goals (SDGs) in India because they promote sustainable practices, environmental literacy, and community well-being. Green libraries are shining examples of success as the nation moves closer to achieving its 2030 Agenda for Sustainable Development. They show that sustainable development is not only feasible but also advantageous for both the environment and local communities. Green libraries are contributing to the creation of a more resilient, egalitarian, and sustainable future for all people by taking a holistic approach to sustainability.

2. Green Library

One institution that is important to a developed society is the library. Shera asserts that the library is a vital component of the social structure that the people have built to meet their needs. In addition, libraries look after environmental groups, which emphasize the importance of libraries in promoting sustainable growth and puts environmental challenges in the context of potential solutions.

In an attempt to write his environmental manual, he sought to build a green library. In the fields of information science and libraries, green libraries are becoming more and more well-liked. A green library, sometimes referred to as a sustainable library, is constructed with environmental concerns in mind. A component of the greater green structure program is green libraries. Libraries, particularly academic and public ones, serve as hubs for lifelong learning for local residents of all ages. The expression 'green library' lacks a precise meaning. The term "green library" refers to the structure and all of its activities. Thus, it appears that the idea is rather wide. A "Green Library" is, to put it simply, an eco- or environment-friendly library. A "green" or "sustainable library" is a building that has been planned, constructed, refurbished, run, or repurposed in a way that is resource- and environmentally-conscious. Libraries should take seriously society's primary issues, which include the role that humans play in climate change and the nation's sustainable development (Varatharajan and Chandrashekara, 2007). An essential part of environmental protection is played by green building. Because of their green architecture and close ties to the community, libraries and librarians offer excellent venues for public education. In addition to giving consumers access to knowledge, a librarian's role includes spearheading the construction of cutting-edge green buildings that preserve the environment. In order to ensure the sustainable growth of libraries in the future, library professionals need to demonstrate their commitment to environmental sustainability by implementing different "greening library" methods and taking proactive measures. Future LIS students and librarians are urged to embrace the "Green Library of Tomorrow" project and grasp the chance to promote libraries in a novel, client-focused, and environmentally friendly manner.

3. Key Features of Green Library

Using a variety of internet resources, including social networking sites and other platforms, librarians should constantly work to advance the cause of green libraries. The most dynamic library is the green one, often known as the eco-library as it requires financial management to be maintained (George, 2013). When it comes to creating robust, thriving libraries, economics and ecology are the two most crucial factors. Basically, the librarians are kept in mind. Nonetheless, the library's contribution to the advancement of social stability must be strengthened. In place of steel, the green library should employ more and more bamboo. The OPAC should be located on the bottom level of the building, together with the reading area, the children's reading room, tale, novel, entertainment books, maps, geographical guides, gazette, online encyclopaedia new arrivals, newspaper, magazine, and so on. Through conferences, seminars, and discussions, they may persuade other librarians to embrace green libraries. A green library combines a number of essential elements to improve community well-being and encourage environmental sustainability. Using eco-friendly building materials, such as recycled components and low-VOC paints, ensures less of an impact on the environment and better interior air quality. The library's carbon footprint is greatly decreased via energy-efficient technologies like LED lighting and sophisticated HVAC systems, as well as alternative sources of energy like solar panels. Water shortage concerns are addressed by water conservation techniques including rainwater collecting, greywater recycling, and landscaping using drought-resistant plants. Programs for comprehensive waste management, such as composting and recycling, reduce the amount of garbage dumped in landfills. Paper and energy use are decreased by using internet resources and green IT techniques. The library acts as a centre for environmental education, providing courses, talks, and neighborhood-based initiatives that support eco-friendly living. Energy efficiency is improved by smart technology like climate control systems and automatic lighting.

4. Review of Literature

Kumar and KD (2015) discovered that the concept and practice of building green libraries is rapidly reaching a tipping point globally and evolving into a library movement. This includes the construction of green library buildings, the greening of current library spaces, the provision of green services for libraries, and the adoption of sustainable and environmentally friendly practices throughout the library. Ecosystem and economics are the two most crucial components of robust, sustainable libraries. This research aims to assess the significance of green libraries and determine the requirements for green libraries in India. to list the main national and worldwide efforts pertaining to green libraries; to describe different approaches and strategies for greening libraries; and to offer recommendations for the construction of green libraries for the sake of environmental sustainability. The study finds that librarians need to be diligent about staying up to date on environmentally conscious developments in the area of librarianship. They also need to raise awareness of these trends and make room in their libraries for libraries to serve as examples of green practices.

According to Kassim *et al.* (2019), there are several reasons why cultural heritage is endangered, including a lack of enthusiasm from the younger generation and short life expectancy, which causes people to pass away before passing on their wisdom to the next generation. This is because it is well acknowledged that cultural heritage was essentially transmitted orally from generation to generation. In order to develop a framework that will explain how information is gathered, transferred, and preserved among conventional baby massage practitioners in Banten, Indonesia, this qualitative study makes use of the company's understanding of conversion theory to examine the acquisition, advancement, and preservation of cultural heritage. The analysis of the practitioner's knowledge transfer in the study used a qualitative methodology. Document analysis, observation, and interviews were used to get the data. The study goals guide the thematic analysis and interpretation of the findings. According to this study, learning traditional massage techniques involves imitation, practice, and observation. Furthermore, interface assistance is emphasized as one of the key factors influencing the successful transmission and acquisition of information. The primary conclusion drawn from this study is that traditional massage expertise is thought to be preserved by the Almighty and passed down to the selected. The study comes to the conclusion that traditional masseurs exchange and acquire expertise through conversations with potential successors, as well as vocally preserving their knowledge.

Sarkar (2020) emphasized the idea of a green library. The goal of a green library is to raise awareness of environmental sustainability at all library levels. Green libraries are eco-friendly and ecologically conscious. Community members are also involved in environmental concerns through Green Library. This essay demonstrates how the Green Library concept may promote environmental awareness and social sustainability. This essay is aware of the seventeen sustainable development objectives of the United Nations and how they affect society. This essay focuses on investigating how librarians and libraries might encourage members of the public to adopt more ecologically friendly lifestyles. Our primary goal in terms of the environment is to use more renewable energy sources and consume less energy in order to preserve environmental sustainability across all levels and leave a more environmentally friendly legacy for future generations.

In an effort to increase user knowledge of green libraries, Gaffar *et al.* (2021) carried out a study to investigate cases of green library building and identification. The function of a librarian, the characteristics of a green library, and the attempts to establish of a green library in India are highlighted in this study by the researcher. Through literature searches, case study analysis, and one-on-one phone calls during the research period, the study discovered that there are some green library projects in India. In India, a good number of libraries have started pursuing green programs. A green library reduces its adverse effects on the environment and improves the condition of the indoor space through site selection that is thoughtful, the use of organic and natural building materials, resource conservation (water, energy, paper), responsible elimination of waste, recycling, etc. The components of creating green libraries, the standards for greening libraries, and the initiatives around green libraries in India.

Mahesh and Babu (2022) discovered that libraries require a lot of natural resources, water, and artificial light for their daily operations. A global movement is underway to expand on the idea of the "green library." The current article goes into great depth on how information centres and libraries around the Indian subcontinent have adopted this project. India ranks among the top nations that contribute to pollution and other environmental issues. Eighty-five information centres and libraries nationwide participated in the study. The study concentrated on the environmental friendliness of the library building, the overall green initiatives implemented in the facilities and library functions, the use of artificial lighting, the proportion of print and e-resources, the use of good lighting and bulbs in the library premises, and many other issues that were addressed appropriately. According to the findings, 60% of institutions have implemented green efforts, 16.77% have an 80% online and 20% print collection, and 7.22% have 100% of their collection available online.

According to Adhikary (2022), preserving the planet and its inhabitants' lives is currently the world's most pressing concern. Libraries are in an ideal position to serve as both an environmentally conscious operator and a promoter of environmental consciousness, so they may play a stronger role in this respect by adopting measures to educate and inspire people to conserve the environment. This brings up the significance of the green library. A "green library" is a library that complies with environmental regulations in a number of ways, such as lowering energy use, acquiring and promoting eco-friendly resources, implementing various green practices, and stocking non-toxic items. Other names for green libraries include sustainable libraries, eco-friendly libraries, and environment-friendly libraries. The paper's primary goal is to discuss the current state of the AKPC Mahavidyalaya library from a green viewpoint. It looks at the ways that the library currently uses its collection, facilities, and services to realistically implement the green library idea. Additionally, it offers recommendations for additional college library resources to support the transition to a greener future.

According to Kunlakrni (2023), the "green concept" is frequently addressed in the context of fundamental library professionals and other areas where humanity faces grave environmental issues. The library is helpful and widely available. Libraries are part of this global initiative. Libraries offer knowledge and information and serve as venues for lifelong learning. Environmental issues like energy conservation and climate change have an impact on how users are ranked when it comes to data resources and software libraries. As libraries' interest in environmental concerns grew, the green library movement got underway in the early 1990s. As a result of the green library movement, there is now more literature on the subject, raising awareness of the subject. The globe at large and today's society are growing increasingly aware of environmental issues. The idea, standards, characteristics, and constituents of a green library are highlighted in this study. The role of green librarians and green library projects in India are also covered in the paper.

Mwanzu *et al.* (2023) state that safeguarding the environment is a global concern that is becoming more and more urgent. Among the growing environmental issues are depletion of ozone layer, the effect of greenhouse gases, global climate change, and global warming. Several Go Green initiatives and a Go Green movement have emerged as a result of the growing number of environmentally concerned customers. These campaigns have offered answers and called on individuals, businesses, and governments to take more initiative in becoming green in all spheres. Libraries are not an exclusion. Despite its significance, not much has been discovered about the green library idea in Kenyan libraries or throughout Africa. Kenya, like a few other developing nations, has reacted to warming temperatures in different ways due to its negative effects. Investigating the strategies libraries are doing to prepare for an environmentally friendly future is crucial. Therefore, in order to provide ideas that might help libraries transition to a greener environment, this study set out to investigate how green library concepts are being used in Kenya for the sustainability of the environment. The research used a qualitative methodology. Out of 227 academic, public, and special libraries in Kenya, a sample of 12 libraries was purposefully selected. Grounded theory was used to analyse the data, which was gathered utilizing semi-structured interview schedules supplemented with observation and a review of the documents. The results demonstrate how Kenyan libraries, operating under the auspices of their parent organizations, have capitalized on their understanding of green principles by embracing and putting into practice green practices and devising plans to improve green libraries and advance environmental sustainability. They have mostly achieved this by remodelling and rebuilding to satisfy the constantly shifting needs of patrons while adhering to green library standards. According to the study's findings, Kenyan library librarians comprehend the principles of green building and have embraced a variety of green projects, if only in part to preserve and improve their standing internationally. Three strategies are suggested by the study to fully implement green initiatives: actively engaging users in green initiatives; enhancing performance and growth through metrics;

and upholding green-library standards in compliance with the Environment, Sustainability and Libraries Section green-library checklist.

According to Solemanpharcy, Gaffar, and Abdul (2023), libraries are essential to the global advancement of environmental education and the sustainable development goals (SDGs). In order to fulfil their function as information centres and agents of good change, libraries have begun to embrace the idea of developing into "green libraries" as society grows more conscious of the pressing need for conservation of the environment and sustainable practices. The purpose of this project is to investigate how green libraries may further environmental education and the SDGs from a global viewpoint. The study examines the many facets of green libraries and the ways they contribute to environmental education and sustainable development via a thorough examination of the literature. It examines several case studies, projects, and best practices from global libraries, showcasing the range of strategies used by these establishments to encourage community involvement, supply access to green information, and raise environmental consciousness. The study's conclusions show that green libraries are crucial venues for promoting sustainability literacy, sharing environmental knowledge, and enabling people to make knowledgeable decisions about environmental concerns. These libraries incorporate sustainable techniques throughout their activities, such as energy-efficient structures, recycling programs, and digital resource optimization, in addition to curating collections of resources with an environmental focus. Additionally, via workshops, seminars, and outreach initiatives, green libraries actively interact with nearby communities, educational institutions, and governmental organizations to promote environmental awareness. This report also lists the difficulties that green libraries confront, such as their lack of funding, technical difficulties, and the requirement for ongoing staff development. It highlights the significance of forming alliances with stakeholders, utilizing digital technology for information sharing, and incorporating sustainability concepts into library practices and marketing strategies in order to address these issues. The report emphasizes how important green libraries are as game-changing tools for accomplishing the SDGs and promoting environmental education globally. It offers insightful analysis and practical suggestions on how scholars, legislators, and library professionals may improve the efficiency and influence of green libraries in advancing a sustainable future.

Sengupta (2024) discovered that the intersection of Open Access concepts and Green Library Initiatives offers a viable path forward for increasing environmental sustainability in academic communication in an era characterized by concerns regarding the environment and the pursuit of knowledge accessible. This research looks at how these two paradigms cross and considers how they may work together to create a knowledge ecosystem that is more inclusive and ecologically conscientious. This integrated strategy holds promise for establishing a more environmentally friendly future for educational institutions and society at large by minimizing the environmental imprint of information transmission while enabling unfettered access to intellectual products. This paper examines the opportunities and difficulties of integrating open access fundamentals alongside Green library initiatives through a review of recent literature and case studies. It provides useful advice on how libraries in general, publishers, and researchers can embrace sustainable practices in the sharing of knowledge. In the end, this combination of green library and open access efforts not only helps to reduce environmental effects but also makes information more accessible to all, which promotes a more inclusive and sustainable academic environment.

According to Patil and Jani (2024), a "green library" is one that was created with ecological sustainability in mind. The research on green libraries in this paper makes use of their programs and resources to educate people about environmental concerns. The concept of the "green library" first gained traction in India in the early 1990s. This paper attempts to define the concept of the "green library." Currently, there is a Digital Knowledge Center with thirty-five Ubuntu PCs linked to the internet. The Delhi University Library System consists of about 37 libraries. Researchers, instructors, and students can access up to 64 top-notch electronic databases via the school network. The Delhi University Library System regularly provides creative information literacy programs for the benefit of instructors, researchers, and students. Calcutta University Library's local network is linked to a server with terminals within the building. Among other resources, the university library's Electronic Public Access Catalogue offers collections of books, periodicals, theses, and dissertations. In addition to the materials listed above, the University provides access to approximately 7,000 electronic journals.

5. India's Adoption of the Green Library

A comprehensive strategy that incorporates resource management, community involvement, sustainable design, and educational programs is needed to establish a green library in India. This would promote environmental stewardship and help the nation achieve its Sustainable Development Goals (SDGs). In order to minimize the need for artificial illumination and climate control, the procedure starts with choosing the best location and orientation for natural light and ventilation. Utilizing environmentally friendly building materials, including recycled materials, low-VOC paints, and timber from sustainable sources, guarantees that the structure is built with less of an adverse effect on the environment and better interior air quality. The museum's carbon footprint is further decreased by including energy-efficient equipment, such as LED lighting and sophisticated HVAC systems, and renewable energy sources, such as solar panels, in line with SDG 13 (Climate change Action). Water conservation strategies help achieve SDG 6 (Clean Water and Sanitation) by addressing the pressing problem of water shortage. Examples of these strategies include rainwater collection, greywater recycling, and using drought-resistant plants in landscaping. Operational procedures concentrate on reducing waste by implementing extensive waste management and recycling programs in addition to sustainable procurement guidelines that give preference to environmentally friendly goods and energy-saving machinery. Ensuring enough ventilation and routine HVAC system monitoring are essential for maintaining a healthy interior air quality, which benefits both employees and guests. Encouraging the use of digital materials minimizes the use of paper, and resource management is made effective by the use of energy and water monitoring systems. The core idea behind the green library notion is community participation. In collaboration with regional environmental groups, educational initiatives focusing on sustainability issues promote lifelong learning and improve environmental literacy, which advances SDG 4 (Quality Education). The library's awareness programs highlight its green activities and instruct patrons on sustainable habits. SDG 11 (Sustainable Cities and Communities) is supported by community-driven initiatives like recycling campaigns, community gardens, and tree planting, which promote social cohesiveness and a feeling of communal responsibility.

The use of intelligent technology, such climate control systems and controlled lighting, improves energy efficiency by adapting to changes in occupancy and environmental factors. Utilizing cloud storage and energy-efficient PCs are two examples of green IT techniques that further lessen their environmental impact. A thorough sustainability strategy that describes the library's commitment to responsible environmental conservation and includes precise goals, tactics, and criteria for monitoring success is used to assure long-term sustainability. Encouraging employees to provide suggestions for improving green projects and routinely assessing and upgrading sustainability procedures are key components of cultivating an environment of continuous improvement. A green library in India may lessen its environmental effect and play a significant role in teaching and inspiring the community to adopt sustainable living habits by including these techniques.

Green libraries provide a viable paradigm for public institutions in India, where considerable issues arise from rising urbanization, resource shortages, and environmental deterioration. These libraries serve as examples of how to incorporate sustainability into regular operations and community activities, which encourages other organizations to do the same. They are vibrant hubs for education, creativity, and community involvement that integrate sustainability into every aspect of their operations and infrastructure. Green libraries are essential to the advancement of the Sustainable Development Goals (SDGs) in India because they promote sustainable practices, environmental literacy, and community well-being. Green libraries are shining examples of success as the nation moves closer to achieving its 2030 Agenda for Sustainable Development. They show that sustainable development is not only feasible but also advantageous for both the environment and local communities. Green libraries are contributing to the creation of a more resilient, egalitarian, and sustainable future for all people by taking a holistic approach to sustainability. India's introduction of green libraries is evidence that public organizations may set the standard for sustainable development and encourage people to live more sustainably. These libraries work as centers of learning, offering materials and activities that advance sustainable living and environmental consciousness. Green libraries play a vital role in empowering and educating communities while also minimizing their environmental effect via the integration of sustainable concepts into their daily operations and outreach initiatives. With this all-encompassing strategy, green libraries are certain to play a crucial role in the country's endeavor to attain the SDGs by fostering a wider cultural transition towards sustainability. Green libraries are a prime example of the vital role that educational organizations can play in promoting a culture that's environmentally friendly and advancing the country's sustainable development strategy as they continue to develop and spread throughout India.

6. Conclusion and Suggestions

There are anticipated hurdles in the process of adopting green initiatives in libraries. The majority of these issues stem from the operations, managerial support, and goodwill that go into a library's ecologically friendly policies. The majority of procedures and choices pertaining to environmental sustainability follow the norm-activation paradigm and value belief model, which are theories that suggest that decisions are typically made not in the best interests of the institution but rather based on the beliefs and actions of managers and employees. The majority of the issues raised are internal, therefore the company can come up with solutions for them. The obstacles encountered had little impact on the greening programs' uptake. With the amount of pollution in the air, water, and soil, deforestation, soil degradation, excessive fertilizer use, increased carbon dioxide emissions and greenhouse effect, unrestricted use of fossil fuels, and other anti-environmental human activities, the world is becoming a less desirable place to live. This makes environmentally friendly projects in the library one of the emerging concepts that are being implemented today. The majority of the reading materials found in libraries are printed items.

Libraries may inform their customers about the qualities of their green architecture and spread awareness of the "green concept" through a number of strategies. It covers topics such as publications, library exhibitions, and pertinent "going green" information in libraries. All colleges and universities should be required by the University Grant Commission to get permission before visiting any green library. Government grants and financial support for the upkeep of these kinds of libraries should be used as a means of promoting green libraries. To guarantee sustainability, the library can set up sessions for strategic planning and thought. Whenever feasible, governments and institutions should work to transform libraries into green structures. Books and financial issues are a constant in each institution's library. The library needs special attention because of issues with dust, dampness, and fungus. The startling rise in greenhouse gas emissions, the retreat of glaciers, and the alteration of weather patterns throughout seasonal events are all signs of the detrimental impacts of global warming. The environmental challenge, which is to protect and sustain life on Earth, is currently a major global concern. The library has the potential to inspire individuals to take proactive steps towards environmental protection. The government need to support the creation of green libraries. It will be necessary for librarians to attend and engage in the green library movement, since several Indian libraries are successfully attempting to build green initiatives. Before visiting any green library, all colleges and institutions must get authorization from the University Grants Commission. Green libraries should be promoted by government grants and other financial aid for the upkeep of these kinds of libraries. The library may host workshops on strategy planning and thinking to guarantee sustainability. Governments and academic institutions should do their hardest to create green library architecture wherever feasible. Every institution's library occasionally has book and budgetary issues. Libraries are susceptible to dust, fungi, and dampness; thus, they need special care. Because of the increase of greenhouse gases, the retreat of glaciers, and the changes in seasonal temperature, climate change is already having negative repercussions. The preservation and maintenance of all living forms on Earth, or the environment, is currently the most pressing issue facing humanity. Libraries may encourage people to contribute toward environmental protection by taking some constructive steps. The establishment of green libraries should have the backing of the government. Numerous Indian libraries and librarians that are attempting to construct green libraries efficiently ought to join the green library movement.

References

Adhikary, S. (2022). Green library initiatives in rural college library-with special reference to AKPC Mahavidyalaya, Hooghly, WB: A study. Revisiting strategies for sustainable development: an eConSus Book Series Vol. 2, 281.

Antonelli, M. (2008). The green library movement: An overview and beyond. Electronic green journal, 1(27).

Asim, M., & Ahmad, P. (2022). Adoption of green practices in university libraries of Pakistan: A qualitative study. Library Philosophy and Practice (e-journal), 1-8.

Aulisio, G. J. (2013). Green libraries are more than just buildings. Electronic green journal, 1(35).

Binks, L., Braithwaite, E., Hogarth, L., Logan, A., & Wilson, S. (2014). Tomorrow's green public library. The Australian Library Journal, 63(4), 301-312.

Gaffar, S. A., Sindhu, P. N., & Kumar, S. K. (2021). The green library initiative in Indian perspective: A study. Library Philosophy and Practice, 1-10.

Gupta, S. (2020). Green library: A strategic approach to environmental sustainability. International journal of information studies and libraries, 5(2).

Hauke, P. (2017). Green Libraries Towards Green Sustainable Development-Best Practice Examples from IFLA Green Library Award 2016–2019.

Hauke, P., Charney, M., &Sahavirta, H. (Eds.) (2018). Going green: implementing sustainable strategies in libraries around the world: buildings, management, programmes and services (Vol. 177). Walter de Gruyter GmbH & Co KG.

Kassim, M. S., Abidin, A. Z., Ab Rahman, A., Sahak, M. D., & Mud, T. N. F. E. (2019). The Green Library Initiative at Universiti Putra Malaysia. Revitalizing the Library for the Nation, 395.

Kulkarni, A. S. (2023). Green Library: Features, elements, standards and green library initiatives in India.

Kumar, P. S., & KD, S. A. (2014). Greening the library for sustainable development. Bangalore: LISCOM. Retrived, 2(04).

Ma, L. F., & Ko, L. Y. (2022). Supporting the sustainable development goals: The role of the Chinese University of Hong Kong Library. The journal of academic librarianship, 48(4), 102562.

Mahesh, V. M., & Babu, R. (2022) Green initiatives: A study on adaption and use among the Indian Libraries and Information Centers.

Maina, J., & Mwiti, F. M. (2023). Analyzing sustainability efforts in supporting the Green Library Initiative by Kenyan University Libraries. African Journal of Science, Technology and Social Sciences, 2(2), 205-213.

Mwanzu, A. (2022). Green initiatives towards a sustainable future: Insights from libraries in Kenya.

Mwanzu, A., Bosire-Ogechi, E., & Odero, D. (2023). Green initiatives towards environmental sustainability: Insights from libraries in Kenya. IFLA journal, 49(2), 298-314.

Omona, W. (2020). The roles of library and information services in achieving Sustainable Development Goals (SDGs) in Uganda. Library Philosophy and Practice, 1-19.

Patil, M. A. M., & Jani, R. (2024). The green library concept and initiatives in India: A new era. Educational Administration: Theory and Practice, 30(5), 3604-3609.

Patil, S. S. (2018). Green Library Initiatives: An overview. In International Conference on Internet of Things and Current Trends in Libraries, 109-116.

Pinto, L. G., &Ochôa, P. (2018). 4 Public Libraries' Contribution to Sustainable Development Goals. Going Green: Implementing Sustainable Strategies in Libraries Around the World: Buildings, Management, Programmes and Services, 177, 46.

PV, V., Chopade, V., Joy, V., & MK, J. (2024). Designing a green library in alignment with the UN's Sustainable Development Goal: A case study of Rajagiri Business School Library, Kerala, India. Electronic Green Journal, (49).

Sarkar, D. (2020). Sustainable development as a goal: Special reference to green library. Int. J. Res. Libr. Sci, 6(1), 182-189.

Sengupta, S. (2024). Towards sustainable knowledge: Integrating open access principles with green library initiatives. College Libraries, 39(I), 11-19.

Solemanpharcy, M., & Gaffar, S. K. (2023). The role of green library for sustainable development goals and environmental education: a study on global perspective. International Journal of Knowledge Management & Practices, 11(1).

Tseng, S. H. (2008). Green library design and evaluation: the Taipei Public Library, Taiwan. New Library World, 109(7/8), 321-336.

Vijesh, P. V., Varsha, C., Joy, V., & Joseph, M. K. (2024). Designing a green library in alignment with the UN's Sustainable Development Goal: A case study of Rajagiri Business School Library, Kerala, India.

20

Digital Engagement in Libraries: Integrating E-Learning Tools

Srikanth H.G.

Librarian, Soundarya Institute of Management & Science - [SIMS] Soundaryanagar Sidedahalli, Nagasandra Post, Bangalore - 560 073, Karnataka, India
E-mail: srikanth.hg@soundaryainstitutions.in

Abstract

The emergence of e-learning has made libraries essential for providing access to e-resources and promoting lifelong learning. Libraries are undergoing a transformation into multifunctional facilities that facilitate learning, creativity, and community involvement, influenced by significant technological advances. This chapter outlines the key aspects of integrating e-learning tools into e-learning tools and technologies in libraries, including e-learning indicators, their benefits, types, implementation strategies, and future trends. By integrating e-learning tools, libraries not only enhance their educational offerings but also position themselves as vital hubs of digital literacy and innovation. This chapter provides practical insights and strategies for library professionals seeking to integrate e-learning tools into their services, ultimately aiming to enrich user experiences and promote lifelong learning in an increasingly digital world. E-learning resources in libraries are cost-effective, personalized, and tailored to cater to diverse learning needs, promoting inclusivity and community building among users. Additionally, e-learning programs in libraries provide opportunities for professional development for librarians, ensuring their relevance in the digital era.

Keywords: *ICT, Digital resources, e-learning, Professional development, Lifelong learning.*

1. Introduction

In the era of digital technology, libraries have expanded their role beyond their initial role as storage facilities for physical material. Libraries have become essential in providing access to electronic resources and promoting lifelong learning due to the emergence of e-learning. E-learning refers to the use of digital technology to access learning materials outside of a traditional classroom setting. E-learning is emerging as a powerful influence in higher education, with a presence on nearly every campus and a growing number of college and university courses. The environment is characterized by growth and dynamism, where fluidity and change are the social, institutional, and technological norms (Mishra & Gupta 2019).

Technology has revolutionized education, turning it into e-learning and internet-based learning. This evolution provides individuals with the opportunity to have more freedom, convenience, and the ability to work virtually. Libraries must modify their resources and services to meet the emergence of e-learners. Academic libraries have now established connections with web-based services and e-resources, which are moving the profession forward at a rate never before seen. (Ramkumar & Kannan, 2023)

2. E-Learning

In e-learning, technological tools aid the learning process. ICT applications facilitate the learning process. E-learning is a form of education that involves the computer, telecommunications, the internet, and storage technologies (Kumbhar, 2009). E-learning generally includes interactive features that encourage learners to actively participate, including tests, discussions, exercises, and presentations with multimedia. These aspects enhance learners' understanding and recall of the subject.

According to CISCO's definition by Jeevan, e-learning is the over reaching umbrella that encompasses education, information, communication, and training. It is the web-enabled system that makes information and knowledge accessible to those who need it, when they need it—anytime, anywhere. In the present context it is a web-based learning. To put it in a more pedagogical frame, e-learning has the following characteristics.

The quality of e-learning is steadily increasing as technology advances. E-learning systems are incorporating virtual reality (VR), augmented reality (AR), artificial intelligence (AI), and other new technologies to improve the quality and engagement of learning experiences. Still, e-learning presents further barriers, including the need for self-control, possible experiences of privacy, and the assurance of the excellence and reliability of online material.

Basically, e-learning is transforming the educational environment by offering flexible, easily available, and interactive methods for people to acquire knowledge and develop in the digital world.

3. History of e-Learning

The history of eLearning is an engaging exploration of technological progress and teaching principles. Here is a concise overview:

In 1728, Caleb Philips invented distance education. Students received weekly course lessons through mail.

In 1840s, Isaac Pitman used written communication to instruct his students in the art of shorthand, now known as stenography. This approach included dispatching tasks by postal mail and then receiving the finished work for evaluation.

In 1924, Sidney Pressey, a professor at Ohio State University, invented the "Automatic Teacher," which was the first electronic learning device. Sidney Pressey specifically designed this machine to provide multiple-choice questions.

In 1954, Harvard Professor B.F. Skinner created the "GLIDER" teaching machine, which made programmed instruction easier.

In 1960s, Donald L. Bitzer created the PLATO (Programmed Logic for Automated Teaching Operations) system, a revolutionary computer-based educational system.

In 1976, Internet-based courses are introduced. The internet's emergence facilitated the creation of web-based courses, indicating a significant transformation in education delivery.

In 1980s, personal computers changed internet-based information access, skill acquisition, and knowledge sharing, allowing people to study from home.

In 1990s, Following the first use of the term "eLearning" in 1999, Learning Management System (LMS) platforms began to emerge, providing well-structured online learning environments.

In 2000s, MOOCs saw a significant increase in popularity. These courses provided free educational opportunities from prestigious institutions to anyone with internet access.

In 2010s, the rise of smart phones and tablets led to the rise of mobile learning, facilitated by social media and collaborative tools.

In 2020s, AI and data analytics revolutionized eLearning, enhancing personalized experiences and advanced analytics. COVID-19 accelerated this adoption, promoting virtual classrooms and hybrid learning models.

Today, eLearning has become an essential component of education and professional growth, as sophisticated platforms provide interactive and customized learning opportunities.

4. E-Learning in Higher Education

In higher education, e-learning utilizes electronic technology to aid and enhance learning. It facilitates the connection between students and instructors, regardless of their physical division, via the use of tools like computers, the internet, and multimedia. This method provides learners with the ability to access educational content at their convenience and from any location. E-learning encompasses several forms, such as online courses, virtual classrooms, and digital materials, hence enhancing the accessibility and adaptability of education to cater to individual needs.

In addition, e-learning minimizes the need for physical facilities, making it an economically efficient resolution for institutions and allowing them to access a wider student base, both nationally and internationally. Considering the limitations of e-learning, such as the need to ensure student participation and provide sufficient technological support, the advantages of e-learning in higher education are unquestionable, and its acceptance is expected to continue expanding in the future.

5. E-Learning in the Libraries

The use of e-learning in libraries has significantly altered the way in which individuals acquire information and instructional materials. Because of the emergence of digital technology, libraries have evolved from conventional storage spaces for physical books to innovative centres of online education. Libraries now provide a wide range of e-learning materials, such as online courses, tutorials, and educational databases, to accommodate various learning requirements and preferences. Users can often access these materials remotely, allowing them to study at their preferred speed and convenience. Furthermore, libraries provide the opportunity to use digital tools and software, allowing individuals to acquire new abilities and improve their proficiency in digital technology. The integration of e-learning in libraries has also enabled cooperation and dissemination of information among users, creating an awareness of community and advancing lifelong learning. Libraries have strengthened their significance in the digital era by adopting e-learning, thus broadening their scope and securing their continued relevance.

Libraries should promote e-learning programs that teach users how to utilize their resources, such as extended lectures series and user education and training programs. It also outlines the duties and obligations of a librarian in meeting the information requirements of users. E-learning refers to the process of transferring up-to-date information among users and using electronic tools to access educational study materials outside of a typical classroom setting (Asif & Singh, 2018).

6. The Importance of E-Learning in Libraries

Today's education relies on e-learning in libraries because it makes learning accessible and adaptable. The digitization method promotes lifelong learning by allowing individuals to continuously acquire relevant information and skills. E-learning plays a crucial role in libraries for numerous reasons.

- **Enhanced Accessibility:** Libraries offer a wide range of e-learning resources, such as online courses, tutorials, and databases that users can access remotely, thereby enhancing the convenience and accessibility of learning.
- **Enhanced Literacy and Digital Skills:** Libraries can offer e-learning resources that help users develop essential digital skills, such as online research, critical thinking, and digital literacy, which are critical in today's information age.
- **Support for Lifelong Learning:** E-learning in libraries supports lifelong learning, enabling users to acquire new skills and knowledge throughout their lives, whether for personal or professional development.
- **Cost-effective:** E-learning resources in libraries are often free or low-cost, making high-quality educational materials accessible to a wider audience, regardless of their financial means.
- **Personalized Learning:** Libraries can tailor e-learning platforms to individual learning needs, offering users an engaging and effective personalized learning experience.
- **Support for Diverse Learning Needs:** Libraries' e-learning resources can cater to a variety of learning needs, including those of users with disabilities, language barriers, or other challenges, promoting inclusivity and equity in education.
- **Collaboration and Community Building:** E-learning in libraries can facilitate collaboration and community building among users, fostering a sense of belonging and promoting social learning.
- **Professional Development for Librarians:** E-learning resources in libraries also provide librarians with opportunities for professional development, enabling them to stay up-to-date with the latest trends and technologies in education and librarianship.

7. Some of the E-Learning Tools and Technologies in Libraries

Libraries are using a variety of e-learning tools and technology to support ongoing education, improve digital skills, and provide access to excellent educational materials. Below are many essential e-learning tools and technologies often used in libraries:

7.1. Learning Management Systems (LMS)

- **Moodle:** It is a free and openly accessible LMS that is widely utilized by educators, administrators, and learners worldwide. The platform offers a flexible, secured, and integrated system for developing customized learning environments. Moodle is a powerful application for blended learning, online education, and other e-learning initiatives. It offers a range of features, including forums, resource sharing, assignments, and grading.
- **Canvas:** It is a stable and flexible LMS used by educational institutions, business organizations, to manage and deliver online courses. The platform provides a wide range of functionalities, including course development, content administration, grading, and communication tools. The application's simple interface and full support make it a preferred choice among both educators and students worldwide.
- **Blackboard**: It is a very popular LMS that offers an excellent foundation for educators to develop, oversee, and deliver online courses. The platform provides a variety of tools and features, such as course material management, exams, grading, and communication tools that enhance an interactive and captivating learning experience. The tool's simple interface and extensive support make it a preferred choice among educational institutions and organizations across the country.

7.2. Massive Open Online Courses (MOOCs)

- **Coursera:** It was created in 2012 by Stanford University academics Andrew Ng and Daphne Koller. It is a worldwide internet-based educational platform that provides a wide range of over 7,000 courses certificates, and degrees in several fields of study. Coursera offers a wide range of learning options for people to enhance their jobs or acquire new skills, with more than 300 colleges and business organizations involved. This makes education accessible, influential, and flexible while still being cost-effective.
- **edX:** Harvard University and MIT created it in 2012 as an online educational platform that offers a wide range of courses, from free options to full-degree programs. It collaborates with more than 90 leading universities and organizations. Offering about 3,000 courses, the objective of this initiative is to ensure that excellent education is within reach for all individuals, irrespective of their geographical location or financial limitations.

- **Khan Academy:** Sal Khan founded this non-profit educational platform in 2006, offering free, high-quality instruction across a wide range of disciplines. The platform offers instructional videos, interactive activities, and customized learning dashboards for learners of every age. Khan Academy further assists teachers and parents by offering resources to monitor student progress and customize educational experiences.

7.3. Gamification and Interactive Tools

- **Kahoot!**: It is a free educational platform that utilizes gaming elements to provide an enjoyable learning experience for any topic, in any language, and on any device. Kahoot! offers a variety of features and tools that enable educators, students, and professionals to design and participate in educational games, quizzes, and surveys. Additionally, it provides a comprehensive collection of materials, such as manuals, templates, and webinars, to facilitate learning and enhance professional growth. In addition, Kahoot! offers solutions tailored for educational institutions, organizations, and people. These include Kahoot! EDU, Kahoot! 360, and Kahoot!+, which give enhanced features and tools for teaching, learning, and evaluation.
- **SMART Boards:** Interactive whiteboards that integrate the user-friendly nature of a conventional whiteboard with the computational capabilities of a computer. They facilitate the participation of instructors and students in interactive classes, presentations, and activities, thereby enhancing the engagement and effectiveness of learning. Digital tools such as ink, shapes, and multimedia resources can seamlessly connect SMART Boards with a range of instructional applications, facilitating collaboration.

7.4. Virtual Reality (VR) and Augmented Reality (AR)

- **Google Expeditions:** It is a technological platform that integrates VR and AR to enhance educational experiences. Teachers may use this technology to provide students with immersive virtual experiences, such as exploring coral reefs, the surface of Mars, and historical sites. By using VR headsets such as Google Cardboard, students have the opportunity to immerse themselves in 360° photospheres and 3D visuals, enhancing the interactive and captivating nature of the learning experience.
- **ARKit:** Apple developed a smart augmented reality framework that seamlessly integrates virtual items into the physical environment of iOS devices. Introduced in 2017, this technology utilizes motion sensors, camera scene capture, and scene processing to provide captivating and

immersive experiences. The primary features include motion tracking, light estimation, people occlusion, 4K video recording, and a depth API using LiDAR Scanner.

8. E-learning has Impacted both Users and Libraries in Various Ways

8.1. Impact on Users

- Accessibility and Flexibility: E-learning enables users to easily access educational materials at any time and from any location, enhancing learning flexibility and accommodating diverse schedules.
- Personalized Learning: Users have the ability to customize their learning experiences according to their own speed and preferences, often using interactive and multimedia materials that improve understanding.
- Cost-effectiveness: E-learning helps to reduce the costs associated with traditional education, such as transportation and lodging.
- Skill Development: It provides opportunities for ongoing professional development and skill improvement, which is particularly beneficial for those who are committed to lifelong learning.

8.2. Impact on Libraries

- Digital Transformation: Libraries have progressively embraced digital technology to facilitate e-learning, establishing digital libraries that include extensive collections of learning resources.
- Enhanced Services: The use of e-learning technology has facilitated libraries' provision of user-centric services, including online tutorials, virtual reference services, and digital lending.
- Challenges: Libraries face obstacles such as the need for skilled personnel, adequate IT infrastructure, and competent leadership to deploy and oversee e-learning technology.
- Role in Lifelong Learning: Libraries are essential for facilitating lifelong learning by providing e-learning materials and assistance in community education programs.

9. Some of the E-Learning Initiatives

- **Digital Library Initiatives at the National Digital Library of India (NDLI):** NDLI is an online collection of educational materials, providing an easily accessible search engine for students, researchers, and everyone interested in continuous learning. The platform integrates information from both domestic and foreign origins, including literature, scholarly papers, dissertations, multimedia lectures, and Open Educational

Resources (OERs). The Indian Institute of Technology Kharagpur is responsible for the development, operation, and maintenance of NDLI. The platform offers unrestricted access to more than 100 million pieces of information in 39 different Indian languages, out of a total of 423 languages.

- **National Programme on Technology Enhanced Learning (NPTEL):** NPTEL is a collaborative effort between the Indian Institutes of Technology (IITs) and the Indian Institute of Science (IISc) with the objective of delivering outstanding learning via online courses. Established in 2003, NPTEL offers a diverse selection of courses in fields including engineering, science, management, and the humanities. NPTEL offers the courses in video and online formats, making them accessible to a diverse range of individuals. NPTEL aims to enhance the learning process by providing comprehensive course materials, assignments, and certification options, thereby bridging the gap between traditional classroom education and modern e-learning.
- **E-Learning in School Libraries: A Case Study from the CBSE Schools:** Implementing e-learning in school libraries is essential for modern education, especially for CBSE schools. These platforms provide 24/7 access to digital materials such as e-books, online tutorials, and instructional videos. They promote collaborative learning and cultivate a sense of community. In addition, they provide educators with resources to develop captivating instructional strategies, thereby enhancing student motivation. Popular e-learning platforms in CBSE schools include Byju's, myCBSEGuide, and Vedantu, which offer a range of digital resources to support student learning.
- **The Public Library of Cincinnati and Hamilton County's Virtual Learning Initiatives:** To promote lifelong learning and digital literacy, the Public Library of Cincinnati and Hamilton County provides many virtual learning programs. These include Professional Online High School, which lets adults get a high school diploma and professional certificate, and Creative bug, which offers over 1,000 arts and crafts video courses. The library also provides LinkedIn Learning for software, creative, and business skills; North star Digital Literacy for fundamental computer and internet abilities; and Universal Class for over 500 online continuing education programs. These efforts allow all ages and ability levels to learn and improve.

10. Challenges and Solutions E-Learning in Libraries

Libraries now consider e-learning to be an essential component of their services since it allows users to easily access a variety of online materials, courses, and tutorials. However, the process of setting up and maintaining e-learning initiatives in libraries might present difficulties. Here are a few common challenges and possible solutions:

10.1. Challenges

- Ensuring that library computers and devices are compatible with e-learning platforms and have sufficient bandwidth to support online learning.
- Patrons may have insufficient digital proficiency to effectively use e-learning materials.
- Ensuring that those with impairments can easily use e-learning materials is known as accessibility.
- Securing enough financing to support e-learning projects is a priority in terms of funding and budget.
- Ensuring that library personnel possess extensive knowledge of e-learning resources and are capable of providing efficient assistance to library users is crucial.
- Evaluation and assessment include quantifying the efficacy and influence of e-learning efforts.
- Establishing collaborations with other organizations to enhance the scope and possibilities of e-learning.

10.2. Solutions

- Ensure that e-learning activities are in line with the library's objectives and goals, and prioritize the needs and preferences of library patrons.
- Ensure that library personnel have the requisite skills and expertise to effectively support e-learning projects.
- Choose systems that have user-friendly interfaces, are widely available and are compatible with library technology.
- Create targeted marketing efforts and promotions to raise awareness of e-learning materials and encourage consumer usage.
- Consistently evaluate the efficiency and influence of e-learning efforts and use data-driven judgments to improve offerings.
- Establish collaborations with other organizations to enhance the range and scope of e-learning programs and possibilities.

- Keep up-to-date with current trends and technology in e-learning, and adjust library services accordingly.

11. Future of E-Learning

The evolution of eLearning in libraries is gradually advancing, with present trends highlighting accessible, collaborative, and engaging varieties. Libraries extensively use AR, VR, and gamification to maintain learners' engagement. Libraries are undergoing a transformation into multifunctional facilities that facilitate learning, creativity, and community involvement, influenced by significant technological advances.

Several key trends that are shaping the future of e-learning in libraries include:

- **Artificial Intelligence:** Libraries are using AI to enhance the educational experience. Learners now receive customized assistance from AI-powered chat bots, and AI-driven adaptive learning systems help tailor learning experiences to individual needs.
- **Augmented & Virtual Reality:** AR and VR are transforming eLearning through the use of realistic 360-degree images, graphical overlays, and an interactive interface. This technology enables educators to completely engage learners in many disciplines, such as mathematics, physics, history, and literature. Due to ongoing technological advancements, the eLearning trend will probably continue for an extended period.
- **Gamification:** Libraries are using gamification to enhance the engagement and enjoyment of studying. Libraries are promoting student engagement by integrating game design aspects into educational experiences, thereby encouraging learners to actively participate in their own learning process.
- **Micro Learning:** Micro learning provides clear and focused learning modules that learners can use at their convenience to acquire specific knowledge. This technique improves the retention of information and enables learners to promptly apply newly acquired abilities, which is highly beneficial for professionals.
- **Collaborative Learning:** Libraries are seeing a growing trend toward collaborative learning. This entails learners collaborating to achieve a shared objective using technology such as video conferencing and online collaboration tools.

12. Conclusion

E-learning, the use of digital technology to access learning materials outside of traditional classroom settings, has become a powerful influence in higher education. It involves computer, telecommunications, the internet, and storage technologies, with interactive features such as tests, discussions, exercises, and multimedia presentations. The quality of e-learning is steadily increasing as technology advances, incorporating virtual reality, augmented reality, and artificial intelligence. E-learning in higher education uses electronic technology to enhance learning, facilitating connections between students and instructors, and minimizing the need for physical facilities. Libraries have evolved from traditional storage spaces to innovative centers of online education, providing a wide range of e-learning materials, such as online courses, tutorials, and educational databases. Libraries should promote e-learning programs that teach users how to utilize their resources and outline their duties and obligations as librarians. E-learning is expected to continue expanding in the future, enhancing the educational environment and promoting lifelong learning.

The integration of e-learning into library services underscores the importance of libraries in modern education. By providing cost-effective and inclusive learning opportunities, libraries support diverse learning needs and contribute to the professional development of both users and librarians. As technology continues to evolve, the role of libraries in e-learning is expected to grow, further solidifying their position as essential hubs of knowledge and learning in the 21st century.

References

Abumandour, E.S.T. (2021). Public libraries' role in supporting e-learning and spreading lifelong education: a case study. Journal of Research in Innovative Teaching & Learning, 14(2), 178-217. https://doi.org/10.1108/JRIT-06-2019-0063

Asif, M., & Singh, K. K. (2018). E-learning, e-searching and e-resource management in the libraries. IP Indian Journal of Library Science and Information Technology, January-June, 3(1), 16-18.

Bajpai, R. P., Hada, K. S., & Bajpai, G. (2016). Academic libraries and e-learning: Initiative and opportunities. International Journal of Library & Information Science (IJLIS), 5(3), 159-165.

Deshpande, B.S., & Sajana, C. (2021). School students' computer skills and the use of online information resources: A case study of CBSE students of Mysore. Library Philosophy and Practice (e-journal). 5052.

Dhamdhere, Sangeeta. (2012). E-learning: Future trends and challenges to academic libraries. International Journal of Information Technology & Computer Science (IJITCS), 2.

Eke-Okpala, Helen. (2010). The perspective of e-learning and libraries in Africa: Challenges and opportunities. Library Review, 59, 274-290. 10.1108/00242531011038587.

E-Student - #1 Source for Online Learning Reviews & Analysis. (2023, November 21). E-Student. https://e-student.org/

Imed Bouchrika. (2024). History of eLearning: Evolution from stenography to modern 2024 LMS platforms. Research.com. Retrieved August 4, 2024, from https://research.com/education/history-of-elearning

Jeevan, V.K.J. Computers @libraries. ESS Publications, New Delhi, 2006, 111.

KS, Abu, & Balasubramani, R. (2019). Developing a web-based e-learning model for Library and Information Science Candidates in India. Library Philosophy and Practice (e-journal), 2202.

Kumbhar, Rajendra (2009). Use of E-learning in library and information science education. DESIDOC Journal of Library & Information Technology, 29, 37-41. 10.14429/djlit.29.228.

Mishra, & Gupta. (2019). E-Learning impact on technological perspectives in library services. Rejuvenating Libraries in Modern Society. In Rejuvenating Libraries in Modern Society (p. 163).

National Digital Library of India (NDLI) (n.d.). National Digital Library of India. Retrieved August 4, 2024, from https://ndl.iitkgp.ac.in/

Ramkumar, & Kannan. (2023). Optimizing e-learning through social media and e-learning tools: Library at the base. Journal of Nonlinear Analysis and Optimization, 14(2).

S, Sivankalai (2021). Academic libraries support e-learning and lifelong learning: A case study. Library Philosophy and Practice (e-journal), 6154.

Shahzad, K., & Khan, S.A. (2023). Effects of e-learning technologies on university librarians and libraries: A systematic literature review. The Electronic Library, 41(4), 528-554. https://doi.org/10.1108/EL-04-2023-0076

Velumani, K. V. (2013). An investigation into the impact of e-resources in modern library and information centers. Retrieved from: http://hdl.handle.net/10603/54204

21

Impact of Village Library for Community Upliftment

Jitamoni Bhattacharyya

Librarian, Biswanath College, Biswanath - 784 176, Assam, India, (Former guest faculty member of Biswanath Agriculture College) E-mail: jitamoni096@gmail.com

Abstract

Village library plays a basic role in raising information literacy among the village community. Village library is one of the effective social institutions that village community improves their skills as well as improves their quality of life and participated in knowledge base society. Information literacy movement is carried out by the help of village library. In this chapter I would like to indicate some major point that barriers of developmental work in village library and at the same time I mention some remarkable work doing by village library mainly teaching-learning process. Assist village community to go ahead by supporting proper document on need baed.

Keyword: *Village library, Village community, Public library, ICT.*

1. Introduction

No doubt Village library is a part of public library. Village library plays an important role for creation knowledge in different field. As we know that maximum people in the village area not aware about their right, even they not know reading and writing. so the rate of literacy is poor. Therefore it can be mentioned that the illiteracy is the main barrier for their development and it will harm their day to day life. Many people in the villages is involve themselves in farming. Agriculture is the main of earning sources. Villagers do not have adequate knowledge in farming and fertilizer that they used for better result even they not know the bad affect of fertilizer that they used. Their life style is not hygienic because of illiteracy.

The importance of village library is one of the effective institutions that village community improved their quality of life and participates in knowledge base society. In rural community get more benefited by the help of rural library services. As library is the institution to spread information, knowledge among the rural community. Village library may afford to improve the literacy rate in rural areas. Information literacy movement is carried out in village areas by the effort of village library services.

Village library can be mentioned that the centre of community development without any barriers such as caste, rich-poor, and all community people can take facility of village library services. All are have equal right to visit the library and acquire information and increase knowledge on their respective areas. It is the service information centre.UNESCO mentioned that every people have the right of literacy skill and it is the basic study for every people which is connected for whole life progress.

Now a day the concept of library has been change. In oldest time people visited to the library and collect books only but now a days this concept totally change. From the oldest period (civilization) to now, the digital environment is able to change the old theme and adopt new ideas new technology. Really, the current century make a drastic change in educational environment and human behavior.

In this age (21st century), where information communication technology has change the picture and has great impact on every human life. It helps human being updated and informative. Village library services and new technology have change the life cycle of human being .Make people knowledgeable one. Village library services and technology support the villagers to increase their knowledge and ideas that they utilized for better result. Original manual technology now changes in to digital technology. For that villagers can know the latest development and adopt strategy for their development and live better way. Village library services provide recent information among its user and the same time collect fruitful information by the use of computer and communication technology. The original manual technology now change in to digital technology

Every citizen needs better quality services provide by village library. Village library adopt various plane and programmed for community users and make them aware about recent development. Now the library is mentioned as knowledge resource centre as the main goal is to promote literacy rate. Village library support village community to collect information and gain knowledge in their won field. Villagers now aware about their right and raise voice against their right. As library is the place here people increase their knowledge and take proper decision in the path of their journey. Library may be the public, may be the academic and may be the special, no problem but main objectives are to promote knowledge among its users group. There is no any barrier all are equal in the eyes of library users'. Village library always tries to create positive environment for the interest of community. Library help to promote digital literacy among the community users as well as innovative ideas, facilitation of all impartial approach to gain knowledge and information in all segments of the people in the society. Village library has playing a crucial role in providing Books, journal, magazine, newspaper, Govt.publications and some others. But in this digital environment village library provided and serve electronic databases-books, e-journals etc.all provide instruction, guidance to simple villagers. For that illiterate people are more benefited and they apply proper method in their field.

2. Some Basic Point Regarding Village Library are Mention as Below

- Services offered for general public, mainly village community having motto to improve the information literacy. increase the rate of literacy
- Various activities such as reader club are open for public to increase their knowledge.
- Increase the reading habit among the children, youth and support them by sharing information.

If people are interested to reading and writing from his childhood, sure they will be the good citizen of the country. Village library open a special division for children and keep material which is effective for them.

- Trying to remove illiteracy and increase the rate of literacy
- Make every segment of people independent reader and motivate them to good educator. So that they stand as good citizen of the society as well as nation.
- Increase sustained multilingual capabilities of children
- The role of little children literacy taken by the village library is a remarkable service and contribution to the society. So that society will be stand as knowledgeable society

3. Rules and Regulation

1. Noise disturbance and unexpected behavior is not allowed within the library premises
2. Proper use of library material as well as well maintenance.
3. Smoking is prohibited in the library campus.
4. Silence should be observed within the library for that many library hanging a notice board where mentioned that "keep silence"

Library is the place of satisfaction, place of relaxation etc.there people visite and makes themselves relaxed also collect valuable information and store knowledge.

- Library is the place where fake information prohibited. Now a day's many fake information published but library is the institution that make people aware about the fake news.
- Village library create social media such as whattsapp, telegram, facebook for spread information among the people. So that public can take proper decision.

4. Basic Element for Village Library

Village library is a branch of public library provides fruitful services among the village community. No doubt it is a noble intuition and noble profession. Some basic elements for village library are:

- Infrastructure
- Computer facility with internet connection
- Books, magazine, newspaper, *etc.*
- Library supported software, *etc.*
- Readers
- Library staff having proper qualification.

Basic areas of village library are:

- Circulation desk
- Reference room
- Computer room with internet connection
- Office room
- Study area i.e. reading facility for reader
- Children reading room
- Playing areas for children
- Book selves area
- Magazine, periodical
- Store room
- Sitting area
- Librarians room, *etc.*

Bibliographic control of material is very important not only for village library but for all libraries. On the other hand, village library staff should create a good environment and a good relation with community and trying to fulfillment of villagers' expectation.

5. Is village Library is a Learning Centre for Village Communities?

Proper infrastructure with proper electronic media village library is one of the most effective institutions that plays a very crucial role in upliftment of villagers and decrease the rate of literacy. Village library supported to teaching learning process so that people acquire knowledge in various field. Maximum people live in village area are engaged themselves on agricultural sector. Maximum are not qualified so they not aware their proper right and their duty to country. Illiteracy is the main barrier for development. People in different segment visited to the library and collect information without pent money. That's why villagers get benefited to know the current information. Village library is known as learning centre because library collect information and stored for future, it is the centre that capture the history of past present and future .It is the carrier of civilization. From the past to present there are many changes found in different fields such as cultural,educational,sports and some others. Our children know about the past event.

Abdul Mukhosis (2021) stated that, "This training is carried out to train the community's skills in the field of entrepreneurship, as well as provide facilities in the development of community skills, especially those in the vicinity of the library, from the manual screen-printing training activity, many of the training participants now have a screen-printing business, such as for product packaging. Some make rice sacks from jute or plastic materials; some make manual screen-printing on soy milk packaging".

As we know that information is one of the requirements for every people. But the people who live in the remote area are far away from attaining of their basic needs. To improve their quality of life style and give support to participate knowledge based society, there is a need of village library. That's why village library plays a remarkable role to improve the quality life of the villagers. On the other hand village library support to activities that performed by simple villagers and uplift for rural community.

Village library creating the reading habit among the village community

Readers Club: village library may create reader club for increasing the value of education. Various activities can take for the benefited of the readers. Village library can take a part of adult education. The main aim is that-increase the rate of literacy.

6. Village Library Face some Barriers to Perform their Duties Such as

1. Shortage of manpower is the main barrier for effective service.
2. Maximum village library in India not have proper qualified person.
3. Main and core element for development of village library is cost. Financially village library is very weak. Infrastructure is not good
4. No enough material such as books, newspaper and magazine etc
5. Not properly implement of electronic device due to lack of technical person.
6. Village library grant provide by Government is very little.

7. Suggestion for Development of Village Library

- Government should release fund from time to time for village library development as well as village community upliftment.
- All the state in India not implemented public library act till now. Act should be implemented for the upliftment of public libraries as well as village libraries.
- Computer with internet connection required for quick information distributed among its user groups.
- User awareness camp/fruitful talk should be arranged for better result.

- Talk with user friendly and take feedback from the users. For that library meet up the demand of the user.
- Sometime arrange quiz, speech competition among the village community, *etc*.

8. Conclusion

Village library always supported to village people to improve their skill as well as increase information literacy rate among the society. Village library is the source of all information/data to rural India. Limited money librarian in village library has to manage all necessary services, teaching skills to its children as well as villagers and make plane for proper utilization of limited money and better services to its users. Now it is a great challenge for librarian to manage all from manual to digital. It has been notice that lack of proper library building in most of rural library in India, with the help of NGO village library can provide better services to its village community. No doubt, library is the only centre that provide best quality services to its reader without any cost, some little amount are taken from the user for enrollment fee.

References

Abiolu, O.A., & Okere, O.O. (2012). Environmental literacy and the emerging roles of information professionals in developing economies. IFLA Journal, 38(1), 53-58.

Akanwa, P.C. (2013). Public library services to children in rural areas. Library Philosophy and Practice (e-Journal), 1–16.

Bhatt, R.K. (2011). Libraries in India. New Delhi: Ane Books.

Bhattacharyya, J. (2019). Library and information science: Diversity of librarianship. Rakesh book service publishers and distributors.

Biradar, B.S., & Kumar, D.P. (2008). Community information needs: A case study of Holehonnur Hobli. SRELS Journal of Information Management, 45(2), 225-232.

Cahyan I.,Titi, & Wicaksono, M.F. (2021). The role of rural library in community empowerment (Leshutama Library Case Study). Library Philosophy and Practice (e-journal).

Fourie, I., & Meyer, A. (2016). Role of libraries in developing an informed and educated nation. Library Hi Tech, 34(3), 422-431.

Kumar, S. (2006). A journey of rural library movements in India: Retrospect and Prospect. SRELS journal of Information Management, 43(3), 295-303.

Nielsen, G.N., & Borlund, P. (2011). Information literacy, learning, and the public library: A study of Danish high school students. Journal of Librarianship and Information Science, 43(2), 106-119.

Samsuddin, S.F., Omar, S.Z., & Shaffril, H.A.M. (2018). Youth development in rural library: ICT gratification as mediating effect. Malaysian Journal of Library and Information Science, 23(2), 111–133. https://doi.org/10.22452/mjlis.vol23no2.7

Sharma, A.K. (2008). Information environment in rural India: Impediments to free flow of Information. Library Herald, 46(1), 42-47.

Suman Rani (2018). The concept of rural library: An overview. JETIR, 5(5), 592-595.

Vashishth, C.P. (2007). Rural information initiatives in India. Library Herald, 45(3), 191-202.

22

Unveiling the Future: Integrating Advanced Marketing Strategies to Foster Inclusivity and Accessibility in Academic Library Services

Dr. Rajinder Kaur Gill

Librarian, Arya Kanya Mahavidyalya, Shahabad Markanda, Kurukshetra - 136 135 Haryana, India, Email: rajinder18gill@gmail.com

Abstract

Libraries, often hailed as the treasure of wisdom, have been instrumental in the acquisition, preservation, and dissemination of knowledge for centuries. With the information explosion, technological revolution, and escalating costs, academic libraries have been expanding their collections to include numerous digital resources alongside print materials and adopting professional marketing approaches to promote their resources and services. Marketing information products and services has become essential for the optimal utilization of libraries, bringing significant changes to traditional practices in library acquisition, organization, processing, storage, retrieval, and reproduction of information. The primary aim of marketing is to create awareness among users regarding library resources and services, ensuring their satisfaction. Libraries now use various media, including print, digital, and human media, to inform users about their resources and services. This paper defines the need, concept, and importance of marketing and promotional activities in libraries. It explores various marketing tools and techniques, including planning, segmentation, and market mix within the field of libraries. Additionally, this paper delves into the integration of advanced marketing techniques in academic libraries to address the growing demand for inclusivity and accessibility. By examining current trends and technological advancements, this study aims to provide a comprehensive framework for libraries seeking to improve their outreach, engagement, and service delivery to all members of the academic community.

Keywords: *Academic libraries, Library marketing, Users' engagement, Technological advancements, Marketing strategies, Outreach efforts.*

1. Introduction

Academic libraries have traditionally been the cornerstones of academic institutions, serving as repositories of knowledge and centers for learning. In the past, libraries were just regarded as physical repositories of books and other reading material, and their role was to acquire, organize, process, store, and provide access to that physical collection. These traditional libraries were characterized by their emphasis on the preservation of physical collections and the provision of a quiet, contemplative space for study and research. However, the role of libraries has expanded significantly in recent years beyond physical collections to include digital resources such as e-books, online journals, databases, and multimedia content. Modern academic libraries are dynamic and multifaceted, providing not only access to information but also a range of services that support teaching, learning, and research. They have become vibrant hubs of activity, offering collaborative spaces, technology-rich environments, and various programs and workshops that enhance the academic experience. The focus has shifted from merely preserving information to facilitating access, promoting information literacy, and supporting the diverse needs of the academic community.

In this fast-paced and information-rich environment, libraries are embracing marketing strategies that have become crucial for libraries to stay relevant and effective. Marketing in libraries involves understanding user needs, creating value propositions, and effectively communicating the library's offerings to its target audience. By adopting data-driven approaches, leveraging digital platforms, and implementing innovative service models, libraries can enhance their outreach efforts and better serve their diverse user base. Marketing helps libraries not only attract more users but also build strong relationships with them, fostering a sense of community and loyalty. It allows libraries to highlight their resources and services, ensuring they are well-used and valued by the academic community. This approach enhances user satisfaction and reinforces the library's position as an essential and integral part of the academic institution.

2. Concept of Marketing

The concept of marketing for non-profit organizations was first introduced by Kotler and Levy in 1969, and later elaborated by Kotler in his 1982 book, "Marketing for Organizations." According to Kotler, non-profit organizations like libraries focus on the production of services rather than goods. In simple terms, marketing for libraries means popularizing the information products and services they offer to users, and involves the exchange of information with other libraries or sister organizations. Marketing of information is a relatively new concept for libraries and information centers, aimed at promoting emerging library products and services. This concept has been transferred from business to public service organizations like libraries. In recent years, marketing has become crucial for libraries due to reasons such as self-protection, competition, increasing customer expectations, and wide access to information. Effective marketing helps libraries improve their image, attract more users, and enhances the reputation of librarians within society. It involves all resources and activities necessary to direct and facilitate the flow of information products and services from producer to consumer. In nutshell, marketing in libraries involves understanding the needs and preferences of patrons and designing services and programs to meet those needs, thereby building strong customer relationships.

3. Definitions of Marketing

Marketing, in general, involves identifying the needs and wants of a target audience and then providing solutions to meet those needs through products or services. It encompasses various activities such as market research, product development, promotion, sales, and customer service. The goal of marketing is to attract and retain customers by delivering value and building strong customer relationships. Various authors have defined the marketing as:

According to the American Marketing Association (AMA) Board of Directors, "Marketing is the activity, set of institutions, and processes for creating, communicating, delivering, and exchanging offerings that have value for customers, clients, partners, and society at large."

Dr. Philip Kotler defines marketing as "the analysis, planning, implementation, and control of carefully formulated programs designed to bring about voluntary exchanges of values with target markets for the purpose of achieving organizational objectives. It relies heavily on designing the organization's offerings in terms of target market needs and desires, and on using effective pricing, communication, and distribution to inform, motivate, and serve the markets."

Mark Burgess describes marketing as "the process by which a firm profitably translates customer needs into revenue."

Folk and Lancaster define marketing as "a systematic process which is a combination of planning, concentrating on customer needs and their satisfaction."

Based on these definitions, we can summarize that marketing is a process that focuses on:

- Understanding users' information needs.
- Providing excellent customer service.
- Ensuring customer satisfaction as the prime objective.
- Aiming to reduce barriers to access and use of information.
- Committing to help users acquire the skills needed to obtain information.
- Identifying the users' market segments and targets.
- Recognizing the organization's strengths as a competitive business.

4. Literature Review

Agyei and Ngulube (2022) examined the marketing strategies employed by selected academic libraries in Ghana. They used a purposive sampling technique to conduct face-to-face interviews with 15 librarians from four academic libraries. The findings revealed that none of the libraries had designated departments solely responsible for marketing their services, nor did they have dedicated budgets for marketing initiatives. Instead, all the libraries relied on informal marketing strategies. This lack of formalized marketing efforts highlights a significant gap in how libraries promote their resources and engage with their users. The study recommends adequately resourcing marketing units within libraries to improve user engagement and justify library expenditures. By adopting more structured and well-resourced marketing strategies, libraries can better communicate their value and services to users.

Uwandu and Osuji (2022) examined the strategies used by librarians to market information resources and library services at the Federal University of Technology, Owerri, and Imo State University, Owerri. They conducted a survey involving 62 librarians from these university libraries. The study revealed that a variety of information resources are available for marketing, including journals, books, newspapers, conference proceedings, e-books, e-journals, e-newspapers, electronic databases, CD-ROM databases, e-theses and dissertations, and internet resources. In addition to these resources, the study identified several library services available for marketing. These services include reference services, circulation services, abstracting and indexing services, Current Awareness Services (CAS), Selective Dissemination of Information (SDI), referral services, inter-library cooperation, internet services, document delivery services, and consultancy services. The study recommended that librarians should focus on employing effective tools and techniques for marketing the resources and services to enhance their utilization and meet user needs effectively.

Maturure and Rakemane (2021) emphasized the critical role of marketing in libraries and information centers, noting a significant evolution from traditional methods to the integration of social media tools. The authors explained that historically, libraries utilized limited approaches to market their collections and services, but in recent years, social media has emerged as a promising and vital channel for this purpose. The study underscored the importance of marketing library services and collections, especially in the dynamic environment of the 21st century. Social media platforms like Facebook, Twitter, Instagram, and LinkedIn are highlighted as effective tools for engaging users and promoting library resources. These platforms offer libraries the opportunity to reach a broader audience, facilitate interactive communication, and provide timely updates on new acquisitions, events, and services.

Agboke and Effiong (2020) explored marketing and promotional techniques within libraries, focusing on creating an environment conducive to accessibility, organizing print and digital materials, and informing users about available resources. This multifaceted approach enhances user engagement and optimizes resource utilization. The strategies employed include producing promotional materials like brochures, flyers, and emails, and maintaining websites. Additionally, libraries compile reading lists, bibliographies, newsletters, and posters to inform and attract users. Special programs and events such as library orientations, workshops, and seminars also play a vital role in promoting library resources. These initiatives not only increase user awareness but also justify the funds expended on the library and enable it to compete with other information stakeholders.

Joshua and Michael (2020) investigated the marketing techniques used to promote library services and resources at the Mustapha Abba Library, Federal College of Education Yola. The study surveyed a population of thirty-six librarians, with twenty-nine participating. The main objectives of study is to focus on need of marketing library services and resources, identify the types of services and resources available for marketing, and examine the current marketing techniques employed by the library. The study examined that most significant reason for marketing library services is to meet the users' information needs. The study highlighted several library services for marketing, including reference services, indexing and abstracting services, and internet access. It also mentioned resources like books, journals, and newspapers being marketed. The findings indicated that the most widely used marketing technique is electronic access to information, followed by exhibitions and displays of newly arrived materials. The present study underscored the importance of effective marketing strategies in enhancing the visibility and utilization of library services and resources, ultimately contributing to the achievement of the library's goals.

Adegoke (2015) discussed the importance of marketing and promotion in academic libraries for attracting and retaining users. His study presented a comprehensive background and current scenario of marketing strategies used in the Abdullahi Fodiyo Library Complex, Sokoto-Nigeria. The research shows that the library employs a variety of marketing strategies, such as publicity advertisement, exhibition and display, library orientation, and public relations to draw more users and encourage them to stay longer in the library. The study concluded that librarians must focus on effective marketing efforts to attract more users to the library.

Kumar *et al.* (2015) conducted a survey to assess the marketing of information products and services at the Central Scientific Instruments Organization under the Council of Scientific and Industrial Research across all disciplines of CSIO, Chandigarh. The findings revealed that most users utilized the library for education, research, knowledge updating, and entertainment. The study suggested that library services and products should be increasingly digitized to reach a wider audience at lower costs and with fewer environmental impacts, enabling faster and broader online access. It also recommended the development of appropriate collections and services to better serve potential users.

Abdul Salami (2013) investigated the marketing of library and information services in six geographical zones in Nigerian Polytechnics. Using a survey method, data was gathered through questionnaires, observations and documentary sources. The sample included librarians and students, selected using purposive sampling techniques. Analysis of the collected data, using frequency and percentages, revealed that the available information resources in most polytechnic libraries included books, journals, pamphlets, and ICT resources. Library orientation and current awareness services were the primary methods used to market these resources. Recommendations from the study included annual displays and exhibitions to attract users to library services.

Bamigbloa (2013) presented a survey of six university libraries in Nigeria including two private and four public libraries. The study examined the seven marketing mix elements and marketing strategies related to library services particularly digital information services. These services and products are available through both open and secured access. The study found that these libraries offer a range of free digital information services and products, such as CD-ROMs, Internet services, OPAC, e-resources, e-books, e-journals institutional repositories, online reference services and computer laboratories. Various marketing strategies, such as library orientation, demonstrations training programs, emails, and library websites, were employed to market these services and justify the expenses. The study also noted that while some services are free, a nominal fee is charged not to recover costs but to maintain the facilities.

Vij (2012) reviewed studies discussing the transformation of libraries in the information society, likening libraries to a service industry where user satisfaction is paramount. The author addressed the marketing of library products and services in India, noting that the pressure from advancing technologies, rising user expectations, high maintenance costs, and decreasing budgets has significantly increased the need for marketing library products and services. Strategic planning is necessary for the marketing process, with the major findings emphasizing activities based on techniques, overcoming obstacles, and the requirement for effective professional skills in marketing library products and services in India.

5. Why Library Marketing

Marketing is crucial for libraries to adapt to the evolving landscape of information access and user expectations. With increasing competition from digital resources and other information providers, libraries have been adopting marketing strategies to promote their unique services and resources to remain relevant. Library survival is dependent on the acceptance of marketing as a fundamental management philosophy (Muema Kavulya, 2004). Social and technological changes of our age are a reason for the need to focus on the customer and argue that marketing and planning are a natural partnership (Budd, 1997). Effective marketing helps libraries raise awareness about their offerings, attract new users, and retain existing ones by highlighting the value and benefits of library services. It also supports the library's mission of community engagement and lifelong learning by ensuring that patrons are informed about available programs, workshops, and events. Library professionals have begun to realize that marketing is an integral part of administration, especially as a means for improving user satisfaction and promoting the use of services by current and potential users due to the information explosion, technology revolution, and escalating library costs (Schmidt, 2006). In a nutshell, marketing empowers libraries to better serve their communities, fulfill their educational roles, and secure their place in an increasingly digital world.

Based on the above philosophy, there are some key points explaining why library marketing is needed:

- **Increasing Awareness**: Many potential users are unaware of the full range of services and resources that libraries offer. Marketing helps to raise awareness among users about various resources and services.

- **Supporting Academic Success**: By promoting resources such as research databases, academic journals, study spaces and workshops, libraries can directly support the academic success of students and faculty.
- **Enhancing User Experience**: Through marketing, libraries can gather feedback and insights about user preferences and needs. This information allows libraries to improve user experience and satisfaction.
- **Promoting Lifelong Learning**: Libraries are key players in promoting lifelong learning. Marketing initiatives can highlight educational programs, workshops, and resources that support personal and professional development.
- **Encourage Usage and Maximizing Resources**: By promoting the full spectrum of available resources and services, libraries can ensure that their collections, programs, and facilities are fully utilized.
- **Enhancing Research Impact**: Academic libraries provide access to a wealth of research materials and tools. Effective marketing of these resources can enhance the research impact of faculty and students.
- **Encouraging Information Literacy**: Academic libraries play a critical role in developing information literacy skills. Promotional activities can highlight workshops, tutorials, and one-on-one support sessions that help users develop essential research and information evaluation skills.
- **Facilitating Collaboration**: Marketing efforts can promote collaborative spaces and services within the library, such as group study rooms, collaborative research tools, and interdisciplinary events.
- **Adapting to Technological Changes**: As academic libraries increasingly offer digital resources and services, marketing helps communicate these changes to users.
- **Justifying Funding and Support**: Effective marketing demonstrates the value and impact of the library to stakeholders, including university administration, donors, and funding agencies. By showcasing usage statistics, success stories, and the library's contribution to academic achievement, libraries can secure necessary funding and support.

6. Key Concepts in Promoting Library Marketing

Before implementing the idea of marketing in libraries, the following key concepts should be kept in mind:

6.1. Marketing Plan

A good marketing plan begins with a mission statement that defines the objectives of the library or information center, including the identification of target market segments. The marketing plan should focus on mission analysis, strategic planning, and the monitoring and evaluation of tasks performed. This ensures that all marketing activities align with the library's goals and are effectively tracked and assessed for improvement.

6.2. Marketing Research

Marketing research is a crucial function that helps understand customer behavior, the relative popularity of services, and the effectiveness of advertising media, pricing policies, and packaging. It involves collecting relevant data to assess market needs by defining research objectives, developing a research strategy, and understanding target market characteristics and demographics. The collected data is then interpreted and analyzed to prepare products and services that meet the library's targets and objectives.

6.3. Market Segmentation

Market segmentation involves dividing the market into groups of customers with similar characteristics and common needs and wants. It is usually categorized into:

- **Demographic Segmentation**: Age, gender, income, education, etc.
- **Socio-Economic Segmentation**: Social status, economic conditions, etc.
- **Geographic Segmentation**: Location-based factors.

Fredrick F. Webster describes market segmentation as a strategy for maximizing market response from limited resources by recognizing differences in customer behavior. It adjusts marketing strategies to cater to the unique needs of different segments.

6.4. Marketing Mix

The marketing mix is a policy framework adopted to achieve success in marketing. It helps in developing specific strategies that serve as a roadmap to achieve targets and goals. In library science, the marketing mix includes the 4Ps:

- **Product**: This includes both tangible and intangible offerings such as books, periodicals, literacy programs, bibliographies, annual reports, statistical surveys, compilations, and services like CD-ROMs, e-books, and digital databases.

- **Price**: This encompasses various pricing strategies such as membership fees, late fines, and service charges, as well as credit and discount policies.
- **Place**: This involves distribution channels, coverage, inventory, locations, and transportation of library resources to ensure accessibility for all users.
- **Promotion**: This includes communication programs aimed at raising awareness and engaging users through advertising, personal selling, public relations, social media campaigns, and community events.

7. Marketing Strategies for Fostering the Inclusivity and Accessibility in Academic Library

Strategy acts as a navigational compass in the evolving landscape of library services, guiding the organization toward its goals amidst complexity and change. Libraries are increasingly adopting marketing strategies to enhance the user experience. A robust marketing strategy involves a multifaceted approach that integrates digital engagement, outreach programs, and community involvement. By leveraging digital platforms like a user-friendly website and active social media channels, libraries can effectively showcase their collections and events. Hosting regular workshops, orientation sessions, and collaborations with faculty further highlights the library's resources and supports academic achievement. Engaging patrons through themed events, interactive competitions, and feedback mechanisms keeps the library responsive and relevant. Creating inviting, tech-savvy spaces maximizes resource utilization and enhances the user experience. Ultimately, a well-crafted marketing strategy boosts visibility, attracts users, and solidifies the library's role in advancing education, research, and lifelong learning. In summary, libraries can implement the following best practices to attract users:

- **Library Website and Web Portal**: Libraries should ensure that the library's website is user-friendly, up-to-date, and optimized for search engines. This helps students and faculty easily find information and resources online.
- **Social Media**: Libraries should utilize social media platforms like Facebook, Twitter, Instagram, and LinkedIn to engage with users, share updates, promote events, and highlight resources. It will help in reaching a broader audience and creating a sense of community.
- **Email**: Libraries should focus on sending regular email to inform patrons about new acquisitions, upcoming events, workshops, and important announcements.

- **Workshops and Seminars**: Libraries should organize workshops, seminars, and webinars on topics like research skills, citation management, digital literacy, and subject-specific databases.
- **Collaborations with Faculty**: Libraries should collaborate with faculty to integrate library resources into the curriculum, providing tailored support for specific courses, and co-hosting events.
- **Orientation Programs**: Libraries should conduct orientation sessions for new students and faculty to introduce them to the library's resources, services, and facilities.
- **Themed Events and Competitions**: Libraries should host themed events, book fairs, reading challenges, and competitions to engage users.
- **Library Week Celebrations**: Libraries should celebrate National Library Week or other similar events to draw attention to the library's role and offerings.
- **Surveys and Feedback Forms**: Libraries should collect feedback regularly from users through surveys and feedback forms to understand their needs and preferences. This will help in improving services and addressing any issues.
- **User Analytics**: From time to time, libraries should analyze usage data to identify trends, popular resources, and areas needing improvement.
- **Highlighting Special Collections**: Libraries should promote unique collections, archives, and special resources that set the library apart from others.
- **Resource Guides and Tutorials**: Libraries should focus on creating and promoting resource guides, and video tutorials to help users navigate library resources effectively.
- **Creating Inviting Spaces**: Libraries should focus on creating comfortable and flexible study spaces, collaborative work areas, and technology-enhanced rooms to attract users.
- **Cultural and Academic Events**: Libraries can host cultural events, author talks, and academic lectures to attract the academic community.

8. Barriers in Marketing of Library and Information Services

- **Limited Budget**: Lack of funds is a major hurdle for marketing the services and products of library.
- **Lack of Expertise**: Lack of knowledge and training in marketing and digital engagement, affecting the effectiveness of promotional efforts.

- **Resistance to Change**: Some library stakeholders may be resistant to adopting new marketing strategies or technologies, preferring traditional methods.
- **Inadequate Technology**: Limited access to modern information and communication technology is another hurdle to marketing.
- **Changing Users' Demands**: Users' needs are increasing day by day rapidly, which makes it difficult to tailor marketing strategies and measure their success.
- **Limited Collaboration**: Ineffective collaboration with other institutions, organizations, or community groups is another hurdle to marketing.

9. Advantages of Marketing of Library and Information Services

- Helps in Popularizing the Library Resources and Services
- Ensure optimum utilization of Library Resources and Services.
- Enhance the status of libraries within the community.
- It attracts financial support from donors, sponsors, and grants.
- Gathers user feedback to ascertain satisfaction and improve experiences.
- Assists in planning for future services and resource acquisitions.
- Builds Positive Connection between Library & Users
- Saves time for both professional and patrons
- Supports broader goals, like promoting literacy, lifelong learning, and community engagement.
- Improves the user experiences and satisfaction.

10. Conclusion

The integration of marketing strategies in academic libraries is essential for fostering inclusivity and accessibility. By leveraging data-driven approaches digital platforms, and innovative service models, libraries can enhance their outreach efforts and better serve the diverse needs of their academic communities. It is an effective technique for fulfilling the satisfaction of users' needs by providing various types of products and services. It is a change in the traditional attitude of libraries toward acquisition, organization, processing, storing, and retrieval of information. A library that recognizes the marketing concept will be closest to its users. As academic libraries continue to evolve embracing these strategies will ensure that they remain relevant and effective in supporting all members of the academic community.

References

Abdulsalami, L. T., & Salami, P. F. (2013). Marketing information services in polytechnics libraries in Nigeria. Journal of Education and Practice, 4(6).

Adegoke, K. A. (2015). Marketing of library and information services in university libraries: A case study of Usmanu Danfodiyo University library, Sokoto-Nigeria. Intel Prop Rights 3(2), 143.

Agboke, A. L., & Effiong, A. E. (2020). Promotional techniques used as marketing strategies for library resources and services. Research Journal of Library and Information Science 4(2), 29-34.

Agyei, D. D., & Ngulube, P. (2023). Exploring strategies for marketing library products and services in university libraries in Ghana. International Information & Library Review 55(2), 143-158. https://doi.org/10.1080/10572317.2022.2100217

Anwar, M., & Zhiwei, T. (2020). What is the relationship between marketing of library sources and services and social media? A literature review paper. Library Hi Tech News, 37(3) 1–5. https://doi.org/10.1108/LHTN-10-2019-0071

Asante, E., Martey, E. M., & Wakye, J. O. (2020). Impact of promotional tools on marketing of library services: An exploratory study of polytechnics in Ghana. International Journal of Technology and Management Research, 2(1), 16–25. https://doi.org/10.47127/ijtmr.v2i1.47

Bamigbola, A. A. (2013). Application of marketing strategies and mix to digital information services (DIS): Nigerian university libraries perspectives.

Bhatt, R. K. (2009). Marketing of LIS products and services for industries by the university libraries: A strategic solution to the problem of financial crunch. Globalizing Academic Libraries, Vision 2020, 449-55.

Budd, J. M. (1997). A critique of customer and commodity. College & Research Libraries 58(4), 309-320.

Dongardive, P. (2013). Marketing library and information services. International Journal of Science and Research, 2(11), 227-234.

Garoufallou, E., Siatri, R., Zafeiriou, G. and Balampanidou, E. (2013). The use of marketing concepts in library services: a literature review. Library Review, 62(4/5), 312-334. https://doi.org/10.1108/LR-06-2012-0061

Jose, A. (1995). Marketing and libraries. Herald of Library Science, 34, 231-234.

Joshua, D., & Michael, D. (2020). Effective marketing techniques for promoting library services and resources in Academic libraries. Library Philosophy and Practice, 1-30.

Kotler, P. (2009). Marketing management. Pearson Education India.

Kumar, A., & Singh, J. (2015). Marketing of Information Products & Services by the Research Scholars & Scientists in Institute of Microbial Technology, Chandigarh: A Study. International Research: Journal of Library and Information Science, 5(1).

Kumar, Anil et al. (2015). Marketing of information product and services by research scholar and scientist in central scientific instruments organization, Chandigarh: a case study. IOSR Journals of Humanities and Social Sciences, 20(3), 138-157.

Madhusudhan, M. (2008). Marketing of library and information services and products in university libraries: a case study of Goa university library. Library Philosophy and Practice, 11.

Maturure, R., & Rakemane, D. (2021). Marketing of library collections and services in the twenty-first century environment: the use of social media technologies. In Examining the impact of industry 4.0 on academic libraries (pp. 191-200). Emerald Publishing Limited.

Muema Kavulya, J. (2004). Marketing of library services: a case study of selected university libraries in Kenya. Library management, 25(3), 118-126.

Rowley, J. (2002). Information marketing in a digital world. Library Hi Tech, 20(3), 352-358.

Schmidt, J. (2006). Marketing library and information services in Australian academic libraries. Marketing library and information services: International perspectives, 120.

Uwandu, L. I., & Osuji, C. E. (2022). Strategies used by librarians in marketing information resources and library services in Federal University of Technology, Owerri and Imo State University, Owerri. Strategies, 8, 1-2022.

Vij, Rajiv (2012). Marketing of library products and services in India. International Indexed and Referred Research Journal, 3(34), 89-91.

23

Marketing of Library and Information Services

Anju Saini

Librarian, SNRL Jairam Girls College, Lohar Majra, Kurukshetra - 136 119, Harayana India, E-mail: anjusaini6080@gmail.com

Abstract

For optimum utilization of information and 24x7 gets to data libraries have changed themselves decisively. Marketing as an idea and as training, actually appears to be obscure to numerous library and data work force. The not-for-profit association like a library is fundamentally taken part in the development of administrations as opposed to products. Administrations are unmistakable in nature principally because of the qualities of being immaterial, indistinguishable, variable and transitory. The marking in library and data setting includes a ton of testing issues. Perhaps of the main thing that a decent scholastic library can offer is a drawn out relationship with individuals who use library administrations. Present days corporate specialists begin marking themselves as their very own. Scholarly custodians also can make best generalization by tweaking library and data administrations with his/her own kind of private brand (bit) of long last relationship.

Keywords: *Academic Libraries, Marketing, Marketing Strategies, Library and Information Services.*

1. Introduction

Libraries and information services have traditionally been seen as quiet sanctuaries of knowledge and learning. However, in today's rapidly evolving digital age, libraries must actively market their services to ensure they remain relevant and accessible to diverse user groups. Marketing these services involves not only promoting the library's offerings but also effectively communicating its value to the community. Marketing in case of libraries is necessary to offer benefits to users' want, reduce barriers to use and access, persuade and inform the users, and carefully plan to satisfy users' needs effectively. "Marketing is so basic that it cannot be considered a separate function within the library. Marketing is a central dimension of the entire library. It is the entirety of the library's operations and services seen from a point of view of its final result, that is, from the customer's point of view" (Gupta and Savard, 2010). Marketing library and information service is an established sub-set of management practices which identifies customer wants and needs and meets those needs within the resources of the organization (Gupta, Koontz & Massisimo, 2013).

2. Importance of Marketing in Libraries

Marketing is crucial for libraries for several reasons:

- **Increasing Visibility**: Libraries offer a wide range of services and resources, from physical books and digital media to community programs and educational workshops. Effective marketing helps increase awareness and ensures that patrons are informed about what is available to them.
- **Attracting New Users:** As libraries face competition from other information sources, including online platforms and digital content providers, marketing helps attract new users and retain existing ones by highlighting the unique benefits of library services.
- **Enhancing Community Engagement**: Marketing initiatives can strengthen the library's relationship with the community by promoting events, partnerships, and programs that cater to local interests and needs.
- **Showcasing Value:** Libraries are valuable community assets, providing access to information, technology, and learning opportunities. Marketing helps demonstrate the library's impact and value, potentially influencing funding decisions and support from stakeholders.

3. Key Strategies for Marketing Library and Information Services

- **Understanding the Audience**: Tailoring marketing efforts to different segments of the community is essential. This involves understanding the needs, preferences, and behaviors of various user groups, including students, researchers, families, and senior citizens.
- **Leveraging Digital Platforms**: Utilizing social media, email newsletters, and the library's website can effectively reach a broad audience. Engaging content, such as blog posts, digital flyers, and interactive features, can attract and inform users.
- **Creating Compelling Content**: Marketing materials should highlight the unique features and benefits of the library's services. Success stories, user testimonials, and case studies can illustrate the library's impact and relevance.

- **Hosting Events and Programs**: Organizing events, workshops, and community outreach programs can generate interest and engage users. Marketing these events through various channels can increase attendance and participation.
- **Building Partnerships**: Collaborating with local businesses, schools, and community organizations can expand the library's reach and create mutual benefits. Partnerships can enhance the library's visibility and provide additional resources and support.
- **Utilizing Data and Feedback**: Collecting and analyzing user feedback and service usage data can inform marketing strategies. Understanding what services are most popular or needed can help tailor marketing efforts and improve offerings.
- **Innovating Services**: Marketing should also focus on promoting new and innovative services, such as digital collections, online databases, and virtual programming. Highlighting these advancements demonstrates the library's commitment to meeting modern needs.

4. Purpose of Marketing in Libraries

Marketing in libraries serves several essential purposes, all of which contribute to enhancing the library's role within the community and ensuring its continued relevance and effectiveness. The strategic promotion of library services is not just about increasing visibility but also about building relationships, meeting user needs, and demonstrating the library's value. Here are the key purposes of marketing in libraries:

4.1. Increasing Awareness and Visibility

- Highlighting Services and Resources: Libraries offer a diverse range of services, including book collections, digital resources, community programs, and educational workshops. Effective marketing ensures that these offerings are well-known and accessible to potential and current users.
- Attracting New Users: By promoting library services and programs through various channels, libraries can attract new patrons who may not be aware of what is available.

4.2. Engaging the Community

- Building Relationships: Marketing fosters a connection between the library and its community. Through targeted communication and outreach efforts, libraries can engage with users, gather feedback, and build stronger relationships with community members.

- Promoting Inclusivity: Marketing helps ensure that the library's services are reaching diverse populations, including underserved or marginalized groups, thereby fostering an inclusive environment.

4.3. Demonstrating Value and Impact

- Showcasing Benefits: Marketing helps illustrate the unique value of library services, such as free access to resources, educational support, and community programs. By demonstrating the library's impact on individuals and the community, libraries can advocate for continued support and funding.
- Providing Evidence of Success: Success stories, user testimonials, and impact reports can be used in marketing efforts to showcase the tangible benefits and positive outcomes of library services.

4.4. Enhancing User Experience

- Informing Users: Effective marketing provides users with timely information about library events, new services, and changes in policies or hours. This keeps patrons informed and helps them make the most of what the library has to offer.
- Improving Satisfaction: By addressing user needs and preferences through targeted marketing, libraries can enhance the overall user experience and satisfaction.

4.5. Supporting Strategic Goal

- Advancing Library Objectives: Marketing aligns with the library's strategic goals by promoting new initiatives, special projects, or changes in service delivery. It helps in achieving broader objectives, such as increasing program participation or expanding digital access.
- Driving Resource Utilization: Strategic marketing can encourage the use of underutilized resources and services, ensuring that the library's offerings are fully leveraged.

4.6. Building Partnerships and Collaborations

- Creating Opportunities: Marketing can facilitate partnerships with local organizations, businesses, and educational institutions. These collaborations can enhance library programs, expand reach, and create additional value for the community.
- Strengthening Community Ties: Through effective marketing, libraries can foster community involvement and support, leading to stronger ties and collaborative efforts.

4.7. Adapting to Changing Needs

- Responding to Trends: Marketing helps libraries stay current with emerging trends and user preferences. By understanding and responding to changing needs, libraries can adapt their services and communication strategies accordingly.
- Innovating Services: Promoting new and innovative services through marketing efforts can ensure that users are aware of and take advantage of advancements in technology and service delivery.

5. Development of Marketing Services in LIS

Marketing as an idea and as training, actually appears to be obscure to numerous library and data staff. A few LIS experts might feel that Marketing is some way or another unseemly for a public help foundation, for example, library and see no space for such practice in a not-for-benefit calling like librarianship. Some information administration supervisors feel that Marketing is hostile to the idea of their exercises. Yet, with expanded contest in the realm of data, promoting is an element for endurance. Simultaneously, there is a typical confusion that limited time exercises alone comprise promoting. Ranganathan's renowned Five Laws of Library Science (Ranganathan1931) was distributed in 1931 and have been talked about in the light of marketing by many creators. Curiously, these regulations can be supposed to be somewhat all around adjusted to promoting hypothesis where the per user (in some cases called the client) is king, and is the point of convergence of the library; in this way the entire association (assets, administrations, offices, rules and guidelines, methodology) becomes users centered. Still in the time of contest Library and information administration experts understand that the conveyance of viable administrations should be founded on essential preparation. The ability of businesses and organizations to promote their services or to make potential users aware of their products can mean the difference between success and extinction (kotler, 1997).

6. Barriers to Marketing of Library and Information Services

Marketing Library and Information Services (LIS) can face several barriers, which may impact the effectiveness of promotional efforts. Here are some common challenges:

6.1. Limited Budget

Libraries often operate with constrained budgets, making it difficult to invest in comprehensive marketing campaigns, advertising, or professional marketing expertise.

6.2. Lack of Marketing Expertise

Library staff may not have formal training in marketing, leading to less effective strategies and a reliance on basic or ad-hoc methods.

6.3. Resistance to Change

Traditional library environments may be resistant to adopting modern marketing practices or technologies, preferring to stick with established methods.

6.4. Underestimating the Need for Marketing

There can be a perception that marketing is less important than the core mission of providing information and resources, leading to minimal investment in marketing efforts.

6.5. Diverse Audience Needs

Libraries serve a wide range of users with different needs and interests, making it challenging to create targeted marketing messages that resonate with everyone.

6.6. Competition for Attention

Libraries compete with numerous other information sources and entertainment options, making it difficult to capture and maintain users' attention.

6.7. Limited Technological Resources

Not all libraries have access to advanced marketing tools or platforms, which can limit their ability to reach users effectively through digital channels.

6.8. Staffing Constraints: Many libraries operate with a small staff, and existing staff may be stretched thin, making it hard to dedicate time and effort to marketing initiatives.

6.9. Measuring Impact

Libraries may struggle to measure the impact of their marketing efforts due to limited tools or methods for tracking engagement and outcomes effectively.

6.10. Community Perceptions

Some communities may have outdated perceptions of libraries, viewing them only as places for borrowing books rather than as dynamic centers for learning and community engagement.

6.11. Inclusivity and Accessibility

Ensuring that marketing materials are accessible and inclusive for all community members, including those with disabilities or non-native speakers, can be challenging.

6.12. Changing Information Landscapes

The rapid evolution of information technology and user behavior requires libraries to continually adapt their marketing strategies, which can be resource-intensive and complex.

Addressing these barriers often involves a combination of strategic planning, staff training, community engagement, and leveraging low-cost or no-cost marketing tools and partnerships. By overcoming these challenges, libraries can more effectively promote their services and enhance their impact on the communities they serve.

7. Conclusion

Effective marketing of library and information services is essential for ensuring that libraries remain integral parts of their communities. By adopting strategic marketing practices, libraries can enhance their visibility, attract and retain users, engage with the community, and showcase their value. In an ever-changing information landscape, proactive marketing helps libraries continue to serve as vital resources for learning, discovery, and connection. The purpose of marketing in libraries extends beyond mere promotion; it is integral to the library's ability to fulfill its mission and serve its community effectively. By increasing awareness, engaging users, demonstrating value, supporting strategic goals, and adapting to change, marketing plays a crucial role in enhancing the library's impact and ensuring its ongoing relevance in an evolving information landscape.

References

AlAwadhi, S., & Al-Daihani, S. M. (2019). Marketing academic library information services using social media. Library Management, 40(3/4), 228-239.

Alkindi, S. S., & Al-Suqri, M. N. (2013). Social networking sites as marketing and outreach tools of library and information services. Global Journal of Human-Social Science Research, 13(2), 1-14.

Das, B. K., & Karn, S. K. (2008). Marketing of library and information services in global era: a current approach. Webology, 5(2), 56.

Garoufallou, E., Siatri, R., Zafeiriou, G., & Balampanidou, E. (2013). The use of marketing concepts in library services: a literature review. Library Review, 62(4/5), 312-334.

Gupta, D. K., & Savard R. (2010) Marketing Library and Information Services. Encyclopedia of Library and Information Sciences, available at DOI: 10.1081/E-ELIS3-120044552.

Gupta, D.K., Koontz, C., & Massisimo, A. (2013). Marketing Library and Information Service 11: A Global Outlook. Munich: K.G. Saur Verlag Publishers.

Horsfall, M. N. (2020). Marketing of library and information services in the digital age: challenges, strategies and implications. Information Impact: Journal of Information and Knowledge Management, 11(2), 13-20.

Jain, P. (2014). Application of social media in marketing library & information services: a global perspective. International journal of academic research and reflection, 2(2), 25.

Jose, A., & Bhat, I. (2007). Marketing of library and information services: a strategic perspective. Vision, 11(2), 23-28.

Kanaujia, S. (2004). Marketing of information products and services in Indian R&D library and information centres. Library Management, 25(8/9), 350-360.

Kotler, P. (1988). Marketing management: Analysis, planning, implementation, and control (6th edition). New Jersey, Englewood Cliffs: Prentice-Hall.

Rao, S. S. (1999). Introduction to marketing of information products and services. In Marketing information products and services: a primer for librarians and information professionals. IDRC, Ottawa, ON, CA.

Weingand, D. E. (1999). Marketing/planning library and information services. Libraries Unlimited, Inc., PO Box 6633, Englewood, CO 80155-6633.

24

Accessibility and Affordability: Open Source Solutions for Rural Library Management

Archanagita Saikia

M. Lib. I. Sc., Gauhati University, Guwahati - 781 014, Assam, India
Email: fftmc.archana@gmail.com

Abstract

Rural libraries frequently struggle to provide the same quality of service as their metropolitan counterparts due to issues with tight finances, remote locations, and outdated technology. This chapter is about how open source software (OSS) could enhance access and reduces costs for managing rural libraries. Compared to proprietary software, open source solutions are more versatile and affordable, allowing libraries to adapt tools to their own requirements without having to pay hefty fees. In order to increase user engagement, boost operational efficiency, and improve library services in remote areas, this study looks at the potential of open source solutions. Our goal is to encourage a wider acceptance of these technologies and bridge the digital divide that rural libraries encounter by examining the advantages and difficulties of using OSS in these settings. In the end, the study outlines policy recommendations and future initiatives to facilitate the integration of open source technologies, thereby improving rural communities' educational and cultural environments.

Keywords: *Automation, Information technology, Open source, Rural library, Software.*

1. Introduction

Libraries play a crucial role as community centers that promote education, culture, and individual growth at a time when information is the currency of advancement. Rural libraries, however, confront unique difficulties that set them apart from their urban counterparts. Rural libraries often struggle with limited funding, being geographically isolated, and having difficulty obtaining technology infrastructure. Yet, in order to guarantee that rural libraries can provide the same caliber of service as their counterparts in larger cities, pricing and accessibility become vital factors. And in light of all of that, they continue to be vital community hubs that contribute significantly to education, cultural enrichment, and lifelong learning.

Open source solutions hold a lot of promise for addressing these problems.Open source software solutions are more affordable and flexible than proprietary software systems, allowing them to be tailored to the specific needs of any rural library. Among other reasons, the library staff finds it easier to catalog documents when they use these kinds of systems. Because these systems use simpler circulation techniques, they assist them with saving expenses while guaranteeing higher user involvement.

This chapter focuses on the ways that open source software addresses two critical issues related to rural library management: affordability and accessibility. Open source software has significant advantages because of its collaborative development model and free distribution strategy. These attributes make it a good choice, particularly in places with limited resources, such as rural areas. By looking at the function of open source solutions, this chapter also seeks to offer ideas on how rural libraries might increase operational effectiveness, improve services, and better serve their communities. By doing this, we intend to encourage a wider adoption of these technologies, which will eventually help to bridge the digital divide and improve rural communities' educational and cultural matrix.

2. Objectives of the Study

The study is aimed at the following objectives-

i. To explain the idea of open source software and emphasize its salient features.

ii. To discuss the unique problems with affordability and accessibility that rural libraries face.

iii. To study the ways in which open source technologies can help rural library management overcome these obstacles.

iv. To identify suggested courses of action and future directions for utilizing open source technologies in rural library environments.

3. A Brief Introduction to Rural Library and Open Source Software

3.1. Rural Library

It is impossible to exaggerate the importance of libraries in isolated locations for the advancement of culture and educational conditions. Because of this, they serve as crucial hubs for learning and information that would otherwise be difficult to locate in these areas. Government initiatives to support literacy, formal education, and the general development of rural populations frequently serve as the impetus for the building of rural libraries.

Millions of villages in India lack access to basic literacy resources, despite having vital library services. Since the few reading resources available to rural children are typically kept in schools behind locked doors, many of them lack access to large libraries.Moreover, Indian rural libraries fall short of global standards in this regard, particularly in the case of the frequently deficient audio-visual resources. Despite the emphasis on high-quality learning in educational programs, rural libraries typically lack the basic amenities. As a result, it is crucial that these services be available to the entire rural population, regardless of location or time. In order to foster community collaboration for improved learning, village authorities should give supporting structures and technical resources to rural libraries. Village librarians can ensure the quality of teaching materials through their policies regarding service and material provision.

Nonetheless, a variety of programs and services targeted at enhancing the general educational and cultural well-being of the community are among the goals of rural libraries. One of its main objectives is literacy promotion, where rural libraries are essential in raising literacy rates because they provide access to a large range of reading materials, including books, newspapers, and magazines, accommodate different literacy levels, and assist people in developing their comprehension and reading skills. Supporting formal education is another goal, which is accomplished by giving pupils access to textbooks and other study tools. These libraries offer materials to help with homework and test preparation in addition to a peaceful study area. Moreover, rural libraries act as hubs for information dissemination, providing locals with knowledge on a range of subjects including business, technology, health, and agriculture. In order to make wise decisions that can improve social and economic well-being, access to information is essential.

In addition to this, rural libraries support the advancement of the social, cultural, and economic facets of rural areas by disseminating information and instructional resources that foster skill development and increase employment prospects. Socially and culturally speaking, they host activities and programs that respect regional customs and identities while also bringing communities together. These libraries also provide a variety of reading materials, allowing community members to choose from a wide selection of novels, factual works, or references that cater to different age groups and tastes.

Besides helping clients choose books according to their reading levels and interests, rural librarians are essential in developing a love of reading and supporting lifelong learning. Additionally, they offer reference services to help people locate particular resources or information. These services are especially helpful for professionals and students who require in-depth knowledge for their studies or employment. Librarians provide instruction in library use, educating users how to identify materials, use the catalog, and conduct research efficiently in order to maximize the benefits of libraries. Beyond aiding with education, rural libraries frequently provide social and personal advice, arranging activities that tackle social concerns and advance the welfare of the community while offering a secure environment in which people can seek counsel and support.

Thus, rural libraries are vital establishments that have a big impact on the cultural, educational, and economic spheres in rural areas. By giving people access to information and resources, they empower people and support the general development of rural communities. The government's commitment to rural libraries emphasizes how crucial these establishments are to reaching more general social objectives related to development and literacy.

3.2. Open Source Software

Software that is released under a license that allows anyone to view, alter, and distribute its source code is known as open source software, or OSS. The portion of software that most computer users never come across is called "source code." Programmers can change the code in order to change how a program or application functions. When a program's source code is available, programmers can improve it by incorporating new features or repairing malfunctioning parts. This method of developing software is not the same as proprietary software, in which the source code is private and can only be altered by the original developers with legal permission. The collaborative and transparent nature of open source software fosters creative problem-solving and quick technological improvements.

Since the majority of open source software is free to use, it's an affordable option for organizations, individuals, & groups. This affordability, which provides access to reliable software solutions without the financial burden of licensing fees, is especially advantageous for libraries with tight budgets. However, because open source software is constantly vetted and reviewed by a large international development community, it can be more dependable & safe than proprietary equivalents. Rapid vulnerability discovery and repair ensures high performance and security requirements.

Some common open-source software choices are as follows:

a) **Linux**: This open-source operating system is extensively used worldwide, powering a large number of servers, PCs, and mobile devices.

b) **Apache HTTP Server**: The Apache HTTP Server, also known as just "Apache," is one of the most well-known and extensively used open-source web servers worldwide. The web server software is available for free and is licensed under the Apache License 2.0.

c) **Mozilla Firefox:** This open-source web browser is well-known for its security features, speed, and customizability.

d) **LibreOffice**: LibreOffice is an office suite available for free and open-source that includes tools for word processing, spreadsheets, presentations, and more.

e) **WordPress**: Designed for creating and administering websites and blogs, WordPress is an open-source content management system (CMS).

However, open source software (OSS) has also grown in popularity within library settings, providing adaptable and affordable solutions to support a variety of library services and operations. These tools help libraries build a collaborative and community-focused atmosphere while managing their holdings, facilitating digital access, and enhancing user involvement.Some of the commonly used open source software integrated with library environments are Integrated Library Management Systems (like Koha and Evergreen); Digital Library Software (Eg., DSpace); Library Discovery and Access Tools (Eg.,VuFind); and Digital Content Management Systems (Eg., Greenstone).

i. **Koha**: Koha is integrated open source integrated library management software, initially released in January 2000. As per the report of Library Technology (2017), Koha is the most widely implemented open source integrated library management system in the world.

ii. **Evergreen**: Georgia Public Library Service for Public Information Network for Electronic services (PINES) developed the Evergreen software initially in the year 2006, which is open source integrated library management software, like Koha.

iii. **DSpace**: DSpace is an open source repository or digital library software that is used to create open access repositories for scholarly and digital content.

iv. **VuFind**: VuFind is an open source library search engine which provides searching and browsing facility beyond the resources of a traditional Online Public Access Catalog.

v. **Greenstone**: greenstone is a set of software tools designed to create and share digital library collections either online or via CD-ROM. It is open source, supports multiple languages, and is distributed under the GNU General Public License.

Open source software has become a vital part of the technology scene because of its many benefits, including affordability, flexibility, accessibility, and teamwork. Because of the community-based development model, which provides a platform for innovations, users are able to access software applications that are generally more secure and offer more customization choices than their proprietary equivalents. It is obvious that open sources will come to define technology in the near future to a greater extent as long as they continue to expand and evolve.

4. The Role of Open Source Software in Rural Library Management

Managing book and media collections can be a difficult chore for many libraries, especially those in rural regions, especially as they increase. Libraries used to rely on card catalog systems, which were difficult to maintain and labor-intensive despite their usefulness. Modern computer technology has completely changed library organization, improving its efficiency and usability. Digital systems that enable users to find books online and pick them up when they arrive have largely replaced the card catalog, saving time and effort.

All libraries do not, however, have access to these cutting-edge tools. Many rural libraries run on tight budgets, & buying new resources usually takes precedence over paying for pricey software programs. Due to budgetary constraints, many libraries are unable to update to the newest technology which makes it harder for them to stay relevant.

One workable answer to this issue is provided by open source software. Open source software is free and available for everyone to use, alter, and distribute in contrast to proprietary software, which is expensive and has limited access to its code. As a result, communities of developers are encouraged to share and work together to create software that is tailored to their own requirements. The community is then notified of these changes, which promotes accessibility and creativity.

Open-source software options abound that rural libraries can take use of. These include document processing applications, digital library software library management systems (LMS), and basic operating systems.With open-source software, these rural libraries may improve their administration systems and provide better services to their community without having to make large investments. This method not only increases the effectiveness of library operations but also gives rural libraries the ability to develop into active hubs of knowledge exchange and learning, promoting literacy and education in underprivileged communities.

5. Accessibility Considerations in Open Source Solutions

For managing rural libraries, open source technologies have several benefits, especially when it comes to accessibility and diversity. Here are some crucial things to remember:

i. **Creating Usable User Interfaces in Low-Tech Settings**

ii. **Simple and Intuitive Design**: To reduce the need for substantial training, library management software user interfaces should be clear-cut and simple to use. This is especially crucial in rural locations where employees might not be as tech-savvy.

iii. **Low Bandwidth Requirements**: To guarantee that software runs well even with sluggish internet connections, it should be optimized for low-bandwidth situations. Offline functionality and lightweight design can help achieve this.

iv. **Mobile Compatibility**: Software ought to be compatible with mobile devices, enabling access via tablets and smartphones, considering the high number of mobile devices in rural areas. For employees and customers who might not have access to desktop PCs, this improves usability.

v. **Localization & Language Support**: Offering user interfaces in regional tongues and languages can significantly enhance usability, facilitating users' ability to browse and make efficient use of library resources.

vi. **Assistive Technology Integration for Users with Disabilities**: Open source solutions ought to be interoperable with text-to-speech software and screen readers so that visually impaired users can independently access library resources.

vii. **Keyboard Navigation**: It is imperative to make software navigable solely through the keyboard for users with mobility impairments, as they might not be able to utilize a mouse.

   Personalized Display Configurations: Enabling users to modify text sizes contrast, and color schemes can enhance accessibility for people with dyslexia or vision impairments.

viii. **Compatibility with Cochlear Implants and Hearing Aids**: Software should work with assistive listening devices and offer captions or visual clues for users who have hearing loss.

Open source solutions can help make rural libraries inclusive and fair by addressing these accessibility issues and granting all community members, regardless of ability to use technology or physical capabilities, access to resources and information. This strategy supports more general objectives of improving literacy and education in underprivileged communities in addition to improving the user experience.

6. Affordability of Open Source Solutions

Rural libraries often operate on tight budgets, price is an important consideration when selecting software. Due to its affordability, open-source software (OSS) presents a strong substitute for proprietary software, particularly for organizations with tight budgets.

- **Cost Comparison of Initial Acquisition Expenses with Proprietary Software**

  a) **Open Source Software**: Software Available for Free The fact that OSS requires no license fees is one of its main advantages. Open-source software is available for libraries to download, install, and use without any upfront payments. This is in stark contrast to proprietary software, which frequently entails high one-time license costs, particularly for digital content platforms or extensive library management systems.

  b) **Proprietary Software:** Proprietary software can be too expensive for rural libraries to use. Financial resources may be further taxed by licensing fees, which are typically hefty and may also include extra costs for user accounts, feature unlocks, or tiers of service.

- **Personalization and Adaptability**

  a) **Open Source Software**: Open source software provides a wide range of free customization choices. Libraries are able to alter the software to better fit their own requirements, improving both functionality and user experience. The requirement for various software purchases or subscriptions may decrease as a result of this flexibility.

  b) **Proprietary Software**: With proprietary software, customization is frequently restricted and expensive, necessitating further purchases or service contracts and raising overall costs.

- **Initial Setup Considerations**

  a) **Technical Expertise:** Technical skill is necessary for setting up Open Source Software (OSS), even though it can be obtained for free. Libraries may need to cover the early costs of hiring outside specialists or investing in staff training. Nonetheless, a lot of open-source communities offer thorough documentation and discussion boards that help speed up the setup procedure.

b) **Hardware Requirements:** Hardware changes may be required for both proprietary and open-source solutions. But OSS usually has less strict hardware requirements, so libraries don't have to spend money on brand-new computers when using the infrastructure that already exists.

- **Long-term Maintenance and Support**

a) **Open Source Software:** While open source software has lower upfront costs, it still needs to be maintained on a regular basis to guarantee functioning and security. This includes security patches, bug fixes, and updates. It may be necessary for libraries to set aside funds for continuing upkeep, either through hiring outside contractors or providing internal staff training.

b) **Proprietary Software:** Upkeep of proprietary software is frequently covered by the licensing cost, and updates and support are given on a regular basis by the vendor. But doing so may lead to dependence on the seller and eventual price hikes.

- **Support and Community**

a) **Open Source Software**: Community involvement is a key component of the OSS support strategy. Active communities that provide free guidance and troubleshooting help are present in many open-source projects. Paid support services from outside vendors, on the other hand, are offered for libraries in need of specialized assistance and strike a compromise between affordability and expert guidance.

b) **Proprietary Software**: As part of the service package, dedicated assistance is usually offered for proprietary software. This guarantees prompt assistance, but it can also be expensive, particularly if premium support services are needed.

- **Longevity and Sustainability**

a) **Open Source Software:** If a project loses community interest or remains dormant, sustainability may become an issue. On the other hand, because the program is open source, future developers can carry on the job, guaranteeing its long-term sustainability.

b) **Software that is proprietary**: Vendors may stop offering certain goods or services, which forces libraries to switch to other ones often at a considerable expense.

Because they have customization features and don't require licenses, open-source alternatives are therefore less expensive. Before making any decisions, libraries should consider the expenses associated with initial installation, ongoing support, and maintenance. Despite this disadvantage, rural libraries can enhance their operations at a relatively low cost by utilizing open source solutions, as they save money over time owing to reduced costs connected with the upkeep and upgrading of these systems.

7. Challenges and Limitations

Use of open-source software (OSS) in rural libraries presents certain difficulties even if it has several benefits in terms of cost and flexibility. These difficulties may make it more difficult for OSS to be effectively adopted and used, especially in remote locations with possible resource and infrastructural limitations.

Connectivity Issues in Remote Areas

Internet Access: Libraries' capacity to obtain and update Open Source Software (OSS) can be greatly impacted by the restricted or inconsistent internet access that many rural communities experience. Regular software updates are essential for preserving functionality and security, and slow connectivity might cause these updates to take longer than expected, leaving computers susceptible. Bandwidth restrictions can impact the performance of cloud-based OSS applications, even in situations when internet connectivity is accessible. This might cause the apps to become less responsive or even unusable during periods of high demand.

Offline Functionality: For essential operations like data synchronization or gaining access to cloud-based resources, several open-source systems significantly rely on internet connectivity. Libraries with patchy internet service could find it difficult to make good use of these services. Nonetheless, libraries in places where connectivity is a problem may have to give preference to an OSS that provides strong offline capabilities, which may reduce the selection of appropriate apps.

- **Technical Expertise and Training Requirements**

  a) A lot of rural libraries don't have enough employees with the technical know-how to install, set up, and manage OSS. This may result in a dependence on outside consultants, raising expenses and possibly postponing implementation. On the other hand, educating library staff in the skills needed to manage the software is typically important for a successful implementation of open source software. For libraries with tight budgets and staff time constraints, training programs can be costly and time-consuming.

b) **Ongoing Maintenance:** In contrast to proprietary software, which has vendor-managed updates and maintenance, open-source software (OSS) mandates that libraries handle these tasks on their own; this requires a certain level of technical expertise and staff time, which can be difficult for smaller libraries with limited staff resources. In addition, libraries that do not have dedicated vendor support must rely on community forums and documentation for troubleshooting; these resources are rich and varied, but can be challenging to navigate without sufficient technical knowledge.

- **Cultural Perceptions and Resistance to Change**

a) **Perception of Reliability:** The belief that open-source software is less secure or dependable than proprietary software may result in resistance from stakeholders who are used to commercial products. Moreover, open-source software may be perceived as complicated or challenging to use, discouraging libraries from implementing these solutions without obvious benefits.

b) **Change Management:** In order to ensure that staff and patrons are comfortable with new systems and workflows, libraries must implement the OSS with effective change management strategies. In addition, obtaining support from all stakeholders staff, patrons, and library boards is essential for successful adoption. This calls for clear communication and cooperation.

In addition to these difficulties, there are other restrictions on using open source solutions for managing rural libraries. For rural libraries, integrating open-source software (OSS) with pre-existing proprietary systems or databases can be challenging. Custom development work is frequently needed, which adds to the implementation complexity. Compatibility issues arising from other members using different systems further exacerbate this complexity for libraries that are a part of bigger networks or consortia, making resource sharing and collaboration more difficult in the end. Libraries must take into account the large upfront expenditures necessary for infrastructure, training, and customization even though OSS can assist lower software costs. To guarantee a seamless transition, careful planning and resource allocation are crucial. Furthermore, there are sustainability issues because certain open-source projects might not be financially sustainable in the long run if community support wanes or the project's developers decide to give it up. In order to reduce potential hazards, libraries must evaluate the viability of the software solutions they have selected and create backup plans.

8. Future Directions and Recommendations

Enhancing affordability and accessibility through the use of open source technologies in rural library management is a huge opportunity. A number of future directions and policy recommendations should be taken into consideration in order to optimize the potential benefits of these technologies:

8.1. Potential Advancements in Open Source Technology for Rural Libraries

Enhanced User Interfaces and User Experience: The development of more user-friendly interfaces for open source library management systems that accommodate a range of technical proficiency and literacy levels should be the main priority in the future. To improve the accessibility of library systems for all users, voice-assisted technology and streamlined navigation may be implemented.

- **Integration with Mobile Technology**: Creating open-source applications that are optimized for mobile devices can expand library services beyond physical venues, given the increasing use of mobile phones. Features like e-book lending, mobile catalog searches, and alerts for library events can improve accessibility and user engagement.
- **Collaborative Innovation and Development**: Fostering a collaborative environment among developers, librarians, and the communities they serve can result in creative solutions tailored to a particular need. Networks for exchanging best practices, case studies, and open source technologies might help to create this cooperative environment.
- **Data Analytics and Insights:** Libraries can gain a better understanding of user behavior, preferences, and needs by putting open-source data analytics technologies into practice. Libraries can enhance services, allocate resources more effectively, and develop community-focused activities by examining this data.

8.2. Policy Recommendations to Support the Adoption of Open Source Solutions

Funding and Grants for Open Source Initiatives: Funds should be set aside by governments and organizations expressly for the creation and application of open source technologies in rural libraries. Funding is available for technical assistance, training initiatives, and the initial configuration of open-source software.

- **Incentives for Open Source Development**: Developers might be encouraged to contribute to open source library management systems by policymakers. These could be in the form of tax breaks, recognition initiatives, or funding for industry-related R&D.
- **Training and Capacity Building**: Effective adoption requires training staff members and librarians on open source technology. Policies should encourage ongoing professional development initiatives that give library staff members the know-how to properly operate and maintain these systems.
- **Infrastructure Development:** In order to guarantee the successful implementation of open source solutions, attention needs to be directed toward enhancing digital infrastructure in rural regions. This entails making sure gear is accessible, supplying dependable electricity, and improving internet connectivity.
- **Collaboration and relationships**: To promote resource sharing and knowledge exchange, foster relationships amongst libraries, educational institutions, non-profits, and technological firms. Working together can result in the creation of specialized solutions to deal with the particular difficulties that rural libraries face.

8.3. Other Considerations

- **Community Engagement and Feedback**: It is crucial to involve the community in order to get input on library services and open-source solutions. Libraries can better match their services to the needs and interests of their patrons by conducting regular surveys, focus groups, and community gatherings.
- **Sustainability and Maintenance:** Libraries need to think about long-term sustainability and maintenance when implementing open source systems. To guarantee the systems' continued functionality and security, this entails making plans for frequent updates, security patches, and continuous technical support.
- **Advocacy and Awareness Campaigns:** Adoption may be accelerated by increasing knowledge of the advantages of open source library management solutions. Cost savings, enhanced accessibility, and the possibility of community empowerment through better library services should all be highlighted in advocacy campaigns.

9. Conclusion

The incorporation of open source technologies into the management of rural libraries offers a revolutionary chance to tackle the two interrelated problems of affordability and accessibility. Through the utilization of open source software's collaborative and cost-effective characteristics, rural libraries can surmount both budgetary limitations and technological obstacles to enhance their community services. This study demonstrates how open source solutions can boost operational effectiveness, improve library services, and provide rural populations with greater access to resources and information, thereby empowering them.

But in order to successfully deploy open source solutions, issues with infrastructure, technical know-how, and community involvement must be resolved. Many remote libraries confront challenges like poor internet access, a lack of funds for technology improvements, and a staffing shortage. Targeted investments in digital infrastructure and capacity-building programs that give library employees the know-how to properly administer and maintain open source technologies are necessary to overcome these obstacles.

To establish an environment that encourages the use of open source solutions in rural libraries, policymakers, library stakeholders, and technology developers must work together. This could entail making investments in digital infrastructure, offering guidance and assistance, and encouraging collaborations that foster resource sharing and innovation. Moreover, creating incentives for the creation and application of open source software can promote creativity and customization made to meet the particular requirements of rural libraries. To guarantee that the solutions put into place are in line with community needs and preferences, libraries must aggressively include local users in the decision-making process. This is why community involvement is so important.

Besides that, ongoing initiatives and campaigns are required to raise awareness and demonstrate the benefits of open source software for library management. Through showcasing the possible financial benefits, improved accessibility, and improved service provisions, stakeholders can promote a broader acknowledgement and integration of these technologies.

Rural libraries may become dynamic hubs of learning and development with the help of various open source technologies, greatly enhancing the educational and cultural development of their local populations. Open source solutions will play an ever-more-important role in ensuring inclusive access to knowledge in rural areas and bridging the digital gap as they develop. In the end, using open source solutions can turn rural libraries into vibrant information hubs that empower communities, ensuring that these essential resources endure in a constantly evolving digital environment.

References

Babbar, P. (n.d.). Open source library software and applications- information communication technology for libraries. https://ebooks.inflibnet.ac.in/lisp5/chapter/open-source-library-software-and-applications/

Choi, N., & Pruett, J.A. (2015). The characteristics and motivations of library open source software developers: An empirical study. Library & Information Science Research, 37 (2), 109-117. https://doi.org/10.1016/j.lisr.2015.02.007

Lamani, M.G., & Sedam, M.V. (2021). Open source softwares for libraries: An overview. International Journal of Research in Library Science, 7 (1), 176-180.

Randhawa, S. (2008, January). Open source software and libraries, Research Gate.

Rani, B. (2017). Rural Libraries, Jnanada Prakashan, New Delhi.

Rani, S. (2018, May). Concept of rural libraries: An overview. Journal of Emerging Technologies and Innovative Research, 5(5).

Techjockey. (2024, March 28). 10 best open source and free library management software. Techjockey.com blog. https://www.techjockey.com/blog/best-open-source-free-library-management-software

Zargaryan, T., & Garibashvili, I. (2012, December). OpenBiblio: Software to digitize small and rural libraries. EIFL. https://www.eifl.net/resources/webinar-openbiblio-software-digitize-small-and-rural- libraries

25

Unlocking Collaboration: The Benefits of Academic Social Networks for Researchers

Mahalakshmi K.R.[1] and Dr. Rajendra Babu H.[2]

[1]Research Scholar, Department of Studies and Research in Library & Information Science, Tumkur University, Tumakuru - 572 103, Karnataka, India and Librarian Atomic Energy Central School, Kaiga Karwar - 581 400, Karnataka, India
E-mail: mahalakshmikr1993@gmail.com

[2]Assistant Professor, Department of Studies and Research in Library & Information Science, Tumkur University, Tumakuru - 572 103, Karnataka, India
E-mail: hrajendra.babu@gmail.com

Abstract

The investigation aims to find the benefits of academic social networks (ASNs) platforms for the research community. ASNs such as Mendeley, ResearchGate, and Academia.edu are essential tools for research collaboration and dissemination. These platforms facilitate various functionalities such as following search, full-text access, and managing article visibility, thereby fostering collaboration and information sharing among researchers. They offer diverse browsing options and advanced filters, significantly enhancing search precision and efficiency. ResearchGate provides robust impact metrics, including citation counts and aggregate citation data, while Mendeley excels in organizing bibliographic data. Despite some limitations, these platforms prioritize user assistance, ensuring seamless navigation and support options to enhance user experience and engagement. The results show that ResearchGate, Academia.edu, and Mendeley are facilitating global research communication. These platforms offer advanced features such as search results, full-text access requests, and discussion forums. ResearchGate provides the most robust tool for research communication and impact measurement, with features like citation counts and aggregate metrics, while Mendeley is valuable for research tracking.

Keywords: *Academic social networks, ResearchGate, Academia.edu, Mendeley, Citations, Social Networking Sites.*

1. Introduction

The aim of social networking sites is to build and maintain social interactions through content creation and communication within networks. However, academic social networks have a different focus. It prioritizes facilitating research and academic importance by connecting research groups and their scholarly outputs more effectively.These platforms typically offer features such as personal academic profiles, connections based on shared academic interests, and tools for managing and monitoring one's research outputs. Additionally, they enable users to utilize the research outputs of others in their field to advance their work. This study focus on to evaluate the capabilities of prominent academic social networking sites in meeting the essential requirements of researchers and the academic community(Bankar & Lihitkar, 2021)

Academic Social Networks are gaining more vogue now a days, reason behind that is it acts bridge between social media and scholarly publication (Ovadia, 2014). Meishar-Tal and Pieterse (2017) Adding to that Academia. edu and ResearchGate are gaining more popularity through allow uploading research paper, abstracts, and links to published article and also engaged in professional interaction. Earlier days researchers are publishing article in peer reviewed conference proceedings and journal article but know the scenario is entirely changed due to introduction of social media into the market, the way academics research content dissemination is changing (Ali and Richardson, 2017; Ovadia, 2014; Veletsianos, 2013).Each academic social networks offers its own combination of tools and services to promote research excertion, communication, collaboration and networking (Vasquez & Bastidas, 2015a).

2. Review of Literature

Researchers have reviewed previous literature related to academic networking sites to provide background studies for a research topic. Social media's rapid growth within academia raises concerns about its impact on education and research. Publishers' focus on monetizing information hinders the free flow of knowledge and exchange of ideas (Roach, & Gainer, 2013). Academic social networks (ASNs) are the solution to these challenges, providing multiple avenues for researchers to connect and exchange valuable data and publications that would otherwise be inaccessible (Veletsianos & Kimmons, 2013). In predatory open-access journals, authors cover the costs, while ASNSs rely on advertisers and investors. Relevant ASNSs include ResearchGate, Academic. edu, Mendeley, and Zotero (Willinsky, 2006). Study analysed the use of academic social networking sites across different disciplines. Using data from 6132 researchers' profiles, Study found that Academia.edu is popular among humanists and social scientists, while biologists' favour ResearchGate. The study also showed that researchers in humanities, social sciences, and natural resources are more active on these platforms compared to biologists (Ortega, 2015). The study has been reported that ASNSs offer a diverse range of tools to enhance research activities, foster collaboration, and facilitate networking. However, managing multiple profiles across various sites could prove to be a time-consuming endeavour (Jain & Makwana, 2022; Vasquez & Bastidas, 2015b).the study found Mendeley readership counts and citations demonstrate a greater impact in social sciences compared to humanities. Utilizing Mendeley readership data enables the effective monitoring of knowledge dissemination across diverse scientific disciplines (Thelwall & Kousha, 2014). Study explained ResearchGate is a powerful tool for showcasing traditional knowledge capital. While Brazil, India, and several other countries have fully embraced this social media platform, China, South Korea, and Russia are falling behind in harnessing its potential (Sayed Mohamed Jelani *et al.*, 2019)

3. Need for the Study

In an era of rapid digital evolution, it is imperative to understand how these ASNs platforms foster the exchange of scientific ideas and contribute to the advancement of knowledge. The research aims to reveal insights into the functionalities and impacts of these platforms, shedding light on their emergence, ownership, global reach, and key features. Understanding the dynamics of ASNs platforms is paramount for researchers and academics to effectively utilize these tools for collaboration, dissemination, and staying informed about developments in their respective fields. The study seeks to bridge a significant gap in the scholarly literature by providing valuable insights into the pivotal role and effectiveness of ASNs in contemporary research communication practices.

4. Research Question

The research examines the features and services offered by four widely used Academic Social Networks (ASNs), namely ResearchGate, Academia.edu, and Mendeley. The study focuses on to answer the following research questions:

1. What are the features and services provided by academic social networks (ASNs)?
2. What is the role of academic social networks (ASNs) in facilitating research communication?
3. How do academic social networks (ASNs) contribute to the exchange of scientific ideas and the advancement of knowledge among researchers and academics?

5. Methodology

The data were collected from 6th April 2024 to 23rd April 2024. Additionally, experts in the fields of library and information science, who have been using these sites for more than one year, were consulted. The structured checklist was designed to examine various features of ASNs, comprising 89 attributes categorized into twelve broad categories such as basic Attributes, browse function, Metrics, User Involvement and Intelligent Functions, User Interface, Display of Bibliographic Information, Settings for Privacy and Display of Text, Filters for Sessions, Output Features, Capability for Users to Upload Content, Features for Assistance : Provision to Upload the Contents by Users, Assistance Features, and provision to upload contents by users.

6. Data Analysis and Discussion

Parity analysis of the features of all the ASNs under study was conducted with the help of a structured checklist.

Table 1: Background information

Sl. NO	Contextual Insights	Academic Social Networks		
		ResearchGate	**Academia.edu**	**Mendeley**
1	Initiated	2008	2008	2008
2	Uniform Resource Locator	http://www.researchgate.net/	https://www.academia.edu/	mendeley.com
3	Owned By	ResearchGate GmbH	Richard Price	Elsevier
4	Founder	Ijad Madisch, Sören Hofmayer, Horst Fickenscher	Richard Price	Paul Foeckler, Victor Henning, Jan Reichelt
5	Headquarters	Berlin, Germany	San Francisco, California	London
6	User Population	20 million as on date 16/04/2024	255 million (Approximately) as on date 16/04/2024	6 million (Approximately) as on date 16/04/2024
7	Total Uploaded documents	60 million Papers	55million papers	70 million papers
8	Geographical Coverage	Worldwide	Worldwide	Worldwide
9	Subject Converge	Multidisciplinary	Multidisciplinary	Multidisciplinary

The basic information of academic social networks are outlined in table-1. Mendeley was established in 2007 and is a division of Elsevier. ResearchGate and Academia.edu were both founded in 2008. Academia.edu claims over 255 million users, while ResearchGate has about 17 million users. All three sites cover a wide range of subjects and have a global reach.

Table 2: Significant Attributes

Sl. No	Typical Significant Attributes	Academic Social Networks		
		ResearchGate	**Academia.edu**	**Mendeley**
1	Follow the search results	YES	YES	NO
2	Request to author for the full text of search results	YES	YES	NO
3	Questions and answers post	YES	NO	YES
4	Show and hide of the abstract	YES	YES	NO
5	Save the search results into account	YES	NO	NO
6	Download the full text document	YES	YES	NO
7	Post about the recent project	YES	NO	NO
8	Remove the document from timeline	YES	YES	YES
9	User can invite colleague	YES	YES	NO
10	User get update about the reads/ Views	YES	YES	NO
11	User can review the contents	YES	YES	NO
12	User can submit feedback about the research work	YES	YES	NO
13	Creation of discussion forums for interest group	YES	NO	YES
14	User get updates about followers	YES	YES	YES
15	Mail the search results of scholarly literature	NO	NO	NO

The key features and services of academic social networks are examined in Table 2. Users can follow search results, request full-text access from authors, and show/hide article abstracts on ResearchGate and Academia.edu. Dedicated discussion forums for specific interest groups offered by ResearchGate and Mendeley. Additionally, users can track search results, post queries and responses, share information about the latest research, and create separate discussion forums for specific interest groups. ResearchGate also provides features such as the ability to remove documents from the timeline and receive updates on followers, which are common across all three academic social networks.

Table 3: Typical Significant Browse Function

Sl. No	Browse Features	Academic Social Networks		
		ResearchGate	Academia.edu	Mendeley
1	Basic search	YES	YES	YES
2	Advance search	NO	NO	NO
3	Title	YES	YES	YES
4	Author	YES	YES	YES
5	Date of publication	NO	NO	YES
6	Journal title	YES	NO	NO
7	Subject	YES	NO	NO
8	Facility of Boolean operators	YES	YES	YES
9	Keywords	YES	YES	YES
10	Facility of Boolean operators	YES	YES	YES
11	Keywords are hyperlinked	NO	YES	YES

Academic social networking sites have various browsing functions, detailed in Table-3. Users can find relevant scholarly information using specific criteria such as title, author, publication date, journal title, and keywords. Advanced filters and Boolean operators are available. Basic search is on all three sites, with advanced tools only on Mendeley. ResearchGate and Academia.edu offer various search parameters, while Mendeley includes keywords, date range, authors, abstract, and title.

Table 4: Statistics and Altmetrics Features

Sl. No	Statistics and Altmetrics Features	Academic Social Networks		
		ResearchGate	Academia.edu	Mendeley
1	Aggregate count of citations for the publication	YES	NO	NO
2	Citations categorized by year	NO	NO	YES
3	Count of citations for each publication	YES	NO	YES
4	Total reads /view of paper/ Total profile view	YES	YES	YES
5	Statistics country wise reads of papers/ Weekly statistics of reads/Weekly statistics downloads	YES	YES	NO
6	Receive notifications about the name of the author who cited the work	YES	NO	NO
7	The paper's references are linked for easy access	YES	NO	NO
8	Graphical representation of citation data	YES	NO	NO
9	Aggregate number of paper downloads	YES	NO	NO

The table-4, covers analytics and altmetrics functionalities, including accessible metrics and impact-based features. ResearchGate provides the aggregate count of citations for the publication indicator. Additionally, both ResearchGate and Academia.edu offer statistics of country-wise reads of papers, weekly statistics of reads and downloads, as well as the count of citations for each publication. Furthermore, all three academic social networking sites provide the total reads/ views of papers and total profile view facility.

Table 5: User Involvement and Intelligent Functions

Sl. No	**User Involvement and Intelligent Functions**	**Academic Social Networks**		
		ResearchGate	**Academia.edu**	**Mendeley**
1	Suggest users based on their research interests	YES	YES	NO
2	Users have the ability to provide comments on the content	YES	YES	NO
3	Users additionally receive notifications via email	YES	YES	YES
4	Personalization of notifications	YES	YES	YES
5	Capability to subscribe to other users	NO	NO	YES
6	Provide recent updates related to articles	YES	YES	NO
7	Users can identify individuals who have visited their profile	YES	YES	NO
8	The site recommends users with similar interests as you type their names in the search box	NO	YES	YES

The table-5, covers the analytics and altmetrics functionalities, ResearchGate provides an aggregate count of citations for the publication indicator. Additionally, ResearchGate.net and Academia.edu offer statistics on country-wise reads of papers, weekly statistics about reads and downloads, and the count of citations for each publication. All three academic social networking sites provide total reads/views of papers and total profile view facilities.

Table 6: User Involvement and Intelligent Functions

Sl. No	User Interface	Academic Social Networks		
		ResearchGate	**Academia.edu**	**Mendeley**
1	Navigation assistance on the database screen is straightforward, transparent, and welcoming	YES	YES	YES
2	Every field is appropriately labelled	YES	YES	YES
3	All labels are written in complete words rather than abbreviations	YES	YES	YES
4	Statistical data regarding total downloads, reads, citations, etc	YES	YES	NO
5	Follower and following profiles accessible from the homepage	YES	YES	YES
6	Details of the user's profile	YES	YES	YES
7	Link to the statistics of the user's profile	YES	YES	NO

The Table-6, reveals that ResearchGate.net, Academia.edu, and Mendeley prioritize user experience by providing clear navigation, appropriate labelling, and complete words. However, Mendeley lacks statistical data, potentially limiting users understanding of their scholarly impact. Despite this, all three platforms offer easy access to follower profiles and detailed profile information, promoting engagement and networking.

Table 7: User Interface facilities provided by Academic Social Networks

Sl. No	Display of Bibliographic Information	Academic Social Networks		
		ResearchGate	**Academia.edu**	**Mendeley**
1	Displaying only concise bibliographic information	YES	YES	YES
2	Displaying bibliographic information alongside the abstract	YES	YES	NO
3	Alphabetical	NO	NO	YES
4	Date – Ascending	NO	NO	YES
5	Date – Descending order	NO	NO	YES
6	Use of upper and lower case in search	YES	YES	YES
7	Clustering results using bibliography items	YES	NO	NO
8	Hyperlinked of author name, subject, keywords	YES	YES	NO

The table-7, Shows that the bibliographic features of Mendeley, ResearchGate, and Academia.edu offer clear and concise information for easy reading. While

Mendeley lacks this functionality, ResearchGate and Academia.edu enhance bibliographic information with abstracts for better contextual understanding. Mendeley's exceptional organization includes both chronological and alphabetical sorting, making research tracking easier. However, Mendeley's lack of hyperlinks for author names, subjects, or keywords may hinder document navigation and comparison. Nonetheless, Mendeley remains effective in providing well-organized access to references and tracking the chronological development of research.

Table 8: Privacy and Display of Text

Sl. No	Privacy and Display of Text	Academic Social Networks		
		ResearchGate	Academia.edu	Mendeley
1	Configuring privacy settings on posts to limit visibility to specific user groups	YES	NO	NO
2	Enable authors to be aware that you've reviewed their publications	YES	NO	NO
3	Option to toggle between making the profile page public or private	YES	NO	NO
4	Option to reveal or conceal user details	YES	NO	NO
5	Option to showcase all results on a single page	NO	YES	NO

The Table-8, The features offered by ResearchGate, such as notifications of reviews, advanced privacy settings, and user-controlled visibility, empower authors to manage their work effectively. Additionally, Academia. Edu's single-page display of results provides convenience. However, both Mendeley and Academia.edu lack crucial features such as post-specific privacy settings, author review alerts, and user-controlled detail visibility, which are essential for user engagement and privacy.

Table 9: Filters for Sessions

Sl. No	Filters for Sessions	Academic Social Networks		
		ResearchGate	Academia.edu	Mendeley
Sorting Options	Institution	YES	NO	NO
	Department	YES	NO	NO
	Researcher	YES	YES	NO
	Questions	YES	NO	NO
	Review	YES	NO	NO

The table-9, displays the filter features of three academic social networking sites. ResearchGate offers extensive session filters, including department, institution, researcher, questions, and review status. Academia.edu supports filtering by researcher and questions, but lacks options for institution and department. Mendeley provides only basic search functionality without session filters.

Table 10: Capability for Users to Upload Content

Sl. No	Capability for Users to Upload Content	Academic Social Networks		
		ResearchGate	Academia.edu	Mendeley
1	Articles	YES	YES	YES
2	Book	YES	YES	YES
3	Book chapter	YES	YES	YES
4	Conference paper	YES	YES	YES
5	Presentation	YES	YES	YES
6	Dataset	YES	NO	NO
7	Patents	YES	NO	YES
8	Research proposal	YES	NO	NO
9	Thesis and dissertation	YES	YES	YES
10	Working paper	YES	NO	YES

Table-10 represents, Capability for Users to Upload Content, Users of ResearchGate.net, Academia.edu, and Mendeley can share a wide range of academic content, including books, articles, conference papers, presentations, theses, and dissertations. ResearchGate.net and Mendeley allow sharing of additional content such as datasets, patents, research proposals, and working papers, which is not available on Academia.edu.

Table 11: Features for Assistance

Sl. No	Features for Assistance	Academic Social Networks		
		ResearchGate	Academia.edu	Mendeley
1	Guidance presented in clear, jargon-free language	YES	YES	YES
2	Textual information devoid of abbreviations	YES	YES	YES
3	Query submission	YES	YES	YES
4	Frequent ask questions	YES	YES	YES
5	E-mail help	YES	YES	YES
6	Online video tutorials	NO	NO	YES

The Table-11, Discuss Users of ResearchGate.net, Academia.edu, and Mendeley can features for assistance provided by all three platforms. They value user accessibility and provide instructions free of complicated jargon, guaranteeing understanding for users with various educational backgrounds. For prompt assistance, users can submit questions, consult the FAQs, or use email support for personalized help. Mendeley offers online video tutorials, particularly helpful for visual learners, and step-by-step explanations for using its capabilities.

Table 12: Impact Feature

Sl. No	Impact Feature	Academic Social Networks		
		ResearchGate	**Academia.edu**	**Mendeley**
1	Citations	YES	NO	YES
2	H-index	YES	NO	NO
3	i10-index	NO	NO	NO
4	Impact Scores	YES	NO	NO
5	Member Statistics	YES	NO	NO
6	Percentile	YES	NO	NO

ResearchGate.net, Academia.edu, and Mendeley are academic social networking sites with various impact features explores in Table-12. Among them, ResearchGate.net stands out with its extensive array of metrics. It offers the h-index to measure productivity and citation effect, citation information, impact scores to evaluate a researcher's importance in their field, percentile rankings to show a researcher's relative standing in terms of citations or effect, and member data that sheds light on its user population. On the other hand, Academia.edu and Mendeley do not offer the h-index, i10-index, impact scores, member data, or percentile rankings, although Mendeley does provide citation information.

7. Conclusion

In conclusion, the study explored how ResearchGate, Academia.edu, and Mendeley facilitate scholarly communication. It explores the roles of ResearchGate, Academia.edu, and Mendeley in facilitating research communication. ResearchGate offers extensive features such as citation counts, advanced privacy settings, and impact metrics like the h-index, enhancing user experience and research impact assessment. Academia.edu provides convenient navigation and discussion forums, while Mendeley excels in research tracking but lacks detailed statistics and hyperlinks. Despite their differences, all three platforms prioritize user accessibility and multidisciplinary coverage supporting millions of global users in their further research.

References

Ali, M. Y., & Richardson, J. (2017). Usage of academic social networking sites by Karachi social science faculty. 12.

Bankar, R. S., & Lihitkar, S. R. (2021). Academic Social-Networking Sites (ASNS) for Research Communication: A Comparative Overview.

Jain, S. K., & Makwana, J. (2022). Use of Academic Social Networking Sites by Academic Community at National Institute of Fashion Technology Centers in India. DESIDOC Journal of Library & Information Technology, 42(4), 253–258. https://doi.org/10.14429/djlit.42.4.17921

Meishar-Tal, H., & Pieterse, E. (2017). Why Do Academics Use Academic Social Networking Sites? 23.

Ortega, J. L. (2015). Disciplinary differences in the use of academic social networking sites. Online Information Review, 39(4), 520–536. https://doi.org/10.1108/OIR-03-2015-0093

Ovadia, S. (2014). ResearchGate and Academia.edu: Academic Social Networks. 6.

Roach, A. K., & Gainer, J. (2013). On open access to research: The green, the gold, and the public good. Journal of Adolescent & Adult Literacy, 56(7), 530–534.

Sayed Mohamed Jelani, A., Ashkar, K., & Sarasu, R. (2019). Research Gate: An Ideal Epitome to Academic Social Networking Sites. Asian Journal of Information Science and Technology, 9(S1), 11–15. https://doi.org/10.51983/ajist-2019.9.S1.230

Thelwall, M., & Kousha, K. (2014). Academia.edu: Social Network or Academic Network? Journal of the Association for Information Science and Technology, 65(4), 721–731. https://doi.org/10.1002/asi.23038

Vasquez, F. K. E., & Bastidas, C. E. C. (2015a). Academic Social Networking Sites: A Comparative Analysis of Their Services and Tools. 6.

Vasquez, F. K. E., & Bastidas, C. E. C. (2015b). Academic Social Networking Sites: A Comparative Analysis of Their Services and Tools.

Veletsianos, G. (2013). Open practices and identity: Evidence from researchers and educators social media participation. 44(4), 13.

Veletsianos, G., & Kimmons, R. (2013). Scholars and faculty members' lived experiences in online social networks. The Internet and Higher Education, 16, 43–50. https://doi.org/10.1016/j.iheduc.2012.01.004

Willinsky, J. (2006). The Access Principle: The Case for Open Access to Research and Scholarship,.

26

A Comprehensive Analysis on the Role of Libraries in Implementing Sustainable Development Goals

Ar. Jagriti Pathak[1] and Ar. Jayantajit Buragohain[2]

[1]*Assistant Professor, Royal School of Architecture, Guwahati-781 035, Assam, India*
E-mail: jagriti108@gmail.com

[2]*Junior Planner, Town and Country Planning, Guwahati-781 035, Assam, India*
E-mail: jntburagohain@gmail.com

Abstract

Libraries are crucial tools for accomplishing the objective of giving easily available, sufficient, and pertinent information. This research paper aims to discover the contributions libraries provide in accomplishing these objectives. It also presented recommendations and outlined the issues impeding libraries' ability to meet sustainable development goals. Libraries support and foster sustainability research on their campuses while frequently serving as the focal point for numerous faculties, fostering links across various academic fields. Worldwide, knowledge generation, collection, organization, preservation, and dissemination have been the responsibility of libraries and librarians. No institution has ever been recognized by humanity with as many responsibilities for the welfare of humanity as the library.It should be highlighted that the goal and vision of the Sustainable Development Goals are to create inclusive development globally whether it is developed or developing country.
Keywords: *Sustainable development goals, Libraries, Information, Role of libraries.*

1. Introduction

Library enhances socioeconomic development by creating a knowledgeable society, empowering marginalized people, and promoting economic activity, skill learning, cultural identity enrichment, and decision-making participation. These services are supported by infrastructure, ICT, media. Libraries are responsible for knowledge, acquisition, management, fostering mutual relationships between librarians, the publics, and stakeholders. Its role as a storehouse of resources worldwide, including acquiring, organizing, preserving, storing, and making information available. Public libraries are important for public infrastructure, community health, and supporting social development goals. The contributions of the various types of libraries are discussed below:

1.1. National Library

A country's reference library is called its national library. It is the resource that all kinds of libraries utilize since it has a collection of all books published by Academic Library nationals in every subject area, as well as writings about the nation written by both nationals and non-citizens and by people in need of materials made in the nation. As a result, it is responsible for maintaining the country's cultural legacy and making the national bibliography available each year.

1.2. Academic Library

Since the academic library is the beating heart of every academic institution, any problems there have an adverse effect on other areas of the organization. As a result, their primary goal is to supply materials to aid in research, teaching, and learning within the programs that the parent school specializes in. These libraries serve academic staff, researchers, students, and members of the general public in need of specialized knowledge. These libraries aid the institution's function in producing human resources, which are the hub of a country's development. The research findings kept in these libraries help shape policies that are implemented in different economic sectors to enhance them.

1.3. Special Libraries

These libraries, as their name suggests, offer materials to support parent organizations' goals and objectives in the areas of research, commerce, finance, industry, diplomacy, or affiliation with a government ministry, department, or agency. They cater mostly to a certain demographic and only serve a small number of clients outside of it. Operationally, they preserve and distribute particular, atomic-level knowledge to individuals; to the extent that academic or public libraries are unable to supply it in detail, special libraries are designated for that purpose.

1.4. Public libraries

Public libraries are accessible to all age groups and literacy levels, including illiterate and literate children, teens, young adults, adults, and specially challenged people. The public library's main purpose is to make reading enjoyable rather than a chore. These libraries are relevant because the communities in which they are located are actively involved and view the library as an extension of themselves. In fact, it is frequently referred to as the "poor man's university" due to its abundance of resources for meeting individuals' information needs and providing reading materials for both official and informal education.

1.5. School Library

As children develop into adults, it is expected that they will first encounter books, the culture of reading, social standards, and ultimately, potential geniuses. These kinds of libraries are seen in elementary schools and focus on creative thinking. These libraries play vital roles in society because they provide an environment where learning, teaching, and reading is valued.

2. Aims and Objectives

The aims and objectives are as follows:

a) To identify the challenges and provide strategies on how Sustainable Development Goals can be creatively implemented in Libraries.

b) To understand various Green Building Techniques for Libraries

3. Scope and Limitations

This research is restricted to the Municipal Boundary of Guwahati city using a age group of 18 - 60 years of 50 for the sample studies. This study has been carried out at a single instance i.e., a cross-sectional study, and has the limitation of sample size.

4. Literature Review

The various aspects of library are discussed below:

4.1. Community Engagement and Outreach

The main idea behind community engagement in public libraries are examined in this paper. It provides a fresh conceptual framework for understanding cooperative, fruitful interactions between service users and providers. Two fundamental variables "influence of authority" and "willingness to learn". The study investigates various factors and components that affect community engagement programs' effectiveness in public libraries. The study's first step toward conducting systematic research is developing a model of the key components of community participation in public libraries. The implications for practice are the main emphasis of this research, which also offers a community involvement paradigm based on participant perceptions and community realities.

4.2. Automated/Assisted Data Management

Library digitization and automation have increased access to collections and facilities, with e-library facilities being the most popular among students.

Online tutorials, textbooks, catalogue, and article databases are the most popular e-library facilities. Graduates and postgraduates use e-library facilities the most, with preferred features including reading full-text articles online, selecting research guides, and online tutorials. Students across different demographics use e-library facilities and library website facilities differently, with different majors affecting their satisfaction with automation. The level of education impacts the selective dissemination of information and positively impacts the charging and discharging of books.

4.3. Training Innovation and Infrastructure

By providing students with a wide range of materials, fostering collaboration and interactive learning, fostering the development of digital and information literacy skills, facilitating individualized and distant learning, and stimulating creativity, innovative library services have a significant positive impact on education and creativity. It also encourages lifetime education and community involvement. Cutting-edge library services help to increase the library's visibility and encourage patronage of its resources. Information technology has led to university libraries becoming more intelligent, more productive, and more user-satisfied. University libraries contribute significantly to user satisfaction by fostering scholarly communication, allowing students to use and access information from anywhere, storing and managing vast amounts of digital content, and creating an atmosphere that encourages academic excellence.

5. Methodology

This research is conducted in the city of Guwahati (Municipal Boundary) where a questionnaire of 14 questions are being circulated among a age group of 50 people between the age 18 – 60 years and also data from secondary sources such as government records, books, articles, and online journals are referred to. Therefore,the study's conclusion underscores the requirement to develop ideas to make libraries successfully implement the various Sustainable DevelopmentGoals.

6. Data Analysis and Findings from the Study

It has been identified through the studythat about 27% of people would at least visit the Library once a month for study and research. The respondents stated that the benefits of visiting the Library are mostly the fact that it provides high-quality study materials along with a quiet environment for studying. They were also asked about their awareness regarding the sustainable development goals and the majority of them stated that they got to know it from the libraries. The respondents were asked how in their area the Libraries are helping in achieving the SDG Goals and they highlighted that the SDG Goals 8, 4, 15, and 16 are mostly achieved due to the various contributions of the Library. They also identified that Libraries help in offering access to knowledge, providing opportunities and guidance

The questionnaire findings are as follows:

a) Do you visit library. If yes, how often you do visit the library.

b) What is the purpose of your visit to the library?

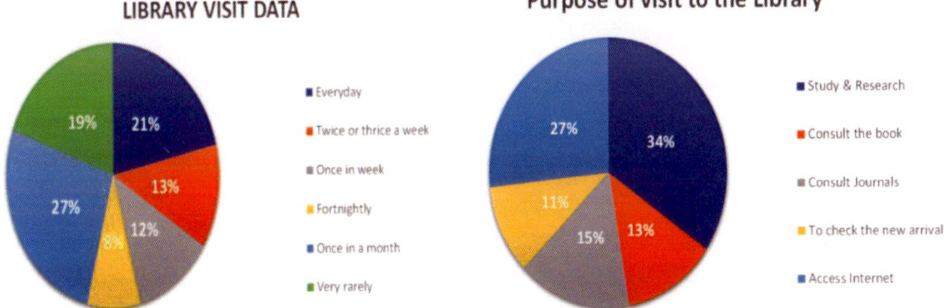

a) According to your opinion the benefits of visiting Library

b) According to your opinion how important is a library for the society

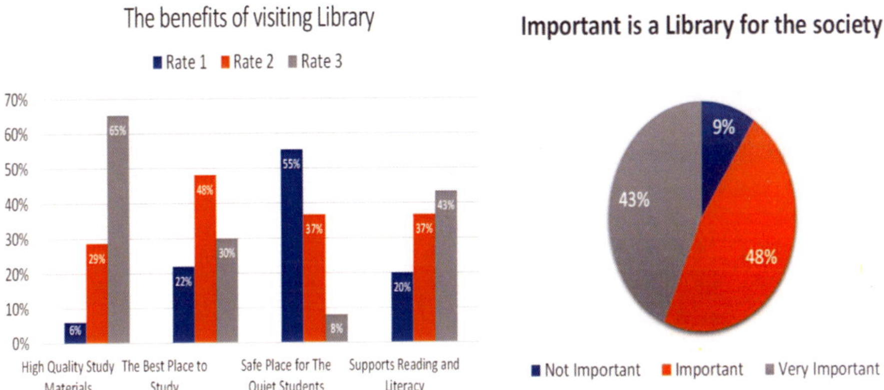

c) Where do you know about sustainable development goals

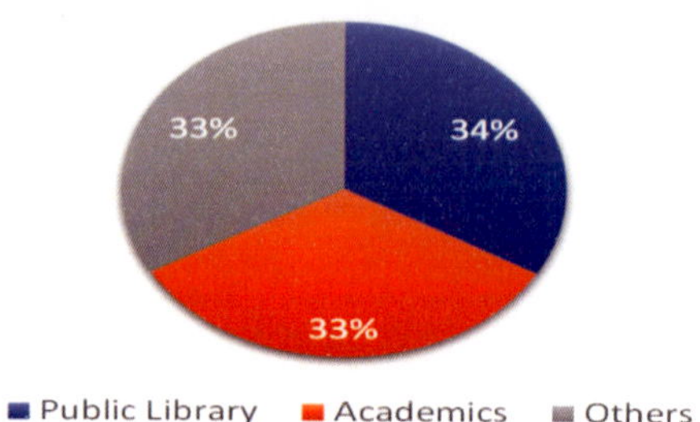

e) In your area how is the library helping to achieve this SDG's Goal

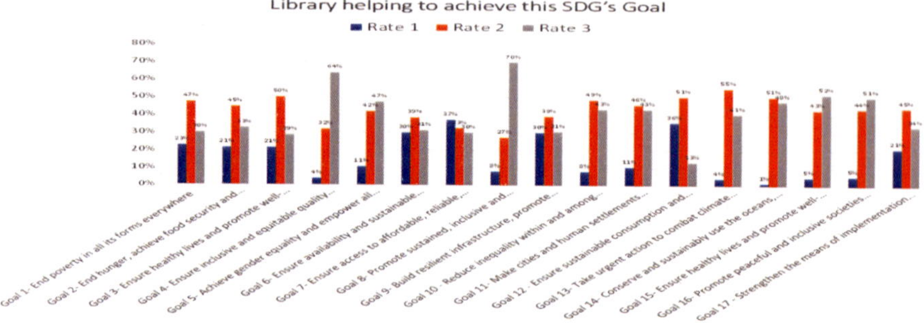

g) Why do you think libraries should support Sustainable Development Goals?

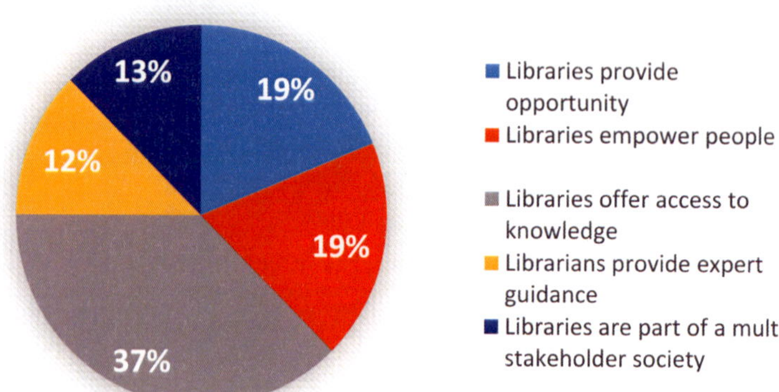

7. Case Studies

The following is a comparative study of India's Important Green Library Buildings:

Table 1 : A Comparative Study of ludia's Importnur Green Library Building:

Institute Name	Certification	Green Initiatives
II T Roorkhee	GRIHA	a) Campus Solarizarion
		b) Produced 42 peak power output (KWP) in 2019 using IITR photovoltaic solar electricity that was installed on the IITR Mahatma Gandhi Library.
		c) Swapping out the resistive fan regulator for an electronic one
		d) Maintaining the air conditioning system's temperature between 25 and *26* degrees Celsius
		e) Installing LED bulbs
		f) Swapping out inefficient tube lights withLED bulbs
		g) Using green building materials
		h) Campus E-Transportation Techniques
Manipal University, Jaipur	GRIHA 5Star Rating for hostels (LEED Indian Platinum) IGBC Certification for academic block	a. Rajasthani architecture
		b) 76% less energy used in air conditioning
		c) Rainwater conservation
		d) Purified water rectification
		e) 50% less fresh water consumption
		f) Complete LED lighting
Anna Centenary Library, Chennai	Certification LEED Gold	a) Eleven 15 Amp outlets in total for electlic charging
		b) Harvesting Rainwater
		c) Albedo plaints are planted on die Library Terrace, and me aluminum terrace has a green roof.
		d) A system af efficient lighting that prevents light pollution
		e) 64% less water used thanks to water fittings that are energy-efficient.
		f) Making use of extremely effective air-cooled chillers
		g) Using wheels with heat recovery
		h) Daylight controls for regions around the periphery
		i) Waste material recycling method
		j) CFC and HCFC Free HBSE locate mechanism
		k) Environment that forbids smoking
		l) The use of CO2 sensors, DCV systems, and densely inhabited areas.
		m) Dust and moisture-resistant techniques

8. Roles of the Library in Achieving Sustainable Development Goals

The following are how Libraries can contribute to achieving Sustainable Development Goals:

8.1. No Poverty

The transmission of information is vital for the fight against poverty.Libraries act as the central hub for this purpose. They offer comprehensive information in a packaged manner about reducing and eliminating poverty. Additionally, ensuring everyone has an equal right to economic resources, including access to basic amenities, land ownership, property, inheritance, new technology, and financial services.

8.2. Zero Hunger

Libraries and information centers should utilize resources to end extreme hunger and guarantee the smooth operation of food commodities by raising awareness and fostering an environment that makes it possible to utilize knowledge on sustainable agriculture practices in all of its forms.

8.3. Good Health & Well-being

Information on eradicating child mortality and enhancing maternal health to the public in print and electronic formats through libraries and information centers. The repercussions of poor diet on health should be highlighted by spreading knowledge about illness causes and natural disease prevention strategies. Libraries should ensure that everyone has access to services related to reproductive health, such as family planning, information and education, and the incorporation of reproductive health care.

8.4. Quality Education

Library and information centre should educate and enlighten adults, children to equip the public with information literacy skills and lifelong learning. Adults and Young Children who have the necessary abilities, both technical and vocational, for employment, respectable jobs, and entrepreneurship. Furthermore, by arming communities with knowledge and the capacity to obtain information for themselves, they must arrange and distribute information that society will use to educate itself on a variety of life concerns.

8.5. Gender Equality

Libraries and information centers should support gender equality by using technologies, particularly information and communications technology, when choosing information resources to aid women's empowerment.

8.6. Clean Water & Sanitation

By creating awareness, organizing campaigns, and hosting symposiums, libraries and information centers could make it easier for people to acquire information about cleanliness, water conservation and sanitation

8.7. Affordable & Clean Energy

Libraries and information centers should make ICT available to the general public, offer study spaces and classes on mobile phone usage. Campaigns should be done to promote reasonable priced energy services.

8.8. Industry, Innovation & Infrastructure

Libraries and information centers should give people access to technology and work to give less developed areas reasonably priced Internet connection. It can be accomplished by having animated discussions, spelling bees and tests via video conferencing project, laptop, and smartphone. These Libraries should also have audio recordings for Visually Impaired, Anti – Plagiarism Tool, Research Information Section.

8.9. Reduce Inequality Goal

Libraries and information centers should support residents' safe and dependable movement by implementing a well-thought-out, well-managed migration strategy. Library services, such as guidance and data about assistance with applications for various sectors like healthcare, education, finance and employment.

8.10. Sustainable Cities & Communities

For future generations, libraries ought to preserve and exhibit the country's natural and cultural history. Libraries can be a source of information storehouse for city planning, building planning strategies, construction and execution. The public library bears the obligation of creating and executing Sustainable design strategies, Disaster Risk Management strategies across all community levels.

8.11. Responsible Consumption

Libraries ought to push businesses to embrace sustainable practices. Libraries may work together to guarantee that information is available in every nation.

8.12. Climate Action

Awareness Programmesaddressing to climate change and related actions, such as disaster effect reduction, early warning, mitigation, and adaptation.

8.13. Life below Water

Scientific data on preservation and conservation methods that can sustainably use for habitat preservation of life under water should be made available by libraries.

9. Challenges

The following are the challenges:

9.1. Inadequate Funding

Libraries lack the necessary funding to fulfill their intended functions, which makes it impossible for them to be both attractive and effective in achieving sustainable development goals.

9.2. Insufficient Staff

The another difficulty of majority of libraries lack the knowledgeable and skilled employees needed to maintain the system which results in delayed accomplishment.

9.3. Inadequate Facilities

Inadequate Facilities are yet another aggravating element affecting information centers and libraries. It has restricted the information and services that libraries and librarians may offer for long-term, sustainable development.

9.4. Poor Infrastructure

Since library resources can get damaged, the proper structures should be constructed so that there is protection against damage to books, computers, or audio-visual items.

10. Strategies

The following are the strategies that can be adopted:

a) The Governing organizations should ensure that libraries and information centers are completely incorporated into the schedule of events both locally and nationally.

b) Libraries and information centers must frequently host symposiums, workshops, and seminars that will act as aneducational platform where people may hear about social, economic, health, and agricultural topics to help people have a new perspective, regardless of their gender, age, or occupation.

c) The parent institution and the government should give financial assistance to the libraries and information centers to address issues with limited facilities and infrastructure, inadequate network distribution, and would also support the publication of research findings carried out by researchers.

d) Sufficient personnel possessing professional qualifications must be offered in order for them to contribute in a high-quality manner to national progress in various areas of life.

e) Building successful advocacy and lobbying campaigns: Directors of libraries should make an effort in long-lasting campaigns for advocacy and lobbying. At all costs, librarians should develop their advocacy and lobbying skills to make sure they do their part to make the SDGs a reality.

10. Conclusion

Hence, it can be concluded that libraries can be seen as role models for instilling environmentally conscious behavior in their users. Libraries provide information in both physical and digital formats. Libraries ought to be planned in a way that they serve as a medium of instruction in all aspects, from creative information preservation to sustainable architectural structure planning to execution and completion. As a result, libraries can act as a catalyst in spreading environmentally responsible habits among its patrons, so generating awareness among its users to prefer a sustainable lifestyle since it would improve the quality of life. Therefore, all countries should be encouraged to devote attention to implanting all the above-discussed strategies into Libraries. This would help in overcoming the obstacles of impending the availability of information and make the country liveable and sustainable.

References

Achieving sustainable development goals: The roles of libraries. (n.d.). International Journal of Applied Technologies in Library and Information Management, 4 (2), 10 - 89-95.

Kishore Kumar, S., & D, V. (2021, February 2). Green libraries- Road map to sustainable development: A Case Study. International Journal of Research in Library Science, 7(1), 85–91. https://doi.org/10.26761/ijrls.7.1.2021.1369

Roles of library in supporting sustainable development goals. (2016, November 8). Paper Presented at the 3rd Conference of Certified Librarians in Nigeria 8th November, 2016 National Space Research and Development Agency, Obasanjo Centre, Airport Road Abuja.

The library as a campus sustainability hub: A case study in increasing community engagement & collaboration in sustainability through academic libraries. (n.d.). International Journal of Librarianship. 5(2), 26-44.

27

Electronic Resource Management Systems: Transforming Libraries in the Digital Age

Mrunalini Dhondiram Gadade[1] and Sagar Shivaji Kumbhar[2]

[1]Librarian, Christ College, Pune - 411 014, Maharashtra, India
E-mail: mrunalini.gadade0706@gmail.com

[2]Librarian, Gokhale Education Society's Arts, Commerce and Science College– Shree-Wardhan, Raigad - 402 110, Maharashtra, India, E-mail: kumbhar020@gmail.com

Abstract

Electronic Resource Management Systems (ERMS) have revolutionized the management of digital resources in libraries, providing essential tools for acquiring, organizing, and maintaining electronic content. This paper explores the complex lifecycle of electronic resources and the pivotal role of ERMS in streamlining workflows, managing licenses, and ensuring efficient access. ERMS integrate seamlessly with library systems, enabling librarians to navigate the challenges of electronic resource management, such as diverse formats, licensing agreements, and evolving user needs. The investigation phase of ERMS involves market analysis and resource evaluation to understand user requirements and assess resource viability. Once resources are identified, ERMS facilitate the acquisition process by negotiating licensing agreements, defining access parameters, and ensuring remote access capabilities. Implementation includes setup and maintenance of access points, cataloging, and troubleshooting, supported by user feedback mechanisms for continuous improvement. Evaluation and review phases utilize usage statistics and user feedback to optimize resource allocation and inform renewal decisions. ERMS provide centralized analytics to track resource utilization and compliance, supporting informed decision-making for resource management. Preservation of electronic resources is ensured through archiving and integration with digital preservation networks. Despite their benefits, ERMS face challenges such as managing the dynamic nature of electronic resources, integrating diverse formats, and providing consistent access across multiple platforms. Integrated Library Systems (ILS) and

Library Services Platforms (LSP) often lack the capability to handle these complexities, necessitating specialized ERMS for effective management. Consortia and individual library approaches pose distinct challenges, requiring ERMS to manage complex licensing agreements and diverse user access requirements while maintaining privacy and independence. The need for efficient management of electronic resources, budgetary constraints, and compliance with licensing agreements underscore the critical role of ERMS in modern libraries. In conclusion, ERMS are indispensable for libraries managing electronic resources, offering centralized management, comprehensive analytics, and streamlined workflows. They ensure efficient resource utilization, compliance with licensing agreements, and improved user experience, making them essential tools for navigating the digital landscape of modern libraries.

Keywords: *Electronic Resource Management Systems (ERMS), Lifecycle of electronic resource management, Transforming libraries, ICT, Digital libraries.*

1. Introduction

Libraries serve as global repositories of knowledge, like to giant treasure boxes. But librarians are faced with a lot of work as more and more things become digital. Electronic Resource Management Systems (ERMS) can help with that (Prajapati, 2016). These are incredibly useful tools that have completely transformed the way libraries handle digital content. The main goal of ERMS is to simplify librarians' lives. They support the acquisition of fresh digital materials, manage subscriptions and licenses, and even provide online resource locators (Anbu *et al.*, 2013). They essentially act as librarians' superhero sidekicks, ensuring that everything works seamlessly in the digital realm. ERMS play a major role in keeping everything organized. They support librarians in maintaining inventory, determining who can use what, and determining when to update it (Mangrum & Ellen Pozzebon, 2012). Libraries are able to ensure that everyone has access to what they need while adhering to all access regulations in this way. However, ERMS aren't limited to library use. They're also excellent for any library user. By making it simple to locate what you're looking for in the library's digital collection, ERMS addresses that. They keep everything organized so you can quickly get to the essentials (Prajapati, 2016).The key component that keeps contemporary libraries running well is ERMS. They make it easier for librarians to handle electronic resources and for users like you and me to find what we're looking for. Thus, keep in mind that ERMS are the unsung heroes who make it all happen the next time you're navigating your library's digital world.

2. What is Electronic Resources Management System

Electronic Resource Management System (ERMS) pertains to the systematic organization, acquisition, and administration of digital resources in libraries and other establishments (Chisman *et al.*, 2007). It includes activities like usage monitoring, licencing, and cataloguing to guarantee effective access to and use of electronic resources. ERMS solutions optimise resource allocation and improve user experiences by streamlining workflows and empowering librarians to manage the lifetime of digital items (Holly & Scott, 2008). Through the centralization of management operations, Electronic Resource Management System (ERMS) enhances the efficiency of information retrieval for users by facilitating the smooth integration and discovery of electronic resources.

"A system that supports management of the information and workflows necessary to efficiently select, evaluate, acquire, maintain, and provide informed access to electronic resources by their business and license terms" (Arms *et al.*, 1999).

"The tools or the system that use to manage the different process associated with the electronic resources is generally called as Electronic Resource Management System (ERMS)" (Patra, 2017).

3. Lifecycle of Electronic Resource Management

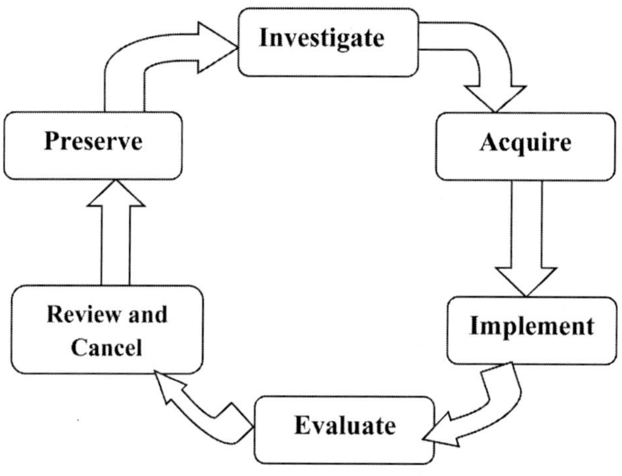

3.1. Investigate

The vital investigative stage, which entails using a methodical technique to comprehend and assess various elements of electronic resources, kicks off the lifespan of electronic resource management. The identification of possible users is the initial step in any investigation into electronic resource management (Hosburgh, 2014). To determine the target audience's demands tastes, and demographics, a thorough market analysis must be carried out. Furthermore, reading through the material that already exists and conversing with providers can offer insightful information about customer expectations and industry trends.

Evaluating the viability of prospective electronic resources is the next stage of the research process. This entails carrying out a sustainability check to assess elements including the dependability of suppliers, the durability of platforms that are available, and the suitability of resources for the institution's long-term objectives.

After that, an overlap analysis is carried out to find any duplicates or redundant information in the electronic resources that are already available (Harrington *et al.*, 2022). This facilitates the process of making well-informed decisions on the platforms to use, including direct resource acquisition from publishers or resource aggregators, in order to guarantee cost-effectiveness and optimal resource utilization. Setting up experiments and creating feedback systems comes next after possible resources have been found. This enables users to engage directly with the materials and offer insightful feedback on their usefulness, applicability, and efficacy. For future reference and well-informed decision-making, trial procedures and feedback must be documented.

In order to guarantee the successful integration & use of electronic resources within the institution's library or information centre, the investigation process in electronic resource management generally entails a thorough exploration of user needs, market trends, resource sustainability, and feedback mechanisms.

3.2. Acquire

The acquisition process of electronic resource management begins by scrutinizing detailed specifications and engaging in negotiations to establish pricing agreements(Harrington *et al.*, 2022). Subsequently, negotiations, reviews, and the signing of licensing agreements are conducted. Derived from these primary processes, additional steps include defining the site and users, specifying where resources will be accessed and determining who will have access to them. Remote access is facilitated through the provision of IP authentication, ensuring secure access from off-site locations.

Article-level linking agreements are established to enable seamless access to specific articles within larger databases or collections, with mutual indemnification clauses providing legal protection for all parties involved (Hosburgh, 2014). Privacy clauses safeguard user data, while provisions for usage statistics offer valuable insights into resource utilization.

Content transfer to third-party discovery tools enhances accessibility, and clauses for perpetual access, such as those provided by preservation networks like LOCKSS, CLOCKSS, or Portico, ensure long-term availability of content.

Negotiations may also encompass clauses for auto-renewal and price cap allowances to provide stability and predictability in pricing over time. Platform fees are negotiated, and considerations for digital rights management (DRM) issues, such as restrictions on copying or downloading content, are addressed(Morales & Beis, 2021).Throughout the acquisition process, careful negotiation and review of licensing agreements are essential to ensure that electronic resources meet the needs of the institution while also being cost-effective and sustainable in the long term. These steps ensure comprehensive access, user protection, and the efficient management and utilization of electronic resources within the institution.

3.3. Implement

The implementation process of electronic resource management involves several key steps. First, access to electronic resources is ensured through the setup of remote login checks and URL maintenance. Cataloging and discovery are facilitated through the creation of A to Z lists, website links, and holdings lists preparation. An admin module is established for efficient management, including trouble shooting and problem logging, along with periodic checks to ensure smooth functioning (McLean & Canham, 2018). Additionally, a feedback mechanism is implemented to gather user input for continuous improvement. Marketing strategies such as email campaigns are utilized to promote awareness and usage of electronic resources. Overall, these steps ensure the effective implementation and maintenance of electronic resource management systems within the institution.

3.4. Evaluate

The evaluation process of electronic resource management involves comprehensive analysis across various dimensions. Periodic access check reports are scrutinized to identify any irregularities or issues, while downtime and problem logs are analyzed to address and mitigate system disruptions effectively. Communication with suppliers is reviewed to ensure seamless collaboration and resolve any potential issues or concerns. Furthermore, users' feedback and evaluations are carefully analyzed to gauge satisfaction levels and identify areas for improvement (Hosburgh, 2014). Usage statistics are examined to assess resource utilization and calculate costs effectively. Additionally, any changes in coverage or platform migration are closely monitored to ensure continuity and optimize resource accessibility. Overall these evaluation processes ensure that electronic resource management systems are continuously refined and optimized to meet the evolving needs of users and maintain the efficiency and effectiveness of resource provision within the institution.

3.5. Review and Cancel/ Renew

The review and renewal/cancel process in electronic resource management involves careful consideration of various factors(Holly & Scott, 2008). If cancellation is contemplated, measures to preserve access to critical resources are prioritized. For renewals, thorough reviews are conducted, examining potential changes in license terms, pricing, DRM policies, and fund availability(McLean & Canham, 2018). Additionally, any alterations in holdings, such as changes in databases or the availability of open access content, are assessed. Based on this evaluation, decisions are made to either renew the resource, initiating a new lifecycle, or cancel, ensuring resources remain aligned with institutional needs and budgetary constraints while maximizing access and value for users.

3.6. Preserve

The preservation process in electronic resource management involves exploring various options to safeguard access to critical resources. This includes examining provisions such as LOCKSS, CLOCKSS, and Portico, which ensure long-term access through digital preservation networks (Harrington *et al.*, 2022). Additionally, resources may be preserved locally through the use of a dedicated server or other means to maintain access in the event of system disruptions or changes in subscription status. By leveraging these preservation strategies, institutions can mitigate the risk of content loss and ensure continued access to essential electronic resources for users.

4. Need of Electronic Resources Management System

A) Nature of e-Resources

Because e-resources are so complex, Electronic Resources Management Systems (ERMS) are essential for efficiently managing digital resources in libraries and other organizations. Different licensing agreements apply to e-resources, making it challenging to track and comply without a framework in place. Access control is managed by ERMS, guaranteeing that content can only be accessed by authorized users in accordance with their roles, IP authentication, and remote access capabilities(Ross & Sutton, 2016). To find resources, precise metadata management is necessary. Administrative technical, and descriptive metadata can all be created, managed, and updated with the use of ERMS technologies. In order to guarantee that e-resources are easily discovered, they also smoothly interface with catalogue systems and library discovery tools. Budget allocation and collection development are informed by usage statistics and analytics gathered by ERMS, demonstrating the value of subscriptions. Financial management and subscription renewals are expedited, avoiding access lapses and supporting budgetary planning. Technical issues are tracked and resolved by ERMS, which also facilitates vendor support interactions(Sadeh & Ellingsen, 2005). Because e-resources are dynamic, they need to be managed and updated often. ERMS automates these tasks so that users always have access to the most recent data.By providing better access and personalized recommendations, ERMS increase user experience by making it simpler for users to locate and utilize e-resources. For the efficient, compliant, and user-friendly management of digital content, a strong ERMS is necessary, especially in light of the complexity of licensing, access, and frequent upgrades.

B) Fragmented Approach to ERM

The necessity of a complete Electronic Resources Management System (ERMS) is underscored by a fragmented approach to Electronic Resources Management (ERM). Tracking license terms, renewal dates, and usage rights becomes difficult without a centralized system, and managing licensing agreements through several spreadsheets and papers causes discrepancies and compliance difficulties(Mangrum & Ellen Pozzebon, 2012). It might be challenging to provide uniform and seamless access for all authorized users when different access control techniques for different resources can perplex users and complicate access management. Furthermore, inconsistent metadata standards impede the integration and discovery of resources, leading to redundant work and possible mistakes in the upkeep of metadata on various platforms(Sadeh & Ellingsen, 2005). It is difficult to obtain a comprehensive picture of resource utilization due to the dispersed consumption data across several systems, which also makes it difficult to make data-driven decisions about the acquisition and renewal of resources. The tedious and error-prone process of manually managing subscriptions and renewals can result in access gaps and ineffective budget management. Resolving technical issues across numerous platforms necessitates corresponding with several vendors, which adds complexity to the support process and delays the resolution of issues. Users encounter a fragmented experience when browsing several platforms and interfaces; erratic interfaces make matters worse by making it more difficult for users to locate and utilize resources effectively. These issues are addressed by implementing a comprehensive ERMS, which guarantees effective, compliant, and user-friendly management of electronic resources by offering centralized licensing management, consistent access control, unified metadata standards, centralized analytics, efficient subscription management, streamlined technical support, and a cohesive user experience.

C) Library Spends Large Sum of Money of Electronic Resources

An efficient Electronic Resources Management System (ERMS) is essential since libraries spend large amounts of money on electronic resources. Investing in these things becomes difficult to manage without a thorough ERMS, which can result in inefficiencies and even financial waste. Centralized licensing management is made possible by ERMS, which also ensures adherence to various agreements and avoids expensive mistakes(Jewell, 2009). Access control is streamlined to enable authorized users to access resources with ease, maximizing efficiency and providing justification for spending. A user's ability to locate and retrieve important content is improved by efficient metadata management in an enterprise resource management system (ERMS). Insights into resource utilization that guide decisions about acquisition and renewal ensure that money is spent on the most useful resources. Centralized analytics provided by ERMS make this process easier. An ERMS makes subscription management more effective, lowering the possibility of access lapses and facilitating improved budgetary planning. In addition, the system streamlines vendor relations and technical assistance, resulting in faster access issue resolution and less downtime(Patra & Jha, 2014). Furthermore, by offering a standardized interface and customized recommendations, a unified ERMS enhances user experience and boosts return on investment. To put it simply, an ERMS is necessary to optimize the return on a library's significant financial investment in electronic resources by guaranteeing effective, compliance, and user-friendly digital content management.

D) Incapability of ILS and LSP in Handling ER Workflow

The inadequacy of Integrated Library Systems (ILS) and Library Services Platforms (LSP) in handling the intricate workflows linked to electronic resources underscores the necessity of an Electronic Resources Management System (ERMS). ILS and LSPs have historically been built with limited capacity to handle the dynamic nature of electronic resources; they are best suited for maintaining physical collections. They have trouble keeping track of various licensing agreements, which calls for careful monitoring of terms, renewal dates, and compliance—tasks that an ERMS can do with ease. Because access control for electronic resources involves a variety of user roles and authentication techniques, it frequently falls outside the purview of ILSs and LSPs, resulting in inconsistent access and other problems that can be fixed by an ERMS. The sophisticated metadata management capabilities required for e-resources are sometimes not offered by ILS and LSPs, but an ERMS offers specialized tools for keeping accurate and up-to-date metadata (Jewell & Mitchell, 2006). Analyses and usage data are essential for assessing the impact and worth of electronic resources. The complex analytics skills required to gather, examine, and report usage data thoroughly are typically absent from ILS and LSPs. On the other hand, an ERMS offers thorough insights into resource usage, assisting in more informed decisions on acquisitions and renewals.ILS and LSPs also struggle with subscription management because they are ill-suited to manage the complex budgeting and tracking needed for electronic resources. Subscription renewals and budget management are made easier with an ERMS, which also lessens administrative work and guards against expensive access lapses. Furthermore, an ERMS facilitates effective tracking and resolution of technical issues, while ILS and LSPs are unable to provide the centralized approach required for the technical support and issue resolution of e-resources. In summary, a specialized ERMS is necessary for effective, efficient, and user-friendly e-resource management due to the shortcomings of ILS and LSPs in handling the complex processes of electronic resources.

E) For Getting Meaningful Metrics

For libraries looking to get meaningful analytics on their electronic resources, an Electronic Resources Management System (ERMS) is essential. Conventional library systems frequently lack the capacity to collect, examine, and comprehend the kind of in-depth usage statistics that are essential for making well-informed decisions. Comprehensive analytics are offered by an ERMS, which tracks parameters including access frequency, download rates, and user engagement. These data are essential for determining the impact and value of e-resources (Prajapati, 2016). By using these measures, libraries may ensure that money is allocated to the most valued resources by making data-driven decisions about acquisitions, renewals, and budgetary considerations. Moreover, an ERMS can combine usage data from several platforms to provide a consistent perspective that makes reporting and assessment easier. This centralization aids in trend identification, license agreement compliance monitoring, and stakeholder return on investment demonstration. All things considered, an ERMS optimizes the administration and use of electronic resources in a library setting by converting unprocessed data into meaningful insights.

F) To Save The Time

Libraries may manage their electronic resources much more quickly by putting in place an Electronic Resources Management System (ERMS). Conventional approaches to managing electronic resources frequently entail laborious and error-prone manual tracking of licenses, renewals, and access concerns. By automating these processes, an ERMS offers centralized control over user access, subscription renewals, and licensing agreements. Staff members can concentrate on more strategic tasks as a result of the automation's reduction of administrative workload. Furthermore, an ERMS simplifies metadata management, guaranteeing that resource data is readily available and regularly updated, improving utilization and discovery without the need for human interaction (Anbu *et al.*, 2013). Additionally, the system combines analytics and usage statistics, doing away with the need for manual data collecting and analysis from several sources. An ERMS speeds up the decision-making process for resource acquisitions and renewals by offering immediate insights and reports. In general, an electronic resource management system (ERMS) expedites operations, decreases human labor, and improves efficiency—all of which result in a substantial time savings.

5. Challenges of Electronic Resource Management System

5.1. Complex Lifecycle of Electronic Resources

Electronic Resources Management Systems face numerous obstacles due to the complex lifecycle of electronic resources (ERMS). From acquisition to disposal, ERMS must handle complex procedures like negotiating multiple license agreements, guaranteeing user access via several authentication methods, and keeping an eye on usage to adhere to licensing requirements (Patra, 2017). For resource discovery and integration, maintaining information uniformly across platforms is essential, as is offering prompt technical help and resolving access problems. Electronic resource archiving and preservation must be managed by ERMS in order to maintain access and comply with legal standards. The process is made more difficult by the need to integrate resources seamlessly into library systems, manage expenses efficiently, and maximize the use of resources.

5.2. Availability in Various Forms and Formats

An essential component of Electronic Resources Management Systems (ERMS) is the difficulty of availability in several forms and formats. A vast variety of electronic resources, each given in a different format and version, must be handled by ERMS. This involves handling databases, multimedia information, e-books, e-journals, and more, each with specialized technological requirements and access requirement (Prajapati, 2016). For the convenience and enjoyment of users, it is imperative to provide seamless integration and access across various platforms and devices. Moreover, ERMS must manage Open Access resources and repositories in addition to supporting a variety of licensing models and access kinds, including subscriptions, perpetual access, and pay-per-view. Due to its complexity, strong metadata management, precise tracking of usage data and efficient coordination with several publishers and vendors are all necessary. In order to meet these issues, an ERMS must be scalable and adaptable enough to keep up with changing user expectations and technological advancements while yet adhering to licensing agreements and industry norms.

5.3. Dynamic Nature of Electronic Resources

Electronic Resources Management Systems (ERMS) face a great deal of difficulties because of the dynamic nature of electronic resources. Electronic resources are always changing—they get new versions, upgrades, and modifications to their licensing and access restrictions on a regular basis. In order to guarantee that users have access to the most recent versions of resources while abiding by licensing agreements, ERMS must effectively manage these modifications. Robust procedures are necessary in this dynamic context to track usage statistics, update metadata, and notify users and stakeholders of any changes.

5.4. ILS and LSP not Able to Manage

The lack of ability of Library Services Platforms (LSP) and Integrated Library Systems (ILS) to efficiently handle the complexity of electronic resources poses a serious threat to Electronic Resources Management Systems (ERMS). Because they are primarily made to manage physical collections, ILS and LSPs find it difficult to meet the special requirements of electronic resources like databases, multimedia material, e-books, and e-journals(Patra & Jha, 2014). These systems frequently don't have the strength to handle various licensing arrangements, handle intricate metadata, or offer thorough usage analytics and statistics. Additionally, they must manage the dynamic nature of electronic resources, which includes upgrades, versioning, and access control, as well as integrate seamlessly with external systems. Inconsistencies and problems with user access may also result from ILS and LSPs' lack of support for the many access models and authentication techniques needed for electronic resources. Libraries need a specialized ERMS that can handle these complexities, offer centralized management, and guarantee effective and compliant access to digital information in order to manage electronic resources.

5.5. Consortia Versus Individual Approach

Navigating the distinctions between consortia and individual library techniques is one of the problems Electronic Resources Management Systems (ERMS) face. When opposed to individual library approaches, consortiums—groups of libraries that collaborate to negotiate and administer electronic resources—present distinct obstacles. Within a consortium, different institutions have different membership arrangements, different needs, and intricate licensing agreements that need to be handled by ERMS(Ross & Sutton, 2016). To manage consortium-specific workflows, shared resource access, and cost allocation, this calls for advanced management skills. Additionally, while preserving the independence and privacy of individual libraries, ERMS must provide centralized reporting and analytics across consortium members. Individual library approaches, on the other hand, call for ERMS to handle resources on an individual basis, which may entail various access models, licensing arrangements, and administrative procedures(Chisman *et al.*, 2007). It is imperative for ERMS to strike a balance between these disparate demands and guarantee fair resource allocation and administration in both group and individual contexts. To efficiently manage electronic resources and maximize resource utilization across all member institutions, a strong ERMS should offer adaptable design choices and thorough support for both individual library models and consortia.

6. Conclusion

In conclusion, Electronic Resources Management Systems (ERMS) are indispensable tools that have revolutionized the management of digital content in libraries and other information institutions. By centralizing and automating workflows related to acquisition, licensing, access control, and usage monitoring, ERMS streamline operations and enhance the user experience. They play a crucial role in ensuring that libraries can effectively manage the complexities of electronic resources, from databases and e-journals to multimedia content and e-books. ERMS also facilitate efficient budget management, support decision-making with comprehensive analytics, and improve the overall accessibility and discoverability of digital collections. Despite the challenges such as the dynamic nature of electronic resources, the limitations of ILS and LSPs, and the differences between consortia and individual library approaches, ERMS provide a robust framework for libraries to maintain organized and compliant access to digital materials. Moving forward, continued advancements in ERMS technologies will be essential to meeting the evolving needs of libraries and users in the digital age.

References

Anbu, J., Kataria, S., & Ram, S. (2013). Dynamics of Managing Electronic Resources: Electronic Resource Management System (ERMS) Initiatives - ProQuest. DESIDOC Journal of Library & Information Technology,, 33(4), 300–305.

Arms, C., Klavans, J., & Waters, D. J. (1999, February 1). Enabling access in digital libraries: A report on a workshop on access management (none) [Electronic book abstract]. None; Digital Library Federation, Washington, D.C. https://old.diglib.org/pubs/dlf102/dlf102.htm

Chisman, J., Matthews, G., & Brady, C. (2007). Electronic resource management. The Serials Librarian, 52(3–4), 297–303. https://doi.org/10.1300/J123v52n03_08

Harrington, C., Haggerty, K. C., & Scott, R. E. (2022). Usability studies in the electronic resource lifecycle. The Serials Librarian, 82(1–4), 164–170. https://doi.org/10.1080/0361526X.2022.2018229

Holly, Y., & Scott, B. (2008). Electronic Resource Management in Libraries: Research and Practice: Research and Practice. IGI Global.

Hosburgh, N. (2014). Managing the electronic resources lifecycle: Creating a comprehensive checklist using Techniques for Electronic Resource Management (TERMS). The Serials Librarian, 66(1–4), 212–219. https://doi.org/10.1080/0361526X.2014.880028

Jewell, T. D. (2009). Electronic resource management in libraries. In Encyclopedia of Library and Information Sciences (3rd ed.). CRC Press.

Jewell, T. D., & Mitchell, A. (2006). electronic resource management: The quest for systems and standards. In Growth, Creativity, and Collaboration. Routledge.

Mangrum, S., & Ellen Pozzebon, M. (2012). Use of collection development policies in electronic resource management. Collection Building, 31(3), 108–114. https://doi.org/10.1108/01604951211243506

McLean, J., & Canham, R. (2018). Managing the electronic resources lifecycle with kanban. Open Information Science, 2(1), 34–43. https://doi.org/10.1515/opis-2018-0003

Morales, J. M., & Beis, C. A. (2021). Communication across the electronic resources lifecycle: A survey of academic libraries. Journal of Electronic Resources Librarianship, 33(2), 75–91. https://doi.org/10.1080/1941126X.2021.1913841

Patra, N. K. (2017). Electronic resource management (ERM) in libraries of management institutes in India. The Electronic Library, 35(5), 1013–1034. https://doi.org/10.1108/EL-01-2016-0020

Patra, N. K., & Jha, S. S. (2014). Comparative analysis of Electronic Resource Management Systems (ERMS): A web study. In Progressive Trends in Electronic Resource Management in Libraries (pp. 118–146). IGI Global. https://doi.org/10.4018/978-1-4666-4761-9.ch007

Prajapati, D. (2016). Electronic resource management in libraries. International Journal of Research in Humanities & Social Sciences, 4(1), 12–16.

Ross, S. V. T., & Sutton, S. W. (2016). Guide to Electronic Resource Management. Bloomsbury Publishing, USA.

Sadeh, T., & Ellingsen, M. (2005). Electronic resource management systems: The need and the realization. New Library World, 106(5/6), 208–218. https://doi.org/10.1108/03074800510595823

28

Institutional Repositories: An Emerging Resource for Scholarly Agricultural Information Communication and Access: A Study

Sanjeev Kumar[1], Dr. Suniti Bala[2] and Amit Kumar[3]

[1]Associate Librarian, M.S. Randhawa Library, Punjab Agricultural University Ludhiana-141 004, Punjab, India, E-mail: srishti88@yahoo.com

[2]Assistant Librarian, M.S. Randhawa Library, Punjab Agricultural University Ludhiana-141 004, Punjab, India, E-mail: sunitipau@gmail.com

[3]Assistant Librarian, M.S. Randhawa Library, Punjab Agricultural University Ludhiana-141 004, Punjab, India, E-mail: 14ak25@gmail.com

Abstract

Open access publishing is gaining importance where individual can make their article accessible to everyone. In the past few years, a boom in open access has led to the growth and development of repositories which allows the researchers to display their work for more distinguishability and visibility. Today librarians are involved to assist the researchers and faculty to know the impact of their research using different modes of metrics. Librarians' focus is shifting from collection development to research evaluation and impact assessment for individual researcher, groups, organizational unit and whole institution. There is remarkable growth in the development of institutional repositories and agricultural institutions are also contributing to the central repository KRISHOKOSH in the field of Agriculture. This article discusses briefly about KRISHIKOSH, its features and contribution of Punjab Agricultural University towards scholarly output. The study uses descriptive research approach to find out how institutional repositories are operating in agricultural institutions. Observation method is used to analyze the data available on the website of Krishikosh. It has been found that theses majorly constitute the total collection of KRISHOKOSH. Only theses submitted to the Punjab Agricultural University constituting the total collection of PAU so there is urgent need to motivate the teachers and scientists to make use of the platform to share their work like research articles, short communications, and chapters for more visibility, growth and development of research.

Keyword: *Open access in agriculture, KRISHIKOSH, Agricultural institutions repository, Agricultural theses database, Online theses in agriculture.*

1. Introduction

Higher educational institutions especially the agricultural universities and institutes have multifaceted mandate which includes teaching, research and extension. Scientific information is generated at rapid pace in the form of research papers, project reports, conference/Seminar proceedings, Theses/ Dissertations, books and so on. It is necessary to manage, preserve and share the intellectual output to the academic community for innovations and betterment of the society. Technological developments, open access and open archive movements have facilitated easy access to information. Open access literature available in public domain removes access as well as price barriers. Realizing the importance of preserving valuable digital and print material of an institution idea of developing institutional repositories was invigorated. It is online digital content free of cost and mostly free from restrictions of copyright and licensing. In other words, it is royalty free literature which can be applied to any digital content that are born digital or the older works digitized later in life. Researchers are not well aware of the various channels to enhance the visibility of their work such as scholarly journals and conferences. Further they don't identify existing library provisions that could help them to share their work on different platforms due to lack of awareness about it. So there is need to create awareness among masses about the institutional repositories and its role in disseminating scientific output of an institution. Institutional repositories are the need of hour for to showcase the research output of an institution.

2. Institutional Repositories and Libraries

Institutional repositories have become common space for community and access to knowledge not only in the work of publication but also on the grey material collection (Salmon & Smart, 2012). "Institutional repositories can be helpful in monitoring the productivity of academic community in relation to scientific output" (Yi-Ping & Tsu-Jui, 2018). "Institutional repositories can be the one of the services organized by the library to manage and serve digital information created by the community to be served" (Oguche, 2018). The main phases in implementation of institutional repositories are (i) Development of scientific work; (ii) Processing of scientific work through self-archiving and digitization (iii) Promotion and dissemination of IRs in increasing its visibility and access. Librarians can serve as strategic information manager who can support the development of intellectual and scientific heritage. Library staff should be encouraged to perform activities like administering, registration and authentication of submission of documents in IR. Further, there is need to develop competences of fresh joiners LIS professionals concerning research processes and its reflection areas (Nicholson & Howard, 2018). Development of institutional repositories goes through different aspects which includes handling, stimulation and dissemination of the scholarly works. Libraries can play an important role in development of institutional repositories which weigh greatly to dissemination of knowledge (Nurdin & Mukhlis, 2019). There should be uniformity in format, capacity and type of file etc. Many of the Indian IRs does not have self archiving facility to the faculties and researchers which is very important in evaluating the progress of IR (Xia & Li, 2007).

3. Objectives of the Study

This study aims:

1. To study the open access initiatives in the field of Agriculture
2. To know the current status of KRISHIKOSH institutional repository
3. To study about the different types' resources hosted in Krishikosh repository
4. To study the contribution of Punjab Agricultural University in supporting and promotion of KRISHIKOSH

4. Methodology

The study applies descriptive research method to discover the operation style of institutional repositories by agricultural institutions. Observation method is used to analyze the data available on the website of krishikosh. Krishikosh repository is accessible at https://krishikosh.egranth.ac.in. The total numbers of different types of documents available in repository were analyzed. Data was analyzed using MS-Excel and percentage method is used to analyze the data.

5. Agricultural Education and Institutional Repositories

Authentic, credible and quick access to information in agriculture sector is crucial for national development. ICAR brought out various digital proposals in upgrading agricultural education and research such as CeRA, Krishikosh, Argoweb, e-krishishiksha etc. Keeping in view, the importance of organizational knowledge for growth of education and research E-Granth subproject was initiated with the support of ICAR under National Agricultural Innovation Project (NAIP to form a platform for providing digital access to huge amount of information available in the National Agricultural Education and Research system (NARES) of India (Jain *et al.*, 2014). Under this initiative digital repository has been figured out to capture, digitize and share the library resources in order to accomplish the information requisites of scientific, teaching and research community.

5.1. KRISHIKOSH

It is a collaborative digital repository for collecting, preserving and disseminating intellectual output of agricultural institutions. It is a distinctive repository of knowledge in agriculture and allied science having collection of theses, previous journals institutional publications, old but valuable books, project reports, technical bulletins, lectures, reprints and other grey literature covering all different state agricultural universities and ICAR institutes. It provides a common platform for implementing open access policy for individual institution's self-managed repository with centralized support. All the issues related to hardware, software and connectivity are administered by Agriculture Knowledge Management Unit at ICAR-IARI and each institute/ University can manage its own repository just as part of KRISHIKOSH. It is jointly run, centrally aggregated with integrated search facility. Krishikosh has enhanced the availability and user-friendliness coupled with preservation of content. Under open access initiatives of ICAR it is mandatory for the all the participating institutions to upload all institutional publications such as journal articles, magazines, brochures, bulletins, project reports, speeches newsletters, annual reports and other grey literature available with the institutions. In addition to it M.Sc and PhD theses (full text) are also submitted in the repository after completion of the work. Access to metadata (Author, Title, Abstract, Publisher etc.) is freely available at the time of deposition whereas unrestricted full text access is available after an embargo period which is not more than twelve months (Sharma *et al.*, 2018).

Currently 103 SAU/ICAR institutes are registered in KRISHIKOSH and 2,75,623 items covering varied area of agriculture in the form of theses, journals, journal articles, book chapters, reports, reprints, etc., are accessible to agricultural fraternity as per Table 1.

Table 1: Content collection in Krishikosh

Sr. No.	Content type	No.	Percentage
1	Theses	200315	73%
2	Journals	17715	6.4%
3	Institutional publications	2173	0.8%
4	Technical Reports	1074	0.4%
5	Books	1074	0.4%
6	Book Chapters	1269	0.5%
7	Articles	17308	6.3%
8	Reports	2247	0.8%
9	Reprints	694	0.25%
10	Training Manual	97	0.03%
11	Proceedings	493	0.1%
12	Other	30869	11%

Different types of documents contributed towards Krishikosh repository are available under Community & Collection of each member institution which is full text searchable. Thus, it has upgraded the accessibility along with maintenance of documents.

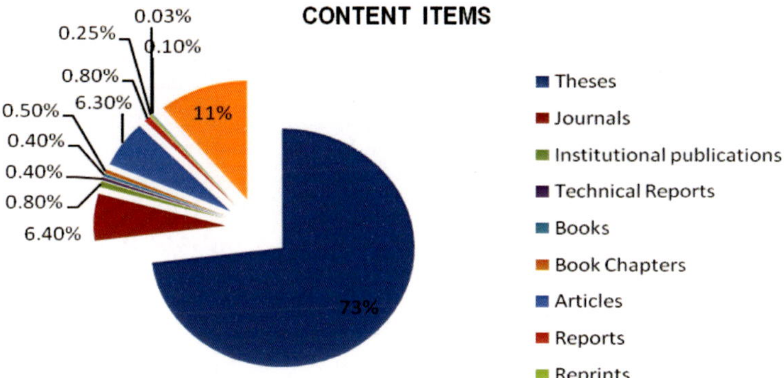

Fig. 1: Content collection

It is clear from the Table 1 that major portion of KRISHIKOSH repository constitutes THESES collection (73%) followed by journals (6.4%) and remaining around 20.6% of the collection comprise of technical reports, book chapters, technical reports and other institutional publications as shown in Figure 1.

Theses collection in KRISHIKOSH: Master level and Doctoral theses covering different area of agriculture and allied sciences are accessible from the repository as shown in Table 2

Table 2: Total No. of Theses in Krishikosh

Sr. No.	Degree	No. of Theses
1	Master level	155083
2	Doctoral Theses	42121
3	Bachelor degree reports	10

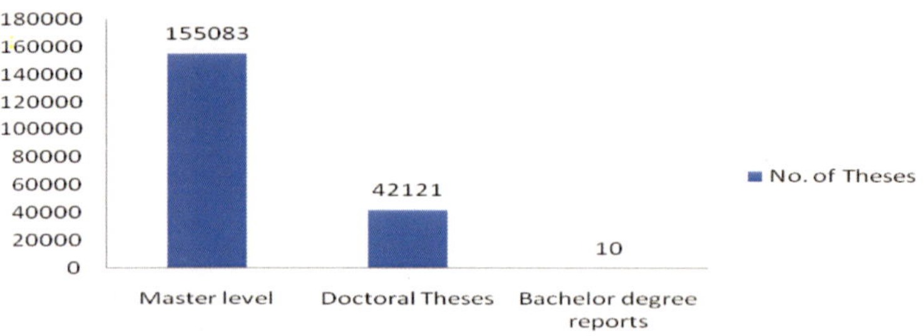

Fig. 2: Total no. of Theses in Krishikosh

Table 3: Subject wise collection of Theses

Sr. No	Degree	Number
1.	M.Sc	120656
2.	M.V.Sc	25091
3.	M.Tech	5730
4.	M.B.A	3561
5.	M.C.A	47
6.	B.Sc	10

Source: krishikosh database (https://krishikosh.egranth.ac.in/)

6. Punjab Agricultural University Library and its Contribution

Punjab Agricultural University is also contributing different types of documents towards Krishikosh repository which is managed by PAU library. PAU library is playing a key role in management, preservation and dissemination of research output of the university to wider audiences. PAU library has well organized collection of e-resources accessible at http://www.pau.edu/msrlibrary/. Library receives one copy of each thesis in CD form which is verified for completeness of the record in the acceptable format to get it uploaded in the repository. After authentication, the record is transmitted for worldwide access. At present Punjab Agricultural University has contributed 6898 theses and 10 Newspaper clippings and 05 reports towards Krishikosh repository.

Library is continuously receiving Master's and PhD theses submitted to the university which are being uploaded in the Krishikosh repository after authentication and proper checking by the library faculty engaged in this work.

Table 4: Collection of PAU in KRISHIKOSH

Type of Documents	**Number**
Theses	6898
Newspaper Clippings	10
Reports	5
Total collection	6913

It is clear from Table 3 that PAU has majorly contributed Theses towards KRISHIKOSH collection which is around 3% of the KRISHIKOSH theses collection as shown in Figure 3

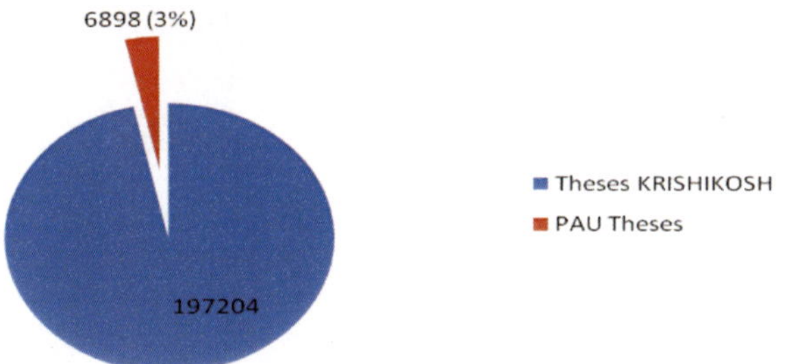

Fig. 3: Contribution of PAU

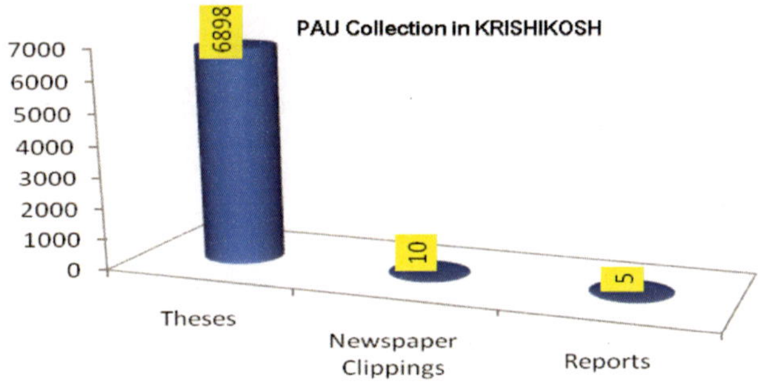

Fig. 4 : PAU collection

7. Conclusion

Institutional repositories are playing key role in preservation and dissemination of institutional output. Krishikosh has become an important platform for sharing information agricultural information. It is providing access to wide variety of documents like Theses, Reports, Articles, Books, Book chapters, Manuals, Lectures etc. Participating institutions are uploading the rare documents and important grey literature which was earlier not available. Each

University and institution should run towards the development of agricultural repository to reach to its end-user so that no one should be deprived of the required information. Librarians should come forward and take lead role in its maintenance and promotion to further strengthen their role as information intermediaries between producer and consumer of information. It has been found that theses majorly constitute the KRISHIKOSH collection which is around 73% and PAU theses collection contributes 3.4% of the total theses collection. Teachers, researchers and scientists should be motivated for self-archiving their work for more visibility and growth.

References

Jain, A. K., Chandrasekharan, H., & Kumar, R. (2014). Final Report of NAIP sub-project 'Digital Library and Information Management under NARS (e-GRANTH)'. Indian Agricultural Research Institute, New Delhi, India

KRISHIKOSH available at https://krishikosh.egranth.ac.in/

Nicholson, J., & Howard, K. (2018). A study of core competencies for supporting roles in engagement and impact assessment in Australia. Journal of the Australian Library and Information Association, 67(2), 131-146.

Nurdin, L., & Mukhlis, S. (2019). Institutional repository as an information infrastructure for scholarly communication: An exploratory study in Indonesia. Malaysian Journal of Library & Information Science, 24(3), 101-114. https://doi.org/10.22452/mjlis.vol24no3.6

Oguche, D. (2018). The state of institutional repositories and scholarly communication in Nigeria. Global Knowledge, Memory and Communication, 67(1/2), 19-33. https://doi.org/10.1108/GKMC-04-2017-003

Punjab Agricultural University https://www.pau.edu/msrlibrary/

Salmon, F., & Smart, C.A. (2012). The role of the academic library in facilitating scholarly communication. Caribbean Quarterly, 58(2/3), 67-78.

Sharma, R., Kumar, A., Pandey, P.S., Himanshu, Kapur, S., & Jain, A.K. (2018). Krishikosh: a digital repository to disseminate agricultural knowledge. Indian Journal of Agricultural Sciences, 88(5), 757-765.

Xia, J., & Sun, L. (2007). Assessment of self-archiving in institutional repositories: Depositorship and full-text availability. Serials Review, 33(1), 14-21.

Yi-Ping, L., & Tsu-Jui, M. 2018. Institutional repositories: A bibliometric study of the social sciences citation index (SSCI). The Electronic Library, 36(3), 504-517. https://doi.org/10.1108/EL-09-2017-0192

29

Reading Drive in Academic Library: An Experimental Study

Yamini P. Galapure

Librarian, Gokhale Education Society's, SMRK-BK-AK Mahila Mahavidyalaya Nashik - 422 005, Maharashtra, India, Email: yaminigalapure.smrk@gmail.com

Abstract

Reading Drive is a unique practice run in our college library. Book Club is established to inculcate reading habit students. This paper describes the activity carried in our college and the efforts taken for successful execution. Book Review system is generated and it shows the output of the book club. Various activities and events to inculcate reading habit are regularly conducted. This resulted in successful outcome through students' email, feedback and book talk etc.. Best reader and Best Book Reviewer appreciation given to the active users of the library. This study is the successful attempt to develop reading habit among students. This paper defines the best practice run in the Academic Library and created Book Lovers in this Digital Age.

Keywords: *Library, Reading habits, Book review.*

1. Introduction

Libraries play an important role in developing reading habit among students. The Academic library specially means to support the teaching learning process and is focus to work on curriculum based education system. Inculcating reading habit among students in academic Library is a challenge. Our College SMRK BK AK Mahila Mahavidyalaya, Nashik is dedicated towards the Women's Education has accepted this challenge and established *Book Club* to inculcate reading habit among students. Reading Drive is one of the unique practices run in our college from 2014. Every year more than 200 students enrolled under the Book Club and read the books other than academics. Under this club, we take various activities such as Book Review Competition, Book talk, Books Exhibitions on Various Occasions, Digital Exhibition of Book Reviews submitted by students, New Arrivals Desk, Ask a Librarian and execute this Reading Drive.

2. Back Ground of the Activity

The College has established *Book Club* to inculcate reading habit among students. Students are encouraged to enroll under *Book Club*. The intention behind it is to develop the reading habit among students and utilize the maximum library resources and facilities.

The details are as below

- **Name of the Activity** :- Book Club (वाचक मंच)
- **Year of Establishment** :- 2014
- **Aims and Objectives:**
  - To cultivate reading habit among students
  - To create inner urge for reading
- **Details of the Activity**
  - Students are enrolled every year for *Book Club*. Library conducts induction program to aware the library services and facilities. During the orientation, students are encouraged to be a member of Book Club. Under this scheme, Library gives additional 01 book other than academics to the members of Book Club. Students can take books of their own choice. Special format is designed for students to give Book reviews. While returning library books, students submit book reviews in the book review form. These students' book reviews are displayed for reading on the notice board and on Library Blog.
- **Enrollment of Students under Book Club**

Table 1: Below table and graph shows the students' enrollment details of last 05 years

Sr. No.	Year	Total Members
1	19-20	256
2	20-21	0
3	21-22	45
4	22-23	131
5	23-24	196

Graph 1: Enrollment of Students during the last 05 years (2019-20 till 2023-24)

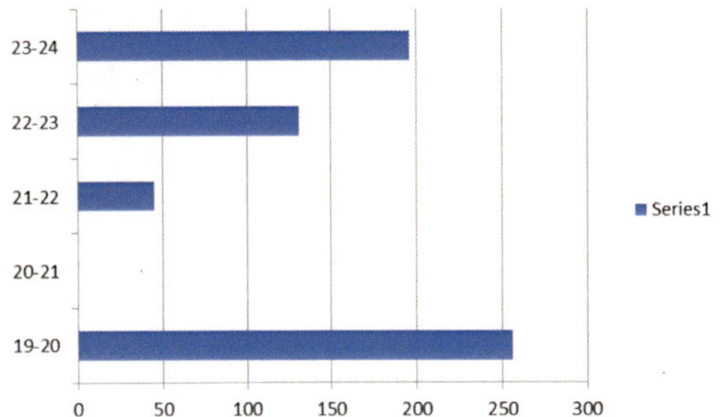

The above graph shows the students enrolled for book club. Due to pandemic in Covid 19, the physical access of students was prohibited hence students were not enrolled for the scheme. But the virtual support was provided for them hence as soon as the physical access was started students started enrolling for book club. This graph shows gradual increase in students' enrollment.

3. Efforts Taken for Enrollment of Book Club Membership and Utilization of Books

- Library User Education Program: Students are oriented to encourage library facilities and services and connect users to library through these programs
- The library's Open Access system provides students with a wide range of exposure and a number of books to choose from.
- Display of Book Reviews published in newspapers on the library notice board for the students.
- Book Review format is designed for students to submit review of the book they read
- Exposure of Book Reviews given by students on Library Blog and on Library Notice Board
- Organise various activities and events for students and appreciate them in the event
- Appreciation of the students as a Best Book Reviewers and Best Reader in the program
- Exposure of Students' Book Reviews presented through Digital form on the Social Media Platforms

4. Library Usage: Observations and Findings

During these last five years, students enjoyed reading several books available in the library. From the year 2018, library has started taking the reviews of the books issued by the students under Book Club. Students are encouraged to read books and give reviews while returning. Till the Academic year 2023-24, we have received more than 100 book reviews of our students.

Active participation of students made library as a vibrant platform for book lovers. Students are now suggesting various books to read. Students connected library through online and offline mode. This has changed the attitude towards the library and resulted in library usage. The level of engagement between the library staff and students increased. Students became comfortable to come in library and became a regular user. Other observed positive changes in the student behavior included added enthusiasm about reading; urge to discuss the read books with classmates; a clearer thought process; and a conviction to give their honest opinions about books.

However, the activity of writing book reviews influenced them so much that by the end of the year, they were deemed ready to give a book-talk based on their written reviews. Such an overall increase in confidence level of students was perceived.

5. Conclusion

This Reading Drive has successfully proved the efforts taken by library. Best readers, Best Book Reviewers were felicitated in the program. This has resulted in maximum utilization of books from library. Activities like Book Review, Book Talk were successful attempt and output of this drive. Books read by students were presented through digital exhibition and made available to ease access for students. Liberty and Choice of reading created Joy of reading among students. Book Reservation system increased. Ask a Librarian service successfully run in our library. This reading drive has successful attempt to develop reading habit among students. Our Library reached to the satisfaction level for successful attempt of this Reading Drive. The five Laws of Library Science are in actual execution and satisfaction in our library.

References

Aina, A. J., Ogungbeni, J. I., Adigun, J.A., & Ogundipe (2011). Poor reading habits among Nigerians: The role of libraries. Library Philosophy and Practice (e-journal), 529.

Green, M.A. (2022). Evolving library services in the digital age. College & Research Libraries, 83(2), 250-265.

J.S. (2019). Digital libraries: Expanding access to reading in the 21st Century. Journal of Library Innovation, 10(2), 75-92.

Johnson, P. (2018). The role of libraries in promoting reading habits. Library Trends, 67(3), 421-438.

L.B. (2020). Community engagement through library program. Public Libraries, 59(4), 201-218.

30

Relevance of E-Learning: A Case Study in Reference to DAV College for Girls, Yamuna Nagar

Dr. Renu Bala

Librarian, PIET Sanskriti Senior Secondary School, NFL Township, Panipat - 132103 Haryana, India, E-mail: renu.bala149@gmail.com

Abstract

Education through Technology and advancement also termed as e-learning has became an integral part for all in today's world involving wide range of in-house trainings, workshops through technological ways, delivery sources and digital approaches. E-learning is transferring of ideas and skills through digitalized and net based system. It includes patterns and process of teaching and making understandable the idea through information Technology. E-learning has brought a lot of change in education, teaching system and all fields of learning including library also. Library and Information science has also gathered revolutionary change due to this advance technology. This paper deals with the impact of e-learning or use of digitalized means on learning and developing aspects of faculty and students of DAV College for Girls, Yamuna Nagar. The Findings focuses on the familiarity of faculty and students with e-learning with the usage of digital resources.

Keywords: *E-learning, ICT, Education, Digital approaches, Digital technologies.*

1. Introduction

The technology is a part of dynamic environment which has brought a lot of change in styles and pattern of various fields. Out of this education is also one of the integral parts. Technology has brought immeasurable effect in the field of education. E-learning consists of various types of facilities such as audio and images delivers text animation *etc*. E-learning process involves many types of local networks or the internet basis learning. E-learning has made learning and developmental activities so frequent which leads to easy access to required information. E-learning is a way of creating knowledge and circulating the same where teacher plays a vital role. The basic objective of e-learning is exploring and adding essential knowledge.

2. E-learning

Learning plays a very significant role in our life. The term learning covers every modification in behavior to meet environmental requirements (Gardner Murphy, 1968). Learning is the acquisition of habits, knowledge and attitudes. It involves new ways of doing things, and it operates in an individual's attempt to overcome obstacles or to adjust new situations. It represents progressive change in behavior. It enables him to satisfy interest to attain goals". Crow and Crow (1973), "our actions are influenced by what we do in past and learn from it". Learning reflects our personality and behavior. By considering all these views we can say learning means the changed and modified behavior of the individual which he undergoes during his/her lifetime. The delivery of learning, training or an education program is by electronic means is added value to the existing learning program.

E-learning involves the use of a computer or electronic device (*e.g.* a mobile phone) in some way to provide training, educational or learning material (Derek Stockley 2003). According to John Chambers (2001, Rosenberg), "the biggest growth in the Internet, and the area that will prove to be one of the biggest agents of change, will be in E-learning." Alvin Toffler (2001) declared, "The illiterate of the 21stcentury will not be those, who cannot read and write but those who cannot learn, unlearn, and relearn." An ancient proverb says: "if we don't change our direction, we'll end up exactly where we are headed". This indicates that everyone will have to constantly change and adapt the new learning styles if they are not to lag behind react to elements of E-learning (since student perception and attitude is critical to motivation and learning) along with how to apply these approaches most effectively to enhance learning.

The process of applying technology in the field of education is called e-learning which includes various sources such as internet, CD ROMS, Computer programs, web quests, online discussion, bulletin board, video *etc*. Media is also one of the parts of encouraging technology based learning. E-learning means different types of learning by a group or class. E-Learning facilities are used in a library where information can be stored in electronic format contains texts, images, sound, video, maps scientific & business data. There will be a vast gathering of users scattered around the globe who are able to assess it easily and conveniently.

3. Pros and Cons of E-Learning

E-learning is ubiquitous in nature, available anytime which comprises traditional system of learning with modern equipments. E-learning provides bulk of material with lesser efforts and works as supportive for all and also increases attention, contribution and attractive. Online availability leads to pollution free environment and support for green environment as saving of papers by non cutting of trees. It is also cost and time effective.

Along with some advantages, sometimes people do not like online information as they are habitual for traditional means. Complex content creates the problems for interaction. It is not easy to measure the effect or results by using of poor and improper technology. Students may feel less interactive with instructor as with e-learning tool. All cannot be contracted. E-learning creates hurdles in public speaking system and group efforts also create hurdles.

4. Sources for E-Learning

Several tools have been employed for e-learning such asBlogs, Twitter, Wikis, Delicious, Specialized software, Slide Share, Emails, Audacity, E-portfolio, Google Search, Animation, Word Press, Messenger, Google reader, Learning management System (LMS), You Tube *etc.*

5. Objectives

- To find out the awareness of e-learning in the users
- To find out the interest of users for e-learning
- To find out the effects of e-learning
- To find out the satisfaction level of users

6. Research Methodology

This paper is based on primary data. Keeping in mind the concepts of e-learning questionnaire has been prepared .The questionnaire contains 26 questions. The questionnaire was distributed to users of DAV College, Yamuna Nagar. Users are divided into three categories i.e. graduate, students, post graduate student and teachers. The No. of users is 160. Secondary sources such as books, research articles, journals, web-portals and internet etc. are also used. Various websites visited to get information related to e-learning. Further various e-journals and other articles are reviewed. Various books and magazines are also reviewed. Some information from proceedings has also been taken.

7. Analysis and Findings

In the present study, as data shown in Table 1, it is observed that out of 160 respondents, 25% were in teaching profession, whereas 53.12% were UG students, and 21.8 % PG Students.

Table 1: Types of Users

Types of Users	No. of the Respondent	Percentage
Teaching Staff	40	25.00
Undergraduate Students	85	53.12
Postgraduate Students	35	21.80
Total	**160**	**100**

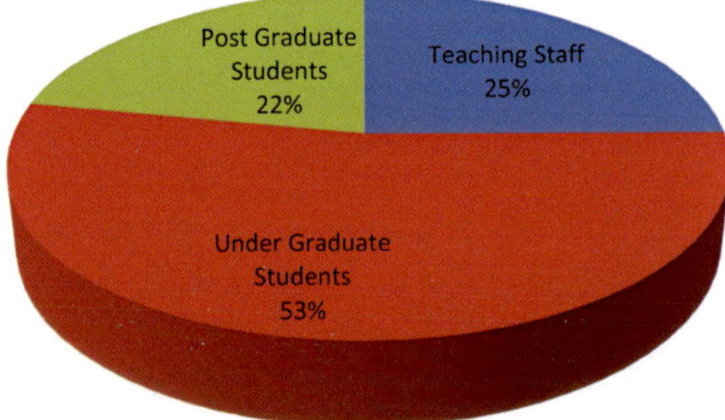

Figure 1: Categories of Users

Table 2: Different Streams of Users

Stream	No. of the Respondent	Percentage
Arts	36	22.5
Commerce	26	16.25
Science	91	56.87
Others	7	4.37
Total	**160**	**100**

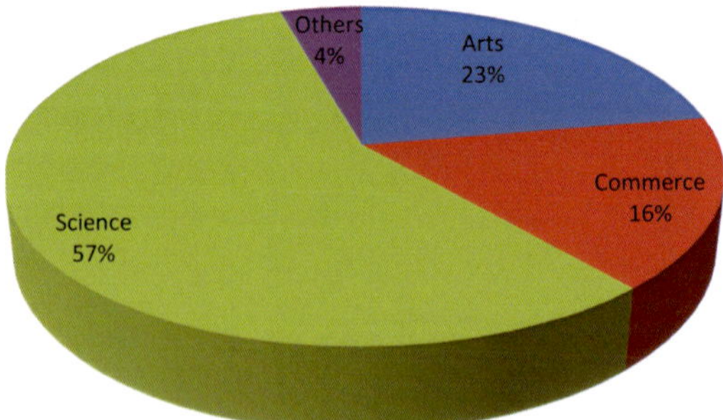

Figure 2: Stream-wise Distribution of Users

Table 3: Usage of computer

Computer Usage	Teacher	Percentage	PG Students	Percentage	UG Students	Percentage
Daily	30	75	7	20	10	11.76
Weekly	7	17.50	14	40	44	51.76
Monthly	1	2.50	1	2.85	6	7.05
Rarely	2	5.0	13	37.14	25	29.41
Total	**40**	**100**	**35**	**100**	**85**	**100**

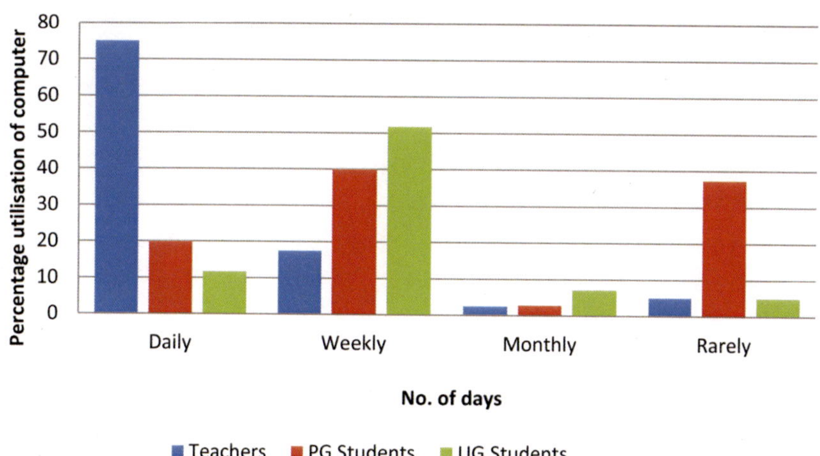

Figure 3: Computer Usage

In the present survey, the data given in Table 3 highlights that 75% teachers utilize computer daily, followed by 17.5% utilize computer for their weekly. Only a small proportion of teachers utilize computer monthly (2.5%) and rarely (5%). But in case of PG Students, maximum utilization of computer has been observed in weekly (40%), followed by rarely (37.14%) and 20% daily basis respectively. The similar trend has been observed in case of UG Students as high percentage in weekly basis (51%) and low percentage observed in monthly basis (7.05%).

Table 4: Duration of time on computer

Hours	Teacher	Percentage	PG students	Percentage	UG Students	Percentage
1 Hrs	28	70	25	71.42	59	69.41
2Hrs	06	15	07	20	18	21.17
3Hrs	04	10	01	2.85	04	4.70
5 Hrs and above	02	5	02	5.71	04	4.70
Total	**40**	**100**	**35**	**100**	**85**	**100**

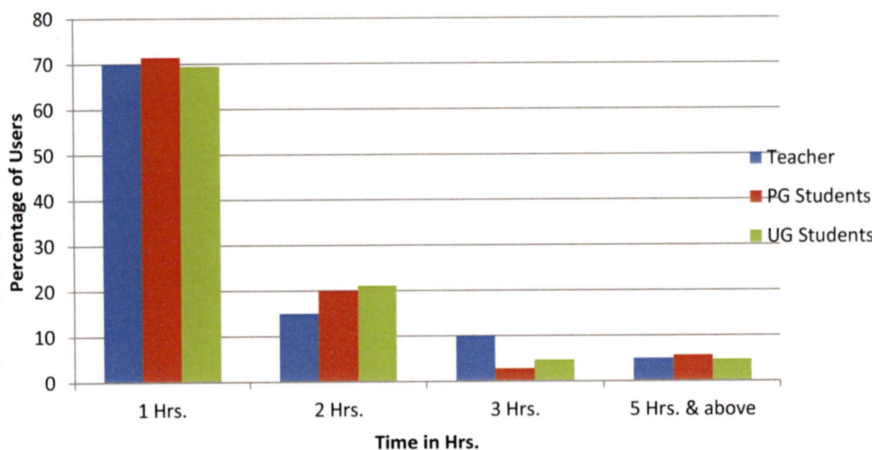

Figure 4: Duration of Time on computer

The data given in Table 4 shows that 70% teachers spent their time on computer for 1 hour, 15% for 2hrs, 10% for 3 hrs, and 5% for 5 hrs. and above. But in case of PG Students, 71.42% spent their time on computer for 1 hrs, 20% for 2 hrs, 2.85% for 3 hrs, and 5.71% for 5 hrs. and above. The proportion of teachers spent 3 hrs. on computer is more than students.

Table 5: Internet usage

Networking Usage	Teacher	Percentage	PG Students	Percentage	UG Students	Percentage
Rarely	04	10	23	65.71	45	5.94
Daily	29	72.5	6	17.14	20	23.52
Weekly	07	17.5	6	17.14	10	11.76
Monthly	0	0	0	0	08	9.41
Addicted	0	0	0	0	02	2.35
Total	**40**	**100**	**35**	**100**	**85**	**100**

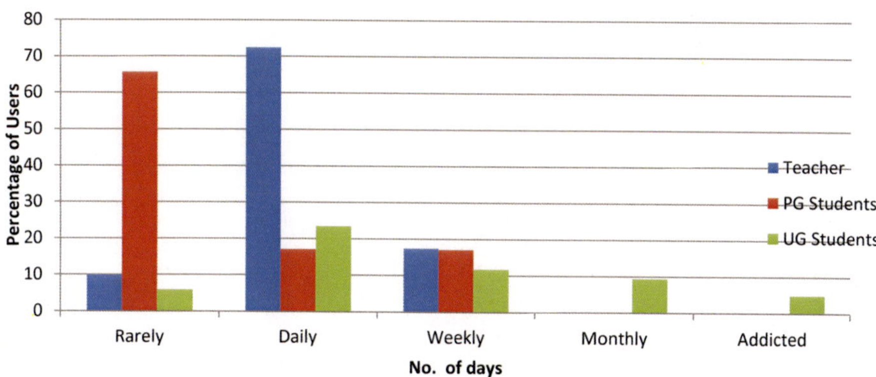

Figure 5: Internet Usage

The presented data in Table 5 highlights that 72.5% teachers utilize Internet daily, followed by 17.5% for their weekly work. Only a small proportion of teachers utilize computer rarely (10%). But in case of PG Students, 23% utilized Internet rarely, 6% equal proportion between daily and weekly basis. The similar trend has been observed in case of UG Students (Table 5).

Table 6: Place of Internet usage

Place of Internet Usage	Teacher	Percentage	PG Students	Percentage	UG Students	Percentage
At Work	05	12.5	1	2.85	01	1.17
At Home	02	5	12	34.28	21	24.70
At College	16	40	19	54.28	58	68.23
At Class	0	0	2	5.71	02	2.35
Everywhere	17	42.5	1	2.85	03	3.52
Total	**40**	**100**	**35**	**100**	**85**	**100**

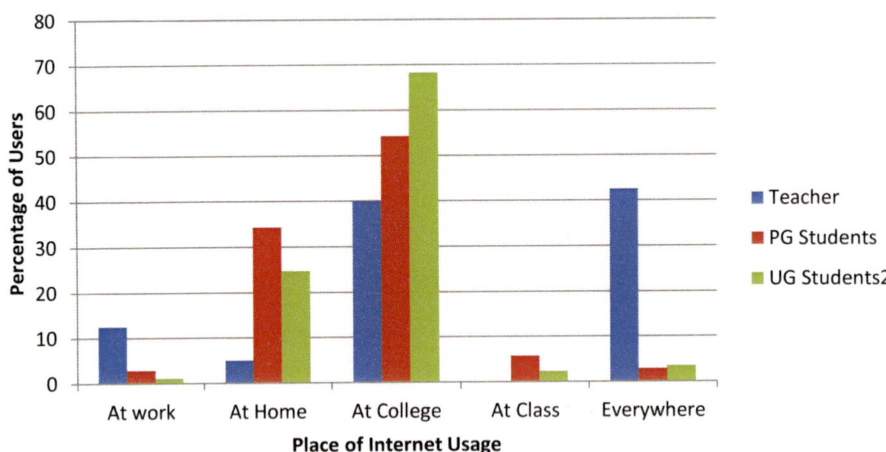

Figure 6: Place of Internet Usage

Table 6 Illustrates that 12.5% teachers utilize computer at their work, 5% at home, 40% at their work and 42.5% at everywhere. Like this 2.85% PG Students utilize computer at their work, 34.28% at home, 54.28% at college, 5.71% in class and 42.5 at everywhere. In the similar way it observed that 1.17% UG Students utilize computer at their work, 24.70% at home, 68.23% at college, 2.35% in class and 3.52 at everywhere. The students of UG & PG utilize their maximum time on internet at college and teachers are use internet at everywhere.

Table 7: Purpose of Internet usage

Purpose of Internet Usage	Teacher	Percent-age	PG Students	Percent-age	UG Students	Percentage
For work	17	42.5	6	17.14	07	8.23
For Studies	5	12.5	22	62.85	55	64.70
For Entertainment	2	5	2	5.71	13	15.29
For Recreational	2	5	1	2.85	04	4.70
Everywhere	14	35	4	11.42	06	7.05
Total	**40**	**100**	**35**	**100**	**85**	**100**

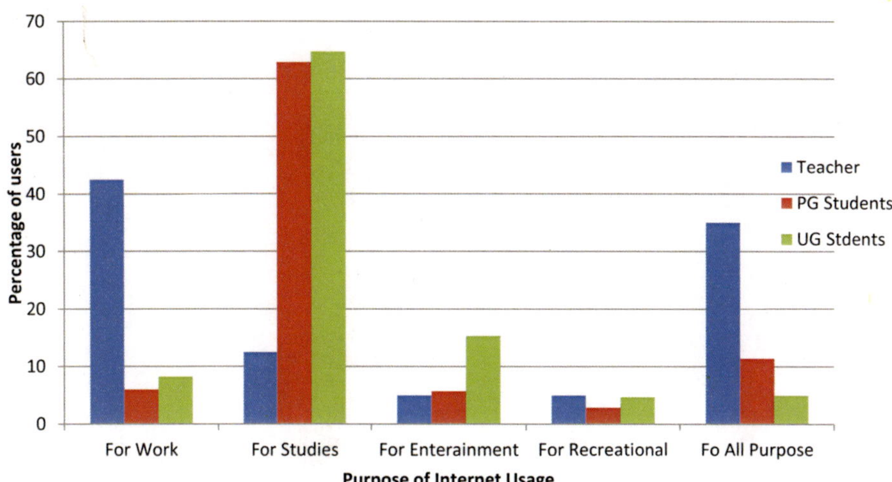

Figure 7: Purpose of Internet Usage

Table 7 shows that 42.5% teachers use the Internet for official or academic work, 12.5% for studies, 5% for entertainment, 5% for recreational work, and 35% teachers use Internet for all other purposes. PG Students use the Internet more mostly for their academic studies 62.85%, 5.71% for entertainment, 2.85% for recreational work and 11.42% for all other purposes. In same way, 64.7% UG students utilize Internet for their academic studies, 15.29% for entertainment, 8.23% use the Internet for work, 4.70% for recreational work and 7.05% for all propose.

Table 8: Usage of Networking Sites

Usage of Networking Sites	Teacher	Percentage	PG Students	Percentage	UG Students	Percentage
Never	09	22.5	21	60	36	42.35
Occasionally	15	37.5	04	11.42	24	28.23
Many times	14	35	10	28.57	23	27.05
Addicted	02	5	-	-	02	2.35
Total	**40**	**100**	**35**	**100**	**85**	**100**

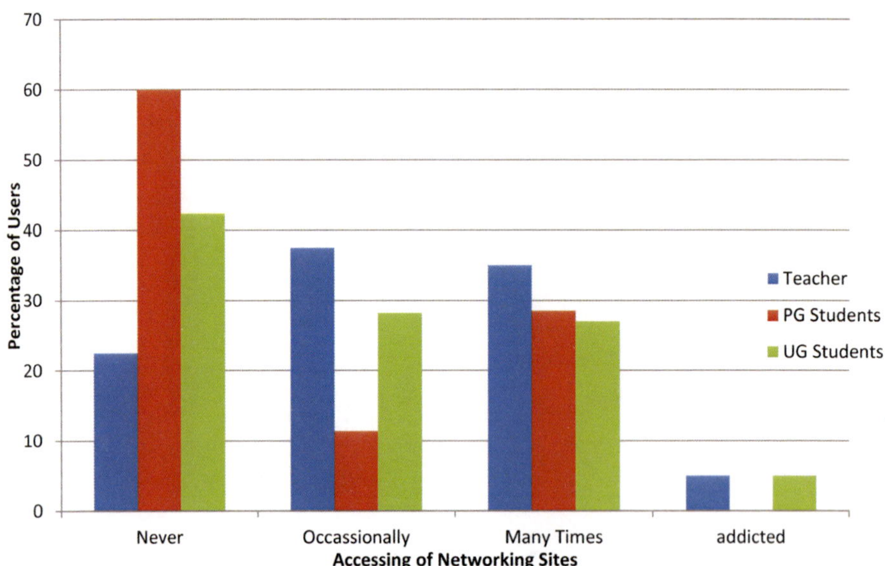

Figure 8: Usage of Networking Sites

The data presented in Table 8 shows that 22.5% teachers never use networking sites, 37.5% occasionally, 35% many times and 5% are addicted. 60% PG students never use networking sites, 11.42% occasionally, and 28.57% many times. 42.85% UG students never use networking sites, 28.23% occasionally, 27.05% many times, and 2.35% were addicted with the use of networking sites.

Table 9: Usage of social networking sites

Social Networking Sites	Teacher	Percentage	PG Students	Percentage	UG Students	Percentage
Email, face book, Chatting etc.	31	77.5	07	20	49	57.6
Online Games	02	5	04	11.4	11	12.9
Online Personal Space like Blogs, Web Pages, *etc.*	01	2.5	01	2.8	0	0

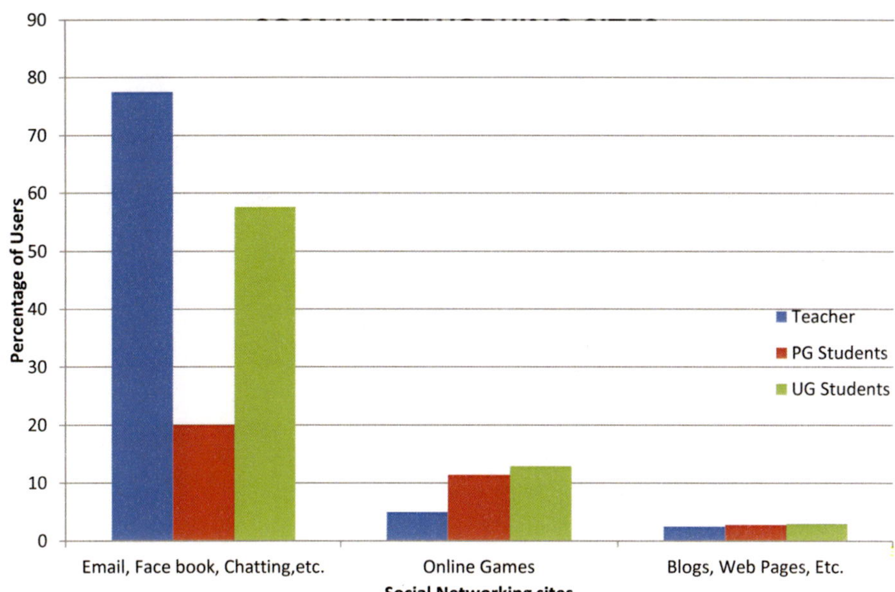

Figure 9: Usage of Social Networking Sites

The present study (Table 9) shows that 77.5% teachers use Internet for email, face book and chatting, 5% for online games, 2.5% for blogs and web pages. On the other hand, 20% PG students use Internet for email, face book and chatting, 11.14% for online games, 2.8% for blogs and web pages respectively. Similarly, 57.6% UG students use Internet for email, face book and chatting, and 12.9% for online games.

Table 10: Usage of messaging and Discussion Tools

Messaging and Discussion Tools	Teacher	Percent-age	PG Students	Percent-age	UG Students	Percentage
Never	06	15	02	5.71	36	42.35
Rarely	06	15	20	57.14	13	15.29
Occasionally	24	60	04	11.42	07	8.23
Daily	03	7.5	09	25.71	21	24.70
Weekly	01	2.5	-	-	05	5.88
Addicted	-	-	-	-	03	3.52
Total	**40**	**100**	**35**	**100**	**85**	**100**

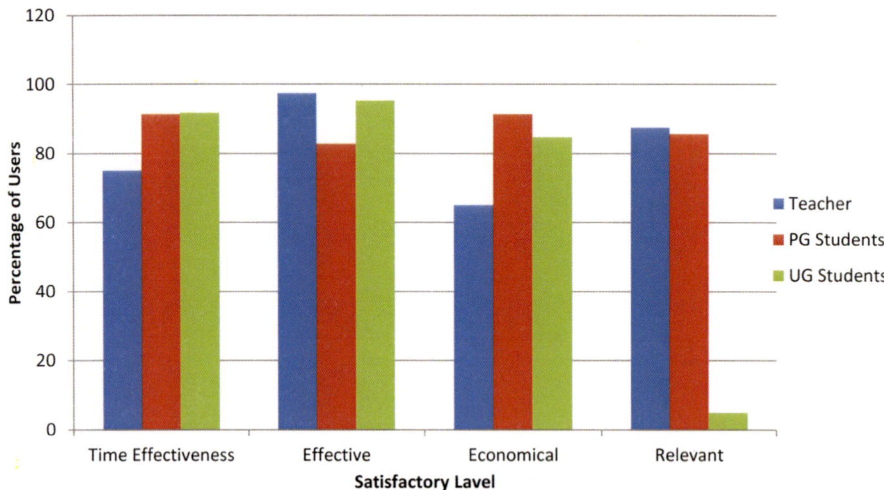

Figure 10: Message & Messaging Tools

The presented data in Table 10 depicts that 15% teachers never use messaging and messaging tools, while 15% teachers use these tools rarely, followed by 60% occasionally, 7.5% daily and 2.5% weekly respectively. Another side, 5.71% PG students never use messaging and messaging tools but 57.14 use these tools rarely, 11.42% occasionally, and 25.71% daily. Similarly, 42.35% UG students never use messaging and messaging tools, while 15.29% use these tools rarely, followed by 8.23% occasionally, 24.70% daily, 5.88 weekly and 3.52% were addicted respectively. The maximum use of these tools was made by the PG students.

Table 11: Accessibility of e-Learning resources

Accessibility	Teacher	Percentage	PG Students	Percentage	UG Students	Percentage
Time Effectiveness	30	75	32	91.42	78	91.76
Effective	39	97.5	29	82.85	81	95.29
Economical	26	65	32	91.42	72	84.70
Relevant	35	87.5	30	85.71	77	90.58

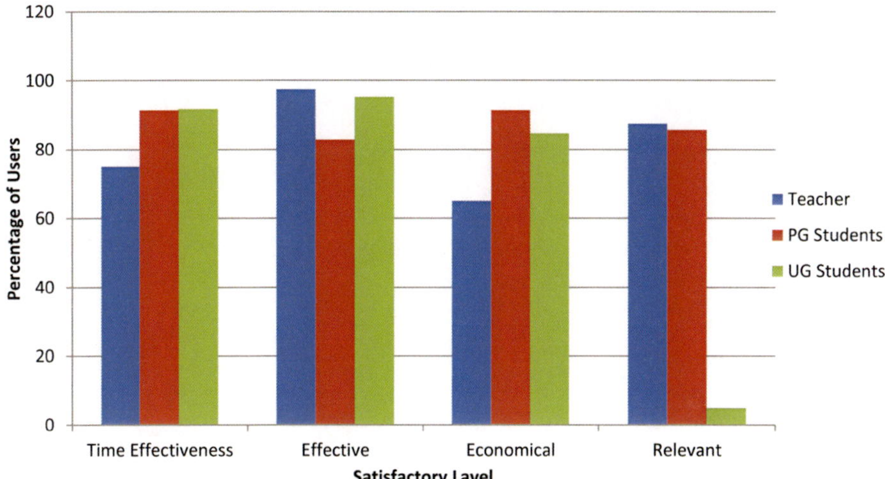

Figure 11: Accessibility of E-Learning resources

The data given in Table 11 reveals that 75% teachers said that e-learning has time effectiveness, 97.5% said effective, 65% said economical and 87.5% said this is relevant. On the other hand, 91.42% PG students said that e-learning is time effective, 95.29% effective, 84.70% economical and 90.58% said it is relevant. This table explains that most of the learners were more satisfied from e-learning.

8. Conclusion

Today computer has become a communication link where the learner and instructor are attached by the computer. The widely enabled technology used by the users has proved very beneficial and useful for every category of users. E-Learning is the leading way where the learner and the instructor interact with each other in a very easy manner. The upgraded and existing technological setup has boosted the interest of users according to advanced environment. E-Learning is based on World Wide Web which has made easy accessibility of information to the users which is shown by this article based on data collected from respondents of DAV College for Girls, Yamuna Nagar (Haryana). Through this survey an effort is made to check that how many people are using Internet facility and e-learning system. This survey shows that time and effort can be saved by using e-learning. The use of E-learning has proved the benefits of self learning for students and faculty. This survey shows the effective utilization of various computer based facilities.

References

Clark, Allan (2004). e-learning Skills. UK Palgrave- MacMillan.

Cross, Jay (2004). An informal history of e-learning on the horizon, 12(3).

Curtin, D.P. *et al.* (1999). Information technology: The breaking wave. Tata Mc Graw-Hill, New Delhi.

Damaraju, S.P., Sai Rani, & Reddy T.N. (2016). Influence of e-learning on the learning and development aspects of IT professional. IUP Journal of Information Technology, XII(4).

Galagan, P. (2000). Getting started with e-learning, training and development. 54(4).

Khanna, Sunaina (2013). E-learning in higher education: Indian perspectives. Libraries in the Digital Era. Chandigarh Librarians' Association. ISBN: 978-81-909943-6-1.

Khanna, Sunaina, & Chohda, Nirupma (2013). Role of libraries in e-learning. Libraries in the Digital Era. Chandigarh Librarians Association. ISBN: 978-81-909943-6-1.

Mahapatra, G. (2006). LIS education in India: Emerging paradigms, challenges and proposition in the digital era. Presented at the Asia-Pacific conference on Library and Information Education and Practice (A-LIEP2006), Singapore 3-6 April.

Mattoo, M.S. (2012). EZPROXY-A powerful platform for advanced e-learning: A phenomenal-case of Kashmir University Library E-resources. In P. Venkata Rao *et al.* (eds.), Libraries and Reading Habit in Electronic Environment, B.P.Publications, Panchkula.

Patnail, Srikant (2001). First book on information technology. Dhanpat Rai and Co., New Delhi.

Pradeep Kumar, & Ranjana (2013). Significance of e-learning in library and information science education through distance mode. Libraries in the Digital Era. Chandigarh Librarians' Association. ISBN: 978-81-909943-6-1.

Wang, M. (n.d.). The strategic role of digital libraries. Library Review, 52(3).

Wasan, Shalini (2012). Paradigm shift from the Internet to the Intranet. In P. Venkata Rao *et al.* (eds.), Libraries and Reading Habit in Electronic Environment. B.P. Publications, Panchkula.

www.elearnspace.org/blog/archives

www.indianjournals.com

www.Internetjournal.com

www.jstor.org

31

Harnessing the Power of Web 2.0: A Comprehensive Review of Tools and Services Transforming Digital Interaction

***Abdul Rajak Ahmed*[1] *and Prince Buragohain*[2]**

[1]*Department of Mass Communication, Mizoram University, Aizawl -796 004 Mizoram, India, E-mail: abdulrajakahmed22@gmail.com*

[2]*Department of Library and Information Science, Mizoram University Aizwal - 796 004, Mizoram, India, E-mail: buragohainprince@gmail.com*

Abstract

The evolution and effects of Web 2.0 tools and services are examined in this study across various sectors, focusing on their applications in education, business, and social interactions. Web 2.0 signifies achange out of static online pages and toward user-generated material that is dynamic, fostering collaboration and interactivity. This review synthesizes findings from numerous studies to highlight the important characteristics of Web 2.0 technologies, including social networking, collaborative tools, and user-generated materials. It discusses the implications of these tools for enhancing learning experiences, improving business practices, and facilitating social engagement. Furthermore, the review addresses challenges such as digital literacy, privacy concerns, and the digital divide. By analyzing case studies and empirical research, this review provides insights into best practices for effectively leveraging Web 2.0 tools and services. The findings underscore the transformative potential of these technologies in reshaping digital interactions and enhancing user engagement.

Keywords: *Web 2.0 tools, User engagement, Academic libraries, Higher-education, Collaboration, 21st-century skills.*

1. Introduction

Web 2.0 speaks about the second version of web advancement and draft that aims to make it easier to communicate and safer to share information, as well as to make the web more interoperable and user-centered. Social networking sites, video-sharing sites, Wikis, blogs, and folksonomies are just a few of the web-based communities, hosted services, and apps that have grown and changed because of Web 2.0 ideas. (King and Brown, 2009; Wikipedia, 2010c). According to Birdsall (2007) web 2.0 is a social revolution since these tools have brought new changes in the society. According to O'Reilly (2005) the advent of Web 2.0 has revolutionized the way individuals and organizations interact with the internet, marking a significant shift from a static web experience to a dynamic, user-driven environment. This transformation is characterized by the emergence of platforms and tools that empower users not only to consume content but also to create, share, and collaborate on it. As a result, Web 2.0 has fostered a more participatory online culture, where users actively engage in the generation and dissemination of information.

An important aspect of Web 2.0 is the emphasis on the content that is produced by users. Unlike the earlier web, where content was primarily produced by a limited number of creators, a Web 2.0 platform for example wikis, blogsand social media networks allows individuals to contribute their perspectives and ideas. This democratization of content creation has led to a rich diversity of voices and a more inclusive digital landscape (Kaplan & Haenlein, 2010). For instance, platforms like Wikipedia exemplify how collaborative efforts can result in comprehensive and constantly updated repositories of knowledge, created and maintained by users from around the world (Bryant *et al.*, 2005).

Collaboration is another cornerstone of Web 2.0 technologies. Tools including Google Docs and Trello permit multiple users to collaborate in real-time, enhancing teamwork and collective problem-solving. This collaborative functionality is particularly beneficial in educational settings, where students can engage in group projects and peer review processes, fostering deeper learning experiences (Harris & Rea, 2009). The ability to collaborate seamlessly across distances has also transformed business practices, enabling teams to operate efficiently regardless of their physical locations (Morrison, 2014).Interactivity is a key aspect of Web 2.0 that distinguishes it from its predecessor. Users can engage with content through comments, likes, shares, and other forms of feedback, creating a two-way communication channel between creators and consumers (Boulos *et al.*, 2011). This interactivity not only enhances user engagement but also allows for the rapid dissemination of information. For example, social media platforms like Twitter enable users to share news and updates in real-time, facilitating discussions and reactions that can shape public opinion and influence events (Bruns & Moe, 2014). Accessibility is another significant advantage of Web 2.0 tools. Many of these platforms are free or low-cost, making them accessible to a wide audience, including individuals and organizations with limited resources (Bennett &

Maton, 2010). This accessibility has led to increased participation in online communities and the democratization of information, as users from diverse backgrounds can contribute to discussions and share their knowledge.

The implications of Web 2.0 tools extend beyond individual users to various sectors, including education and business. In education, the integration of Web 2.0 technologies has transformed traditional teaching methods. Educators are increasingly utilizing platforms like Edmodo and Moodle to create interactive learning environments that facilitate asynchronous learning and collaboration among students (Bates, 2015). These tools support diverse learning styles and promote engagement through multimedia content and interactive activities. Moreover, the flipped classroom model, which leverages Web 2.0 tools to deliver instructional content outside of class, allows for more interactive, hands-on activities during class time, fostering deeper understanding and application of knowledge (Bergmann & Sams, 2012).

2. Problem Statement

This review examines the integration of web 2.0 tools in higher education, libraries and other institutions, including benefits, challenges, and barriers. Despite prior research, this study is crucial in addressing challenges faced by libraries and other institutions during web 2.0 integration, including financial constraints, insufficient personnel, and lack of policies. This study greatly enhances the writing on integrating social media in libraries by giving a complete and up-to-date summary. The review's results will help libraries offer better services and stay relevant in the digital age, which will eventually make a bigger difference in the areas they serve.

3. Objectives

- To evaluate the interest of incorporating Web 2.0 tools in libraries
- To find the challenges and considerations faced while adopting web 2.0 tools
- To figure out the factors that make it easier for libraries and other institutions to adopt Web 2.0 tools

4. Methodology

This literature review employs a comprehensive approach to synthesize findings from various research on the use of Web 2.0 tools in educational settings. Methodology involves the following key steps:

a) Literature Search and Selection

A systematic search was conducted using academic databases such as Semantic Scholar, Google Scholar, and ERIC to find related studies that came out between 2010 and 2023. The search terms were made up of groups of popular phrases like "Web 2.0", "education", "higher education", "student engagement", "collaboration", and "21st-century skills". In order to be included in the review, studies had to meet the following norms:

- Focus on the use of Web 2.0 tools in educational contexts, particularly in libraries at higher education
- Examine the impact of these tools on student engagement, collaboration, and/or the development of 21st-century skills
- Present empirical findings or case studies to support their conclusions
- Be published in peer-reviewed journals or conference proceedings

b) Data Extraction and Synthesis

From each study that was included, important details were taken out, like the research goals, methods, results, and what they signify. The extracted data was organized in a tabular format to facilitate comparison and synthesis across studies.

c) Thematic Analysis

In order to find the common themes and patterns a method called thematic analysis was used to find patterns and themes that went through all of the works that were looked at. The researchers coded the studies and then met to discuss and refine the emerging themes. The final themes were:

1. Enhancing student engagement and learning
2. Fostering collaboration and communication
3. Developing 21st-century skills
4. Addressing challenges in implementation

d) Quality Assessment

The quality of the included studies was assessed using a modified version of the Mixed Methods Appraisal Tool (MMAT) (Hong *et al.*, 2018). This tool evaluates the methodological quality of studies across different study designs, including qualitative, quantitative, and mixed methods. Studies were rated as high, moderate, or low quality based on their adherence to established research standards.

e) Data Synthesis and Reporting

The findings from the thematic analysis and quality assessment were put together to give a full picture of the circumstancesof current state of research on Web 2.0 tools in education. The synthesis focused on identifying key themes, highlighting areas of consensus and disagreement, and discussing the implications for educational practice and future research. The results are presented in a narrative format, with direct quotes and examples from the included studies to illustrate key points.

f) Limitations

There are several limitations of this review. First, only English-language works were included in the search, which might have left out important studies written in other languages. Secondly, this review focused on higher education contexts, it's possible that the results can't be applied to other academic settings. Lastly, the studies that were included were not all of the same quality, with some relying on small sample sizes or lacking rigorous methodological approaches.

This literature review employed a systematic and rigorous methodology to synthesize current research on the application of Web 2.0 tools in higher education. The findings highlight the potential of these tools to enhance student engagement, foster collaboration, and develop 21st-century skills. However, challenges remain in ensuring equitable access and addressing privacy concerns. Future research should continue to explore best practices for integrating Web 2.0 tools into educational settings while addressing these challenges.

5. Important Components of Web 2.0 Tools

Web 2.0 tools are distinguished by several defining characteristics:

- **User-generated content**: Platforms such as blogs, wikis, and social media allow users to create and share their own content, fostering a diverse range of perspectives and ideas (Kaplan & Haenlein, 2010).
- **Collaboration**: Tools enable real-time collaboration among users, enhancing teamwork and collective problem-solving (Harris & Rea, 2009).
- **Interactivity**: Web 2.0 tools facilitate two-way communication, allowing users to engage with content through comments, likes, shares, and other forms of feedback (Boulos *et al.*, 2011).
- **Accessibility**: Many Web 2.0 tools are free or low-cost, making them accessible to a wide audience, including individuals and organizations with limited resources (Bennett & Maton, 2010).

6. The Intersection of Libraries and Web 2.0: Transforming Information Access and User Engagement

Libraries have long been regarded as vital resources for information access and literacy. However, the rise of digital technologies and the internet has prompted libraries to adapt to changing user needs and expectations. The integration of Web 2.0 tools into library services presents opportunities for enhancing user engagement and improving service delivery.

a) Enhancing User Engagement

With Web 2.0 tools, libraries can connect with their users in innovative and useful ways. Social media platforms, blogs, and interactive websites allow libraries to share information, promote events, and engage with patrons. For instance, libraries can use platforms like Facebook and Twitter to provide real-time updates on new acquisitions, upcoming programs, and community events. This interaction not only keeps people up to date, but it also builds community.A study conducted on the use of Web 2.0 tools by library professionals in Lagos State tertiary institution libraries revealed that these technologies significantly enhance communication and collaboration among library staff and patrons (Ogunleye *et al.*, 2015). The findings indicated that while the extent of Web 2.0 tool usage for service delivery was average, there was a clear recognition of their potential in improving library services. Libraries that actively engage with their users through these platforms can create a more vibrant and interactive environment, encouraging greater participation and feedback.

b) Improving Service Delivery

Web 2.0 tools also provide libraries with innovative ways to enhance their services. For example, libraries can utilize collaborative platforms like Google Docs and wikis to create shared resources and knowledge bases. These tools enable librarians to work together on projects, share expertise, and develop comprehensive guides for users. Additionally, libraries can implement online reference services, allowing patrons to ask questions and receive assistance in real-time.Online learning groups have also grown because of Web 2.0 tools used in libraries. These communities provide users with access to a wealth of resources, including tutorials, webinars, and discussion forums. By fostering collaboration and knowledge sharing, libraries can empower users to take charge of their learning and information-seeking processes.

c) The Use of Edmodo in Academic Libraries

A case study conducted on the use of Edmodo, a Web 2.0 tool, in higher education libraries demonstrated its effectiveness in facilitating communication and collaboration between students and librarians (Öztürk & Akgün, 2012). The study revealed that while teacher-initiated communications were more frequent than student-initiated ones, Edmodo provided a platform for students to engage with library resources and seek assistance. This case highlights the potential of Web 2.0 tools to bridge the gap between library services and user needs, fostering a collaborative learning environment.

d) Library 2.0 Initiatives

Integration of libraries isn't happening everywhere in the world at the same rate or in the same way because of different factors. For example, Pienaar and Smith (2008) wrote about how they created a Library 2.0 service model for the academic library at the University of Pretoria in South Africa. They discovered that the library is already very close to becoming a Library 2.0 library. They hope that their case study will inspire other university libraries to follow the same for their users. According to Baker (2008) many libraries have embraced the concept of Library 2.0, which emphasizes user participation, collaboration and the integration of Web 2.0 technologies into library services. For example, the San Jose Public Library implemented a Library 2.0 initiative that included the use of blogs, social networking sites, and interactive online tools. This initiative aimed to engage users in the library's decision-making processes and gather feedback on services and programs. The library's efforts resulted in increased user satisfaction and a stronger sense of community among patrons.

e) Web 2.0 Tools and Libraries

According to Lwoga (2014), adding Web 2.0 tools to the MUHAS library improved the quality of its services in Tanzania. The purpose of the exploratory survey with a questionnaire was to find out how the students saw things and what they liked best about Web 2.0. Their results show that adding Web 2.0 tools to the university library has helped people get better service and encourage users and libraries to talk to each othera lot more. People are more interested in using libraries when they can give and receive comments more easily.

A study investigating the use of social media by academic libraries found that these platforms significantly enhance user engagement and information dissemination (Morris & Stommel, 2016). Libraries that actively use social media to promote events, share resources, and interact with patrons reported higher levels of user satisfaction and participation. This case underscores the importance of leveraging Web 2.0 technologies to create a more connected and engaged library community.

After looking at the websites of Indian university libraries and talking to library staff, Santosh (2017) looked into how Web 2.0 tools are used in these places. The research showed that Indian libraries are way behind developed countries when it comes to using Web 2.0 tools. There aren't enough funds, buildings, or systems for the library staff to use Web 2.0 tools in their work, even though they'd like to. It is also important to have internal (willingness, personal interest) and external (ICT training and help) motivators for integration. It follows that the amount of growth in libraries' use of social media is affected by the parenting nation's abilities and support. Another important factor is the type of school and how the managers run things. We now have a lot of high-tech tools that can help with the merging process thanks to improvements in web technology.

f) Applications in Education

Web 2.0 tools have significantly impacted educational practices, enabling innovative teaching and learning methods.

- **E-learning:** Platforms like Moodle and Edmodo leverage Web 2.0 technologies to create interactive learning environments that facilitate asynchronous learning and collaboration among students (Bates, 2015). These tools support diverse learning styles and promote engagement through multimedia content and interactive activities.
- **Flipped Classroom:** Educators increasingly use Web 2.0 tools to implement flipped classroom models, where instructional content is delivered online outside of class, allowing for more interactive, hands-on activities during class time (Bergmann & Sams, 2012). This approach fosters deeper understanding and application of knowledge.
- **Digital Literacy:** The integration of Web 2.0 tools in education enhances students' digital literacy skills, preparing them for the demands of the modern workforce (Hague & Payton, 2010). Students learn the critical thinking, communication and teamwork skills they need to be successful in the 21st century by using a variety of digital tools.

g) Challenges and Considerations

While Web 2.0 tools offer numerous benefits, they also present challenges that users and organizations must navigate.

- **Digital Divide:**Warschauer (2004) mentioned that despite the numerous benefits associated with Web 2.0 tools, challenges remain. The digital divide continues to be a significant problem, as everyone does not have the same level of access to technology and the internet. This disparity creates barriers for individuals and communities that lack the necessary resources to engage with Web 2.0 technologies effectively. Addressing the digital divide is essential to ensure equitable access to information and opportunities.
- **Privacy and Security Concerns:**Privacy concerns arise as users share personal information on social media platforms, necessitating awareness and proactive measures to protect user data (Tufekci, 2015). To lower

the risks, organizations need to have strong privacy rules and teach their employees about data security. The sharing of personal information on social media raises privacy issues, necessitating awareness and proactive measures to protect user data.

- **Information Overload:**Information overload is another challenge posed by the vast amount of content generated through Web 2.0. Users may struggle to discern credible sources amidst the noise, making it essential to develop critical information literacy skills. Educators and organizations must prioritize teaching these skills to empower users to navigate the complex digital landscape effectively.The vast amount of content generated through Web 2.0 can lead to information overload, making it challenging for users to discern credible sources. Developing critical information literacy skills is crucial for navigating this complex landscape (Bawden & Robinson, 2009).

7. Case Studies Related to the Application of Web 2.0 Tools

a) Case Study 1: Scholarly Communication

Gu and Widen-Wulf (2011) did a case study to look into how the use of social media and Web 2.0 tools might change the way scholars communicate. They learned that scholars are familiar with Web 2.0 tools and use social media to easily share material and talk to each other as scholars. The study helps librarians figure out how to improve scholarly services and training in the social media world, taking into account how people use information and communicate scholarly ideas differently now that we live in the social media age.

b) Case Study 2: Educational Institutions

A study conducted on the use of Web 2.0 tools in higher education revealed that institutions integrating these technologies into their curricula experienced enhanced student engagement and collaboration (JISC, 2009). Faculty members reported improved communication with students and greater opportunities for interactive learning.

c) Case Study 3: Web 2.0 Tools in STEM Education

A study examining the use of Web 2.0 tools to motivate STEM students found that incorporating these technologies into engineering courses can significantly enhance student engagement and learning outcomes (Šćepanović *et al.*, 2018). Researchers implemented an Educational Recommender System called ELARS, which suggested relevant Web 2.0 tools to students based on their learning activities and preferences. The study involved two consecutive generations of STEM students and the results showed that most of them were positive and eager to use Web 2.0 tools to help them learn. However, the researchers also noted that students tended to work on assignments intermittently, suggesting that future studies should address this issue of non-continuous work habits (Šćepanović *et al.*, 2018).

d) Case Study 4: Web 2.0 Tools in Science Education throughout the COVID-19 Pandemic

The COVID-19 pandemic has speed up the need for innovative teaching approaches, including the use of Web 2.0 tools. A study conducted during this period investigated the effectiveness of an educational website enriched with Web 2.0 tools for science education (Çakır& Yaman, 2022). A quasi-experimental design with a pretest-posttest control group was applied to find the impact of the website on student achievement and motivation. The study focused on the unit of "Force and Energy" and found that the students did better in school and were more motivated when they used Web 2.0 tools. These findings suggest that Web 2.0-enhanced educational websites can serve as an effective alternative or supplement to online learning during the COVID-19 pandemic and beyond (Çakır & Yaman, 2022).

e) Case Study 5: Nurturing 21st Century Values through Web 2.0 Tools in ESL Classrooms

Building 21st century skills, like basic reading and writing skills, as well as character traits, is important for students getting ready to be responsible and useful global citizens. But most schools systems put most of their attention on basic skills like reading, writing and math. Srinivas (2021) did a study on how ESL teachers can help students improve their skills and character by giving the students language-learning projects using Web 2.0 tools for communication, collaboration, critical thinkingand creativity made it easier for people to develop personality traits like curiosity, persistence, adaptability and social and cultural awareness. The research found that Web 2.0 tools with features like asynchronous environments, easy content creation and interactive environments gave students many chances to have deep, long and reflective talks. Through observational data, the researcher identified the acquisition of character qualities like adaptability, social awareness, and cultural sensitivity as students engaged in Web 2.0-mediated activities.

f) Case Study 6: Factors Influencing Web 2.0 Tool Usage among Academics in South Africa

Despite the benefits of Web 2.0 tools in higher education, their adoption by academics in South Africa has been limited. A study investigated the factors influencing the usage of Web 2.0 tools among academics at two South African tertiary institutions. Using a case study research strategy and a quantitative approach, the researchers used a questionnaire to get information from the target population. Individual factors, group factors, perceived value and perceived quality are all strong predictors of Web 2.0 tool use, according to the findings. The study has practical implications for academic stakeholders in private higher education institutions, emphasizing the importance of integrating Web 2.0 technologies into teaching practices (Makhathini *et al.*, 2022)

g) Case Study 7: Adopting Edmodo as a Web 2.0 Tool in Higher Education

The advent of new internet technologies has not only altered teaching and learning but also completely changed the way people communicate, with Web 2.0 tools playing a significant role in this transformation. A case study examined the use of Edmodo, a Web 2.0 tool, in higher education to teach classical Turkish literature to third-year undergraduate students. To gather information for the study, the researchers used a mixed research technique, which included qualitative and quantitative approaches of data collection. In addition open ended survey was carried out in order to determine the students' opinion on Edmodo. According to the findings, the frequency of contacts launched by the teachers through Edmodo was much higher than the frequency of communications made by students. This case study provides insights into how Web 2.0 tools can be effectively integrated into higher education curricula to enhance teaching and learning experiences (Öztürk & Akgün, 2012).

8. Discussion

The integration of Web 2.0 tools in educational settings has become increasingly significant as educators and institutions seek to enhance teaching and learning experiences. This discussion synthesizes insights from various studies on the use of Web 2.0 tools, focusing on their impact on student engagement, collaboration, and the development of 21st-century skills. The findings underscore the transformative potential of these technologies in fostering a more interactive and participatory learning environment.

One of the primary benefits of Web 2.0 tools is their ability to enhance student engagement. According to research conducted by Šćepanović *et al.* (2021), the use of digital tools in English as a Second Language (ESL) classrooms facilitated the development of competencies and character qualities among students. The study highlighted that features of Web 2.0 tools, such as participatory environments and asynchronous communication, provided opportunities for meaningful conversations and collaboration. This finding aligns with the broader literature indicating that interactive technologies can increase student motivation and involvement in the learning process (Baker, 2008).Moreover, the study by Öztürk and Akgün (2012) on the use of Edmodo, a social learning platform, revealed that while teacher-initiated communications were more frequent, the platform still encouraged student participation. This dynamic suggests that even when educators lead discussions, Web 2.0 technologies can help pupils to communicate their ideas and opinions in an open and interesting surroundings. Using these tools improves not only learning but also helps students acquire critical thinking and problem solving techniques by means of communication and teamwork.

Collaboration is a key component of effective learning, and Web 2.0 tools facilitate this process by enabling students to work together on projects and share resources. The research conducted by Morris and Stommel (2016) emphasizes the importance of social media in promoting collaborative learning experiences. Libraries and educational institutions that leverage these toolsmay build online communities where students may interact, exchange ideas and connect, thereby facilitating knowledge sharing among them. This collaborative approach aligns with constructivist and connectives learning theories, which emphasize the importance of social interaction in the learning process (Siemens, 2005). The importance of developing 21st-century skills, such as critical thinking, creativity, and collaboration, cannot be overstated. According to the World Economic Forum (2015), these are the kind of talents that are absolutely necessary for students to have in order to flourish in a world that is becoming increasingly complicated and interconnected.Web 2.0 tools provide a unique opportunity for educators to nurture these competencies through interactive and collaborative learning experiences.

The study by Makhathini *et al.* (2022) highlights the factors influencing the use of Web 2.0 technologies among academics in South Africa. The findings indicate that perceived usefulness and ease of use significantly impact the adoption of these tools. This suggests that educators must be provided with adequate training and support to effectively integrate Web 2.0 technologies into their teaching practices. By fostering a culture of collaboration and communication, educational institutions can empower both teachers and students to harness the full potential of these tools.

Research by Šćepanović *et al.* (2021) demonstrates how Web 2.0 tools can facilitate the acquisition of character qualities such as adaptability, social awareness, and curiosity. By engaging in collaborative projects and discussions, students can develop these essential skills in a supportive and interactive environment. Furthermore, the study by Öztürk and Akgün (2012) highlights the role of Edmodo in fostering communication and collaboration among students, reinforcing the idea that Web 2.0 tools can serve as powerful catalysts for skill development.

Despite the numerous benefits associated with Web 2.0 tools, challenges remain in their effective implementation within educational settings. The digital-divide, which refers to the discrepancies in access to technology and the internet is one of the important challenges that we face today. As noted by Warschauer (2004), not all students have equal access to digital resources, which can hinder their ability to engage fully with Web 2.0 tools. Educational institutions must tackle this problem by giving equal access to technology and by making sure that all students are benefitted from these innovative tools.

Additionally, concerns regarding privacy and data security must be addressed. The use of social media and online collaboration tools raises important questions about user privacy and the protection of personal information. Libraries and educational institutions must establish clear policies and guidelines to safeguard user data while promoting engagement (Tufekci, 2015). In order to maintain a respectful and secure atmosphere for learning, teachers should also teach pupils about the importance of digital citizenship and responsible behavior when using the internet.

9. Conclusion

For the purpose of increasing student involvement, cooperation, and the development of vital skills for the 21st century, the incorporation of Web 2.0 tools into educational settings has the potential to revolutionize the teaching and learning experiences experienced by students. According to the findings of a number of studies, these technologies provide educators with exceptional chances to develop dynamic and participatory learning environments that enable students to take an active role in their own education. However, in order to ensure that all students are able to avail themselves of the benefits that Web 2.0 technologies have to offer, it is necessary to address obstacles such as the digital divide and concerns around privacy. The provisions of training and assistance for instructors, the establishment of explicit regulations to safeguard user privacy and the prioritization of fair access to technology are all requirements that educational institutions must fulfill. Students will need to be prepared to flourish in a world that is becoming increasingly complicated and interconnected. By the utilization of Web 2.0 technologies, which will become increasingly important as the educational environment continues to undergo transformation. By adopting these technologies, educators have the ability to establish dynamic learning communities that encourage collaboration, critical thinking and creativity this in turn will provide students with the skills they need to be successful in the 21st century.

References

Baker, M. (2008). Library 2.0: The future of libraries in the digital age. Library Journal, 133(12), 30-33.

Bates, A. W. (2015). Teaching in a digital age: Guidelines for designing teaching and learning. Tony Bates Associates Ltd.

Bawden, D., & Robinson, L. (2009). The dark side of information: overload, anxiety and other paradoxes and pathologies.Journal of Information Science, 35(2), 180-191. https://doi.org/10.1177/0165551508095781

Bennett, S., & Maton, K. (2010). Beyond the 'digital natives' debate: Towards a more nuanced understanding of students' technology experiences. Journal of Computer Assisted Learning, 26(5), 321-331. https://doi.org/10.1111/j.1365-2729.2010.00360.x

Bergmann, J., & Sams, A. (2012). Flip your classroom: Reach every student in every class every day. International Society for Technology in Education.

Birdsall, F.W. (2007). Web 2.0 as a social movement. Webology, 4 (2). Retrieved January 10, 2010, from.www.webology.ir/2007/ v4n2/a40.html

Boulos, M. N. K., Maramba, I., & Wheeler, S. (2011). A review of the online social networking impact on health education: Implications for health education. Health Information & Libraries Journal, 28(1), 1-23. https://doi.org/10.1111 /j.1471-1842.2011.00973.

Bruns, A., & Moe, H. (2014). Structural communication in social media: The role of Twitter in the Arab Spring. Social Media Society, 1(1), 1-12. https://doi.org/10.1177/2056305114558452

Bryant, S. L., Forte, A., & Bruckman, A. (2005). Becoming Wikipedian: Transformation of participation in a collaborative online encyclopedia. Proceedings of the 2005 International Symposium on Wikis, 1-10. https://doi.org/10.1145/1109530.1109532

Çakır, R., & Yaman, S. (2022). A Teaching Suggestion in the COVID-19 Disease Pandemic Period: The Educational Website Enriched by Web 2.0 Tools. In Cases on Smart Learning Environments (pp. 1-16). IGI Global.

Chapman, A., & Russell, R. (2009). Shared Infrastructure Services Landscape Study: A survey of the use of Web 2.0 tools and services in the UK HE sector.

Durovic, G., Dlab, M. H., & Hoic-Bozic, N. (2018). Motivating STEM students to use Web 2.0 tools for learning: A case study. Proceedings of the 2018 International Conference on Information and Digital Technologies. https://doi.org/10.1109/icimp1.2018.8325856

Garcia, D.L. (2003). Technology and Social Inclusion: Rethinking the Digital Divide. By Mark Warschauer. Cambridge, Mass.: MIT Press, 2003. Pp. xii+260. $32.95. American Journal of Sociology, 109, 767-768.

Gu, F., & Widén-Wulff, G. (2011) Scholarly communication and possible changes in the context of social media. The Electronic Library 29(6), 762–776.

Hague, C., & Payton, S. (2010). Digital literacy across the curriculum. Futurelab. Retrieved from https://www.nfer.ac.uk/publications/FUTL50/FUTL50.pdf

Harris, R., & Rea, A. (2009). The role of Web 2.0 in the development of collaborative learning. In Emerging technologies for the classroom: A learning sciences perspective (pp. 13-29). Springer.

Hoffman, D. L., & Fodor, M. (2010). Can you measure the ROI of your social media marketing? MIT Sloan Management Review, 52(1), 41-49.

Kaplan, A. M., & Haenlein, M. (2010). Users of the world, unite! The challenges and opportunities of social media. Business Horizons, 53(1), 59-68. https://doi.org/10.1016/j.bushor.2009.09.003

Keshavarz, M., & Mirmoghtadaie, Z. (2023). Book Review: Teaching in a Digital Age: Guidelines for Designing Teaching and Learning–Third Edition, authored by Anthony

William (Tony) Bates (Tony Bates Associates Ltd., 2022). The International Review of Research in Open and Distributed Learning.

King, D. L., & Brown, S. W. (2009). Emerging trends, 2.0, and libraries. The Serials Librarian, 56(1e4), 32e43.

Makhathini, N. S., Zuva, T., & Zuva, K. (2022). Factors Influencing Web 2.0 Technology Usage Among Academics: A Case Study of Two South African Tertiary Institutions. (1-18). IGI Global.

Malureanu, C., Malureanu, A., & Lazar, G. (2021). Motivational Factors of the Usage of Web 2.0 Tools for E-Learning in Business.

Mangold, W. G., & Faulds, D. J. (2009). Social media: The new hybrid element of the promotion mix. Business Horizons, 52(4), 357-365. https://doi.org/10.1016/j.bushor.2009.03.002

Morris, S., & Stommel, J. (2016). Social media and the academic library: A case study of the University of Mary Washington. Journal of Library Administration, 56(1), 1-10. https://doi.org/10.1080/01930826.2015.1071131

Morrison, D. (2014). Remote working: A guide to working from home. Routledge.

Ogunleye, O. S., Ojo, J. A., & Ojo, A. A. (2015). Use of Web 2.0 tools and services by the library professionals in Lagos State tertiary institution libraries: A study. Library Philosophy and Practice. Retrieved from https://digitalcommons.unl.edu/libphilprac/1230

O'Reilly, T. (2005). What is Web 2.0: Design patterns and business models for the next generation of software. O'Reilly Media. Retrieved from https://www.oreilly.com/pub/a/oreilly/tim/news/2005/09/30/what-is-web-20.html

Öztürk, E., & Akgün, Ö. E. (2012). Using Web 2.0 Tools in and beyond the University Classrooms: A Case Study of Edmodo. In Proceedings of the 4th International Conference on New Horizons in Education (pp. 1-7).

Padayachee, I., & Moodley, K. (2022). Factors influencing Web 2.0 technology usage among academics. International Journal of Technology and Human Interaction, 18(1), 1–22. https://doi.org/10.4018/ijthi.293189

Pienaar, H., & Smith, I. (2008). Development of a library 2.0 service model for an African library. Library Hi Tech News 25(5), 7–10.

Santosh, S., (2017). Adoption of web 2.0 applications in academic libraries in India. DESIDOC Journal of Library & Information Technology 37(3), 192–198

Šćepanović, S., Žižović, M., & Radojičić, Z. (2018). Motivating STEM students to use Web 2.0 tools for learning: A case study. In Proceedings of the 2018 International Conference on Information and Digital Technologies (pp. 1-6). IEEE.

Siemens, G. (2005). Connectivism: A learning theory for the digital age. International Journal of Instructional Technology and Distance Learning, 2(1), 3-10. Retrieved from: http://www.idtl.org/Journal/Jam_05/article01.html

Srinivas, R. (2021). Conversations through Web 2.0 tools: Nurturing 21st century Values in the ESL Classroom. TESOL International Journal, 16(2), 1-15.

Tufekci, Z. (2015). Algorithmic harms beyond Facebook and Google: A research agenda for social media. Proceedings of the 2015 ACM Conference on Computer Supported Cooperative Work and Social Computing, 1-10. https://doi.org/10.1145/2675133.2675148

Warschauer, M. (2004). Technology and social inclusion: Rethinking the digital divide. MIT Press.

Wikipedia (2010c). Definition of Web 2.0. Retrieved June 10, 2010, from. http://en.wikipedia.org/wiki/Web_2.0.

World Economic Forum. (2015). New Vision for Education: Unlocking the Potential of Technology. Retrieved from https://www.weforum.org/reports/

32

Scope of Academic Research in Big Data Environment: Study the Role of Library Analytics

Dr. Swapan Khan

Librarian, Narasinha Dutt College, Howrah - 711 101, West Bengal, India
E-mail: khanswapan@gmail.com

Abstract

In the era of social media, the constant increase in data generation presents a number of difficulties for academic study. Big Data research methodologies present a variety of challenges because of factors such as difficult data generalization, data linkage issues, ethical concerns, and more. Some huge data handling issues have been resolved by the development of new technologies. These days, data can be presented in a machine-readable manner and altered by researchers utilizing a variety of software and analytical tools. The present study focuses on various aspects of Big Data and finds the scope of academic research activities in Big Data environment.

Keywords: *Big data, Data analytics, Academic research, Library science, Library analytics.*

1. Introduction

Today, all the organisations rely on assessments of data provided by their users, consumers, internal procedures, and day to day operations to find new avenues for betterment and expansion of the product and services. These insights provide a large and intricate set of professionally generated, maintained, analyzed, and modified facts. "Big Data" refers to the vast amount of data (Bansal, 2022, April 6). For instance, social media platforms facilitate human contact every day and millions of people connect and share a vast array of diverse types of information through these platforms. Big Data is viewed as "incredibly significant" in academic research. In the Big Data context, data visualization tools and technologies are essential for analyzing large amounts of data and making decisions based on facts (Adilin Beatrice, 2021, March 2). Information and data are represented graphically in data visualization.

By using visual components like charts, graphs, and maps, data visualization tools provide an intuitive way to observe and understand trends, outliers, and patterns in data. The process is called Big Data Analytics that describes the process of uncovering trends, patterns, and correlations in large amounts of raw data to help make data-informed decisions in various research works.

2. Objectives

The main objectives of the study are as follows:

i. To state an overview of the big data and big data analytics
ii. To show the scope of use of big data in library and information science
iii. To understand the library analytics
iv. To discuss the areas of academic research in big data environments

3. History of Big Data

According to Gartner, "Big data is high-volume, high-velocity and high-variety information assets that demand cost-effective, innovative forms of information processing for enhanced insight and decision making." The Wikipedia defines "Big Data is a broad term for data sets so large or complex that they are difficult to process using traditional data processing applications. Challenges include analysis, capture, curation, search, sharing, storage, transfer, visualization, and information privacy".

The idea of a big volume of data is not entirely new in its fundamental sense. By 300 BC, the Egyptians made an attempt to compile all of the "data" into the Alexandrian library. Furthermore, in order to establish the best army deployment, the Roman Empire used to meticulously examine its military data. However, the first known example of large-scale data analysis dates back to 1663, when John Graunt, the first demographer and maybe the first epidemiologist, had to cope with a tonne of data while researching the bubonic plague that was wreaking havoc on Europe at the time. In 1965, the US government built the first data centre, with the intention of storing millions of fingerprint sets and tax returns. The term 'Big Data' has been in use since the early 1990s. Most people credit American computer scientist John R. Mashey, whose work on Silicon Graphics popularized the word, however it is unclear who used it originally. Human comprehension is being pushed to the limit by the enormous changes in data volume and generating speed during the past 20 years.

4. Big Data Analytics

Artificial intelligence (AI), mobile devices, social media, and the Internet of Things (IoT) are all becoming increasingly popular. As a result, data sources have become more sophisticated than those for traditional data. An organisation collects structured, semi-structured, and unstructured data in real time and on a wide scale from sensors, devices, video/audio, networks, log files, transactional applications, online, and social media (Simplilearn, 2022, April 3). The technique can work with a variety of structured and unstructured datasets.

Analyzing data from multiple sources is known as big data analytics. It explains how to find patterns, correlations, and trends in vast amounts of unprocessed data to support data-driven decision-making. It provides an almost limitless supply of commercial and informational wisdom. In order to generate data-informed judgments from very large, diverse big data sets which can range in size from terabytes to zettabytes and contain structured, semi-structured, and unstructured data the analytics leverages traditional statistical approaches.

5. Big Data Analytics in Libraries

To fulfill the needs of the researchers & the community as a whole, libraries & information centers are essential for organizing knowledge & disseminating information. More online resources and services are being offered by libraries, and librarians are able to get more online data from social networking sites like Facebook, LinkedIn, Instagram, and others by using analytics tools (Garoufallou & Gaitanou, 2021). Library professionals have acknowledged the significance of Big Data in library services within this framework. Certain individuals have designated the library as a "Smart Library" or a "Intelligent Library" by utilizing and implementing the advantages of a substantial amount of user data produced from diverse sources (Ahmad, JianMing & Rafi, 2019). Library and information centers are facing a big challenge to provide specialized services in a in a highly competitive environment due to change in the requisition patterns of the user communities, their interaction in of social medias, application of personalized devices, open source movements etc. Different types of information are collected from the users in social medias, internet and from other AI system etc.

These days, literary discussion clubs, bookstores, writers, and other information providers and suppliers compete with libraries (Hamad *et al.*, 2020). Big Data is essential for libraries to comprehend the evolving needs of their patrons and to reorganize and reshape their services and procedures in response. It holds the answers to numerous important questions about patterns, trends, and associations of user behavior (Gibbons, 2013).

6. Collaborative Initiatives on Big Data

Libraries play significant roles in a major part in many collaborative initiatives on Big Data projects that have been found. Here are a few of the initiatives:

i. **Data Conservancy:** Under the National Science Foundation's DataNet program, the Data Conservancy (DC) was established. The DC's primary goals are to improve data infrastructure and to better understand the difficulties associated with gathering, storing, and curating various kinds of research data.

ii. **Data To Insight Center:** At Indiana University Bloomington (IUB), the Data To Insight Center (D2I) is a research center dedicated to data science and data-driven computing. D2I has made groundbreaking contributions to the field of data provenance and its role in fostering more dependable, credible, and publicly accessible research outcomes.

iii. **GeoMAPP:** In order to enable the retrieval of historical data for use and study in the future, the GeoMAPP project aims to regularly archive geographic digital content about land parcels, zoning, roadways, and jurisdictional boundaries. Partners in GeoMAPP include the Library of Congress, the North Carolina State Archives, the Kentucky Department for Libraries and Archives (KDLA), the North Carolina Center for Geographic Information and Analysis (CGIA), Montana State Library, Utah Automated Geographic Reference Center (AGRC), and so forth.

iv. **DataONE:** DataONE, led by PI William Michener, is an initiative driven by the community which was funded in August 1, 2009 (dataone.org). Access to data from various member repositories is made possible by it, facilitating better Earth and environmental data search and discovery. It encourages the use of best practices for instructional materials and resources. Gaining a deeper understanding of and protecting the environment that supports life on Earth is the primary goal of DataONE.

v. **DataCite:** Numerous national libraries throughout Europe, as well as a few in North America, formed it. It gives research data and other outputs from the member organizations persistent identifiers (e.g DOIs). Making research output discoverable and facilitating members' connections and sharing of persistent identifiers with the larger research ecosystem are the primary goals. Through outreach initiatives and community-building projects, it encourages data sharing and citation.

vi. **Meta Archive Cooperative:** In 2004, as part of a project funded by the Library of Congress National Digital Information Infrastructure & Preservation Program, six research libraries launched it. It is an international network for digital preservation made up of archives, libraries, and other memory-related organizations. Promoting and assisting in the long-term preservation of digital content are the primary goals.

7. Challenges with Big Data

It is very difficult for libraries and information centers to offer specialized services in a setting where competition is fierce. Due to shifts in user groups' demand patterns, their interactions on social media, the use of customized devices, open source movements, etc. Users of social media, the internet, and other AI systems, among other places, provide various forms of information that are gathered. To provide the best possible service to their patrons, libraries and information centers must analyze this kind of big data. Most important challengers are as follows:

7.1. Sharing and Accessing Data

i. Access to data from open repositories can occasionally lead to significant issues.

ii. The majority of industries experience large data from heterogeneous sources being inaccessible.

iii. Retrieved data can occasionally be imprecise and lacking. As a result, it causes bad decisions to be made and has a negative impact on all information systems.

iv. Data interchange presents a number of issues, including universal formats, data types, and structure, among others.

v. Legal issues pertaining to private or public data present a significant obstacle.

7.2. Privacy and Security

i. There is a possibility that personal information will be combined with externally available massive data, which could reveal details about a person that they would prefer to keep private.

ii. This challenge has legal, intellectual, technical, and delicate implications.

iii. Information gathered from people can occasionally raise a company's worth. By gaining insights into their life that they are ignorant of, this is accomplished.

iv. It is quite challenging to identify sensitive information from a vast amount of data on a daily basis. However, the checking is essential, particularly when done in real time.

7.3. Analytical Challenges

i. When an organization has too much data, it raises important questions about data analytics.

ii. The vast volumes of data that need to be analyzed for this kind of work can be classified as unstructured (unorganized data), semi-structured (semi-organized data), or structured (organized data).
iii. Making the best use of data for analytics is essential.
iv. It's quite difficult to identify the key information.

7.4. Technical Challenges

i. **Fault tolerance**
   - Fault tolerance computing is very difficult and involves complex algorithms; it is another significant technical difficulty.
   - Emerging technologies such as big data and cloud computing have always meant that in the event of a failure, the damage should always be contained within an acceptable bound, meaning that the task should never have to start from scratch.

ii. **Quality of data**
   - The quality of data storage is always crucial in order to extract better findings and draw conclusions from big data. It reduces unnecessary data and concentrates on high-quality output.
   - Big Data needs to be stored in a huge store. Large data storage is always required by the majority of businesses and IT executives in order to organize data in a way that maintains data quality on a systemic level.
   - Making sure that the data is relevant presents another challenge. It also begs the questions of how much data would be sufficient for making decisions and how accurate the data that is saved is.

iii. **Scalability**
   - It addresses several issues related to managing and carrying out different tasks so that any workload objective can be met economically.
   - Big data projects that businesses deploy today can develop and grow quickly. Therefore, cloud computing has emerged as a solution to the Big Data scalability challenge.
   - Additionally, it calls for effectively resolving system malfunctions. It raises another concern regarding the proper use of king storage devices.

8. Big Data in Academic Research

The field of social science research especially academic research could be significantly altered by big data, and administrative social science data could be especially advantageous. It is emphasized that big data is a diverse collection of many data kinds rather than a single, cohesive sort of data resource, despite the fact that it is currently a trendy catchphrase in social science (Connelly *et al.*, 2016). The last several decades have seen the development of new methods for gathering, storing, and analyzing data. This has led to the area of data science emerging, which combines statistical, mathematical, computational, and algorithmic techniques to extract knowledge from large data sets (Leonelli, 2020).

Researchers play a crucial role in the knowledge society, which is made possible by big data. It is true that researchers have been slothful to embrace big data as a source for their research study. This could be partly due to the fact that big data are not curated by scholars, but rather by governmental or commercial entities (Raubenheimer, 2021). Therefore, problems are encountered when dealing with large amounts of data handling by the researchers such as huge data volume, continuous development of data sources and also the problem of the time variations in large data sets.

9. Conclusion

The big data set is growing larger every day as several exabytes are added at an ever-increasing pace. Aside from this, heterogeneous datasets, including both structured and unstructured ones, as well as a variety of data types and file formats, are typically included in large data sets. Finding and retrieving pertinent information in real time from the large data cosmos is a highly challenging task (Singh & Singh, 2015). It is intended that librarians from the walled library will contribute to this effort by virtue of their knowledge of the structure and management of information. The demands of library patrons will only increase in the future due to the digital environment (Showers, 2014). According to Semeler *et al.* (2017), librarians ought to be able to relate to the development, management, and preservation of data. Because of their crucial role in Big Data analytics, librarians' skills and expertise must be improved in order to execute Big Data analytics.

References

Adilin Beatrice, A. (2021, March 2). All about the basics of big data: History, types and applications. Big Data Latest News. Retrieved April 2, 2022 from https://www.analyticsinsight.net/all-about-the-basics-of-big-data-history-types-and-applications

Ahmad, K., Jian Ming, Z., & Rafi, M. (2019). An analysis of academic librarians' competencies and skills for implementation of Big Data analytics in libraries: A corelational study. Data Technologies and Applications, 53(2), 201-216. DOI: 10.1108/DTA-09-2018-0085

Alaisawi, S., & Khalifa, S. (2020). Artificial Intelligence in data mining and big data. Retrieved April 23, 2022 from https://www.analyticsinsight.net/the-difference-between-artificial-intelligence-and-machine-learning/

Bansal, S. (2022, April 6). What are the key characteristics of big data? Retrieved April 22, 2022 from https://www.analytixlabs.co.in/blog/characteristics-of-big-data/

Big Data Analytics: What It Is, How It Works, Benefits, and Challenges (n.d.). Retrieved April 13, 2022 from https://www.tableau.com/learn/articles/big-data-analytics

Connelly, R., Playford, C. J., Gayle, V., & Dibben, C. (2016). The role of administrative data in the big data revolution in social science research. Social Science Research, 59, 1-12. DOI: https://doi.org/10.1016/j.ssresearch.2016.04.015.

Garoufallou, E., & Gaitanou, P. (2021). Big data: Opportunities and challenges in libraries- A systematic literature review. College & Research Libraries, 82(3).

Gaur, C. (2020, December 11). Top 6 big data challenges and solutions to overcome. Retrieved April 13, 2022 from https://www.xenonstack.com/insights/big-data-challenges

Ghorui, K. (2019, August 26). Applications of big data. Retrieved April 1, 2022 from https://www.geeksforgeeks.org/applications-of-big-data/

Gibbons, S. (2013). Techniques to understand the changing needs of library users. IFLA Journal, 39(2), 162–167. DOI: https://doi.org/10.1177/0340035212472846

Hamad, F., Fakhuri, H., & Abdel Jabbar, S. (2020). Big data opportunities and challenges for analytics strategies in Jordanian Academic Libraries. New Review of Academic Librarianship, 1–24. DOI: https://doi.org/10.1080/13614533.2020.1764071

Lawton, G. (2022, January 5). 10 big data challenges and how to address them. Retrieved April 12, 2022 from https://www.techtarget.com/searchdatamanagement/tip/10-big-data-challenges-and-how-to-address-them

Leonelli, S. (2020). Scientific research and big data. The Stanford Encyclopedia of Philosophy. Retrieved March 21, 2024 from https://plato.stanford.edu/archives/sum2020/entries/science-big-data/

Raubenheimer, J. (2021). Big data in academic research: Challenges, pitfalls, and opportunities. In: Prodromou, T. (eds), Big Data in Education: Pedagogy and Research. Policy Implications of Research in Education, vol. 13. Springer, Cham. DOI: https://doi.org/10.1007/978-3-030-76841-6_1

Semeler, A. R., Pinto, A. L., & Rozados, H. B. F. (2017). Data science in data librarianship: Core competencies of a data librarian. Journal of Librarianship and Information Science. DOI: https://doi.org/10.1177/0961000617742465

Showers, B. (2014). Developing a shared analytics service for academic libraries. Insights: The UKSG Journal, 27(2), 139-146.

Simplilearn (2022, April 3). What is big data analytics and why it is important? Retrieved April 15, 2022 from https://www.simplilearn.com/what-is-big-data-analytics-article

Singh, M.K., & Singh, D.K. (2015). Cataloguing to facilitate big data analytics. 10th International CALIBER-2015.

33

Green Libraries

Dr. Suman Singh

Assistant Professor, (SOH) Library Information Science, Uttar Pradesh Rajashri Tandon Open University, Prayagraj - 211 013, Uttar Pradesh, India
E-mail: drsumansingh33333@gmail.com

Abstract

In the 21st century, sustainability has become a critical focal point for various industries and institutions, as society seeks to mitigate the effects of climate change and environmental degradation. Among the numerous sectors embracing this green revolution, libraries have emerged as key players. The concept of "Green Libraries" or "Sustainable Libraries" refers to institutions that integrate environmental responsibility into their operations, services, and infrastructure. These libraries are not only centers of knowledge and learning but also serve as role models in promoting sustainability, reducing carbon footprints, and fostering environmental awareness among the public.

Keywords: *Green libraries, Infrastructure, Design, Environmental library.*

1. Introduction

The idea of Green Libraries transcends mere energy conservation or waste reduction. It encapsulates a broader commitment to sustainability, encompassing green building practices, sustainable operations, environmentally friendly collections, and programs that educate and engage communities on sustainability issues. This essay explores the multifaceted nature of Green Libraries, examining their significance, the principles that guide them, and their impact on communities and the environment.

2. The Emergence of Green Libraries

The Green Library movement is relatively recent, gaining momentum in the late 20th and early 21st centuries. This movement arose from a growing recognition of the library's role in society as a community hub and educational institution. Libraries, traditionally seen as repositories of knowledge and culture, began to evolve in response to societal shifts towards environmental consciousness.

One of the earliest and most significant steps in the Green Library movement was the construction of environmentally friendly library buildings. The Leadership in Energy and Environmental Design (LEED) certification, established by the U.S. Green Building Council in 1993, provided a framework for assessing the sustainability of buildings, including libraries. The LEED certification became a benchmark for Green Libraries, with numerous institutions striving to meet its standards in areas such as energy efficiency, water conservation, indoor environmental quality, and the use of sustainable materials.

In addition to LEED certification, the concept of Green Libraries has been influenced by global environmental initiatives such as the United Nations' Sustainable Development Goals (SDGs), particularly Goal 11, which focuses on making cities and human settlements inclusive, safe, resilient, and sustainable. Libraries have embraced these goals, recognizing their potential to contribute to a sustainable future through education, community engagement, and resource management.

3. Green Building Design and Infrastructure

One of the most visible aspects of Green Libraries is their physical infrastructure. Green Library buildings are designed to minimize environmental impact and maximize energy efficiency. This begins with the site selection, where considerations include the use of previously developed land, proximity to public transportation, and the potential for natural lighting and ventilation.

4. Sustainable Design Features Commonly Found in Green Libraries Include

Energy Efficiency: Green Libraries often utilize renewable energy sources such as solar panels, wind turbines, or geothermal systems to power their operations. They incorporate energy-efficient lighting, heating, ventilation, and air conditioning (HVAC) systems. For example, the Varina Area Library in Henrico, Virginia, boasts a geothermal HVAC system that significantly reduces energy consumption.

Water Conservation: Many Green Libraries implement water-saving technologies such as low-flow faucets, toilets, and irrigation systems. Rainwater harvesting and greywater recycling are also common practices. The Vancouver Island Regional Library in British Columbia features a rainwater harvesting system that supplies water for landscape irrigation.

Sustainable Materials: Green Libraries prioritize the use of eco-friendly building materials, including recycled, reclaimed, or rapidly renewable resources. The Cedar Rapids Public Library in Iowa, for instance, incorporated reclaimed wood from a nearby tornado-damaged warehouse into its design.

Indoor Environmental Quality: To ensure a healthy indoor environment, Green Libraries focus on improving air quality, maximizing natural light, and using non-toxic materials. The San Diego Central Library in California features large windows that allow natural light to flood the interior, reducing the need for artificial lighting and creating a more pleasant atmosphere for patrons.

Landscape Design: Green Libraries often incorporate native plants and drought-resistant landscaping to minimize water use and support local ecosystems. The National Library of Singapore is surrounded by green spaces and gardens that not only enhance the building's aesthetic appeal but also contribute to urban biodiversity.

5. Operational Sustainability

Beyond their physical infrastructure, Green Libraries are committed to sustainable operations. This involves adopting practices that reduce waste, conserve resources, and promote environmentally responsible behavior among staff and patrons.

Waste Reduction: Green Libraries prioritize waste reduction through recycling programs, digital services, and responsible procurement. Many libraries have reduced their reliance on paper by digitizing catalogs, records, and communication. The Chicago Public Library, for example, has implemented a comprehensive recycling program that includes paper, plastic, glass, and electronic waste.

Resource Conservation: In addition to energy and water conservation, Green Libraries strive to reduce their consumption of other resources, such as paper, ink, and plastics. The Boston Public Library's "BYOB (Bring Your Own Bag)" program encourages patrons to use reusable bags instead of plastic ones.

Green Procurement: Sustainable procurement practices are another hallmark of Green Libraries. This includes purchasing eco-friendly products, such as recycled paper, energy-efficient appliances, and non-toxic cleaning supplies. The University of California, Davis, Shields Library, for instance, has committed to purchasing environmentally responsible office supplies and equipment.

Sustainable Transportation: Green Libraries encourage sustainable transportation options for both staff and patrons. This may involve providing bike racks, offering electric vehicle charging stations, or promoting public transit use. The Boulder Public Library in Colorado offers bike-sharing stations and ample bike parking to support sustainable commuting.

6. Green Collections and Programs

Green Libraries extend their commitment to sustainability through the development of green collections and educational programs that promote environmental awareness and action.

Green Collections: Green Libraries curate collections that focus on environmental issues, sustainability, and climate change. These collections include books, magazines, digital resources, and multimedia materials that educate patrons on topics such as renewable energy, conservation, and sustainable living. The New York Public Library, for example, offers a "Green Living" collection that provides resources on topics ranging from organic gardening to energy-efficient home design.

Sustainability Programming: Green Libraries often host programs and events that educate the community about environmental issues and sustainable practices. These programs may include workshops on topics like composting, energy conservation, or sustainable gardening. The Los Angeles Public Library's "Sustainable Sundays" series offers workshops on topics such as zero-waste living, urban farming, and sustainable fashion.

Partnerships and Collaborations: Many Green Libraries collaborate with local environmental organizations, government agencies, and schools to promote sustainability. These partnerships can lead to joint initiatives such as community clean-up events, tree planting drives, or environmental film screenings. The Vancouver Public Library, for instance, partners with the City of Vancouver and local environmental groups to offer programming that supports the city's Greenest City 2020 Action Plan.

Digital Services: To reduce their carbon footprint, Green Libraries increasingly offer digital services that allow patrons to access resources online. E-books, audiobooks, and streaming services reduce the need for physical materials and associated environmental costs. The San Francisco Public Library's robust digital collections and online services make it easier for patrons to access information without the need for physical travel or resources.

7. Impact on Communities and the Environment

The impact of Green Libraries extends beyond their immediate environmental benefits. These institutions play a crucial role in shaping sustainable communities, influencing public behavior, and fostering a culture of environmental stewardship.

Community Engagement: Green Libraries serve as community hubs where individuals can learn about sustainability and take part in environmental initiatives. By hosting workshops, lectures, and events, these libraries empower community members to make informed decisions about their environmental impact. The Brooklyn Public Library, for example, has become a center for environmental education, offering programs on topics such as sustainable urban farming and climate change.

Environmental Advocacy: Green Libraries advocate for environmental responsibility and climate action within their communities. Through their programs and collections, they raise awareness about pressing environmental issues and inspire collective action. The Seattle Public Library, for instance, plays an active role in the city's climate action efforts by hosting educational programs and collaborating with local environmental organizations.

Cultural Shift: Green Libraries contribute to a broader cultural shift towards sustainability. By demonstrating environmentally responsible practices, these libraries set an example for other institutions and individuals to follow. The success of Green Libraries has inspired similar initiatives in schools, businesses, and other public spaces, contributing to a more sustainable society.

Environmental Impact: The environmental impact of Green Libraries is significant. By reducing energy consumption, conserving water, and minimizing waste, these libraries contribute to the reduction of greenhouse gas emissions and the preservation of natural resources. The cumulative effect of these efforts can be substantial, particularly as more libraries adopt green practices.

8. Challenges and Future Directions

While the Green Library movement has made significant strides, it also faces challenges that must be addressed to ensure its continued success and growth.

Financial Constraints: Implementing green practices and building sustainable infrastructure can be costly. Many libraries, particularly those in underfunded or rural areas, may struggle to afford the initial investment required for green initiatives. However, the long-term savings from energy efficiency and resource conservation can offset these costs. Innovative funding mechanisms, such as grants, donations, and partnerships, can also help libraries overcome financial barriers.

Awareness and Education: Despite the growing popularity of the Green Library movement, not all library staff and patrons are aware of its importance or how to participate. Ongoing education and training are essential to ensure that everyone involved understands the principles of sustainability and how to apply them in the library setting.

Technological Challenges: As libraries increasingly rely on digital services to reduce their environmental impact, they must navigate the challenges associated with technology, such as digital equity, cybersecurity, and the environmental impact of electronic waste. Ensuring that all patrons have access to digital resources without exacerbating the digital divide is crucial.

Evolving Standards: The field of sustainability is constantly evolving, with new technologies, practices, and standards emerging regularly. Green Libraries must stay informed about these developments and be willing to adapt their practices to maintain their commitment to sustainability.

Community Resistance: In some cases, communities may resist the changes associated with the Green Library movement, particularly if they perceive these changes as inconvenient or unnecessary. Libraries must engage with their communities to address concerns, educate the public about the benefits of green practices, and build support for their initiatives.

Looking ahead, the future of Green Libraries appears promising. As environmental concerns become more pressing, the demand for sustainable practices in all sectors, including libraries, is likely to grow. Green Libraries are well-positioned to lead the way in this regard, not only by continuing to implement sustainable practices but also by serving as advocates for environmental responsibility within their communities.

One potential area of growth for Green Libraries is in the realm of digital sustainability. As libraries increasingly rely on digital platforms to deliver services, they must consider the environmental impact of their digital infrastructure, including data centers, servers, and electronic devices. By adopting green IT practices, such as using energy-efficient hardware, optimizing server usage, and promoting digital literacy, libraries can reduce the environmental footprint of their digital services.

Another promising avenue for Green Libraries is the expansion of their role in community resilience. As climate change continues to impact communities around the world, libraries can serve as critical resources for disaster preparedness, response, and recovery. By offering information on climate adaptation, emergency planning, and community resilience, Green Libraries can help communities build the knowledge and skills needed to navigate the challenges of a changing climate.

9. Conclusion

Green Libraries represent a powerful convergence of environmental stewardship and knowledge dissemination. By integrating sustainability into their buildings, operations, collections, and programs, these institutions are redefining what it means to be a library in the 21st century. They are not only reducing their own environmental impact but also inspiring and educating their communities to do the same.

The Green Library movement is a testament to the idea that libraries can be more than just repositories of books; they can be active agents of change, promoting a more sustainable and equitable world. As the movement continues to grow and evolve, Green Libraries will undoubtedly play an increasingly important role in shaping the future of our communities and our planet.

While challenges remain, the success of Green Libraries demonstrates that sustainability and knowledge are not mutually exclusive. Instead, they are complementary forces that, when combined, can drive meaningful progress towards a more sustainable and informed society. As we look to the future, the Green Library movement offers a model for how institutions can adapt to the challenges of the 21st century while staying true to their core mission of serving the public good.

References

Choudhury, S. (2015). Green library: scope and implications at the national and international level.

Hauke, P., Grunwald, M., & Wilde, A. (2014). Green Libraries Coming Up! National and international initiatives fostering environmental sustainable libraries and library services. In BOBCATSSS 2014 (pp. 1–7).

S P, R., & M, Em. K. (2013). Energy efficiency in green buildings– Indian concept. International Journal of Emerging Technology and Advanced Engineering, 3(3), 329–336.

Shah, L., Kumar, S., & Shah, M. K. (2015). Green libraries in academic institutions: Need of the hour- Social issues and environmental problems. Journal of Research-GRANTHAALAYAH, 39035(2008), 1–5.

Tseng, S. (2008). Green library design and evaluation: the Taipei Public Library, Taiwan. New Library World, 109(7/8), 321–336.

34

Mapping the Scholarly Productivity of Assam University: A Bibliometric Analysis Based on Shodhganga During 2001-2023

Tarali Barman[1], Mathias Swargiary[2] and Sagarika Medhi[3]

[1]Guest Faculty, DLIS, Cotton University, Guwahati - 781 001, Assam, India
E-mail: barman.tarali@rediffmail.com

[2]PhD Research Scholar, DLIS, Cotton University, Guwahati - 781 001, Assam, India
E-mail: mathias95912@gmail.com

[3]PhD Research Scholar, DLIS, Cotton University, Guwahati - 781 001, Assam, India
E-mail: samudrika1230@gmail.com

Abstract

The present study provides a mapping of the scholarly productivity of Assam University by doing bibliometrics analysis based on the theses uploaded in online ETDs database hosted by INFLIBNET Centre, Shodhganga. The study is arranged in such a way that, categorization is done based on chronological wise and department wise submission of theses. The time period of 22 years has been set from 2001-2023 to evaluate the growth and expansion of theses uploaded. There is total 2035 nos. of these uploaded till date consisting of various disciplines, which includes Natural Sciences, Arts and Humanities, Engineering, Business Administration, Commerce and Law and Language and Literature. The highest number of theses was completed and uploaded in Shodhganga in 2014, with 240 theses (11.79%). The lowest number of theses was in 2002, with 14 theses (0.69%). The Ecology and Environmental Science, Life Science, and Bengali Departments appear to be the most active in terms of research output, as reflected by the high number of theses uploaded. The following study is important to understand the progress of research and its areas of Assam University, as it will reflect the various research outputs and valuable contributions of the institute in academic integrity.

Keywords: *Assam University, Bibliometric, Research, Shodhganga, Thesis.*

1. Introduction

Electronic theses and dissertations (ETDs) are a great resource for scholars in academic institutions. They assist academics in comprehending the study carried out in prior years. When study results are made accessible on an open access platform, they are extensively utilized for further inquiry and teaching purposes. Theses and dissertations need to be made publicly available so that the users can get access to it. So, the emergence of Shodhganga proves to be very useful to the students, researchers, faculties and all other users (Vijayakumar and Vijayakumar, 2007). A significant number of doctorate research projects are now being conducted worldwide in several fields within the natural sciences. Bibliometrics focusses on analyzing the attributes of published literature, such as the distribution of literature on different subjects, statistical examination of content kinds, references, citations, and co-citations within and across fields (Pritchard, 1969). Davarpanah and Aslekia (2008) conducted bibliometric research on PhD theses, however these investigations were often limited to a certain university or department. Shodhganga serves as a digital archive for freely accessible whole texts of Indian electronic theses.

According to the UGC PhD rule 2009, it is now required to upload PhD theses on Shodhganga repositories. This allows for a better understanding of the University of Northeast's contributions in the field of Natural Science (UGC, 2009). This study will be invaluable for institutions and research centers engaged in scientific research, as it will aid in the creation of their collection development policy and in minimizing duplication when receiving collections free of charge.

2. Literature Survey

Dutta and Balaraman (2024) evaluated the research contribution of Presidency University, Kolkata accessible through Shodhganga: An Electronic Theses and Dissertations (ETD) database of INFLIBNET. There were 97 theses submitted to the database and found that department of Life Sciences holds the top position with total 28 (28.86%) theses. They also got that majority of theses were uploaded in English followed by Bengali and Hindi. Suryawanshi (2024) conducted a bibliometric study to examine the availability and legibility of Law Faculty PhD program thesesuploaded on the Shodhganga repository. They retrieved the data from Shodhganga ETD Repository and Google Scholar. The study found that 149 Indian universities have deposited 3441 PhD theses in Shodhganga Repository till December 2021.They also got that Punjab University has uploaded the maximum number of PhD theses on Shodhganga & Kuvempu University has the maximum number of citations received.

Similar bibliometric analysis was carried to determine the research output of Tezpur University during 1995-2023(Daimary, 2024). The study covered four academic disciplines: Management Science, Humanities & Social Science, Engineering, and Science and the data was segmented into three periods: 1995–2004, 2005–2014, and 2015–2023. The study highlights diverse growth patterns across disciplines and evaluates research impact through citation quantity and Average Citation Per Paper (ACPP) metrics, alongside a Relative Specialization Index comparing Tezpur University's global and national contributions in each field. The study found that Science and Engineering departments were the primary contributors and Management Science departments experienced the most substantial growth expansion.

Mondal and Khan (2021) analyzed a bibliometric mapping of Library and Information Science recorded through the Shodhganga reservoir of theses with special reference to West Bengal. The research found that total number of research paper recorded in Shodhganga repository on subject Library Science is Two 2143 up to February 2021 in India. They also found that the subsequent number of research paper recorded in Shodhganga repository from West Bengal is 124 which was submitted by 1 Central University and 5 State Universities. Highest number of research paper submitted in Shodhganaga repository by University of Burdwan is 37.

Hazarika and Sudhier (2022) conducted a bibliometric analysis of theses uploaded in Shodhganga related to the domain of Assamese Language and Literature. They found that 445 theses were submitted with the related domain among four Universities, viz.,Gauhati University, Dibrugarh University, Mahapurusha Srimanta Sankaradeva Viswavidyalaya and the Bodoland University. The study also found that the 2017-2018 period was the most productive year of theses submitted with the total number of 100 PhD theses. The research also found that Gauhati University and Dibrugarh were the highest contributor with 389 and 46 PhD theses uploaded in the database. Also, Dr. Umesh Deka of Gauhati University is identified as the Research Supervisor to guide the highest number (24) of theses followed by Dr. Dipti Phukan Patgiri of the same university. In case of reference these from Dibrugarh University received the highest number of 160 references.

3. Objectives

The objectives of the study are:

a) To find out the total uploaded theses by Assam University

b) To study chronological wise distribution of PhD awarded theses

c) To find out Subject and Year wise distribution of thesis during 2001-2023

d) Identify which subject area contribute most numbers of thesis

4. Methodology

The present work is an assessment study under natural research. It basically follows the survey method in general and document analysis method by using bibliometric techniques in specific. The study is based on the electronic theses available in the Shodhganga from 2001 to 2023. Data for the study were collected from the INFLIBNET website at Shodhganga@INFLIBNET Centre (http://shodhganga.inflibnet.ac.in/) and tabulated using MS -Excel and then analyzed by quantitative methods. The analysis was conducted for Assam University, Silchar which is situated in Assam, North-East to know and the year of dissertation publication, and major subjects of published dissertations.

5. Scope and Limitation

This study covers the theses that uploaded on Shodhganga, during the period of 22 years from 2001 to 2023, submitted by one of the Central Universities of Assam- Assam University. This period is selected with the purpose of a comprehensive study covering each subject. Electronic theses in each department submitted on Shodhgangaup to 2023 will be studied in present research. There are 2035 nos. of thesis submitted by this University on Shodhganga. The study is also limited to the theses published during the period of 2001-2023.

Table 1: Year wise distribution of thesis by Assam University

Sl. No	Year	Nos. of Thesis	%
1	2001	23	1.13
2	2002	14	0.69
3	2003	17	0.84
4	2004	15	0.74
5	2005	31	1.52
6	2006	26	1.28
7	2007	34	1.67
8	2008	48	2.36
9	2009	59	2.9
10	2010	76	3.73
11	2011	74	3.64
12	2012	99	4.86
13	2013	173	8.5
14	2014	240	11.79
15	2015	185	9.09
16	2016	94	4.61
17	2017	175	8.6
18	2018	148	7.27
19	2019	95	4.67
20	2020	98	4.82
21	2021	163	8
22	2022	108	5.31
23	2023	40	1.97
Total		**2035**	**100**

Table 2: Thesis Uploaded by Assam University Under Natural Science

Year	Chemistry	Biotech-nology	Computer Science	Ecology & Environmental Science	Earth Science	Information Technology	Life Science	Mathe-matics	Micro-biology	Physics	Statistics
2001	0	0	0	1	0	0	1	2	0	0	0
2002	0	0	0	0	0	0	0	0	0	0	0
2003	0	0	0	1	0	0	1	2	0	0	0
2004	0	0	0	0	0	0	0	0	0	3	0
2005	1	0	0	3	0	0	0	0	0	0	0
2006	1	0	0	6	0	0	3	0	0	1	0
2007	3	0	1	1	0	0	4	3	0	0	0
2008	1	1	1	14	0	0	2	2	0	1	0
2009	1	0	1	12	0	0	5	1	0	2	0
2010	6	0	4	8	0	1	7	1	0	2	0
2011	4	2	0	6	0	0	7	2	0	2	0
2012	3	3	0	13	0	0	13	0	0	6	0
2013	7	6	4	13	0	0	25	7	0	1	0
2014	5	13	3	12	0	0	7	2	0	3	0
2015	3	7	6	17	0	3	15	0	3	3	0
2016	6	6	4	20	0	2	15	1	2	3	0
2017	1	3	5	14	0	0	15	1	2	6	0
2018	6	5	3	12	0	1	12	1	2	7	0
2019	2	6	5	10	7	0	25	2	3	6	0
2020	1	5	6	5	3	0	3	0	4	5	1
2021	2	7	1	8	1	0	9	1	1	3	2
2022	4	4	1	3	1	0	6	0	3	2	2
2023	3	1	2	1	0	0	2	3	0	0	0
Total	**60**	**69**	**47**	**180**	**12**	**7**	**177**	**31**	**20**	**56**	**5**

Table 3: Thesis Uploaded by Assam University under Engineering Departments

Year	Agricultural Engineering	Computer Science & Engineering
2001	0	0
2002	0	0
2003	0	0
2004	0	0
2005	0	0
2006	0	0
2007	0	0
2008	0	0
2009	0	0
2010	0	0
2011	0	0
2012	0	0
2013	0	0
2014	0	0
2015	1	0
2016	0	0
2017	0	3
2018	2	3
2019	1	10
2020	1	0
2021	1	2
2022	0	1
2023	0	1
Total	**6**	**20**

Table 4: Thesis Uploaded by Assam University Under Arts & Humanities

Year	Anthro-pology	Eco-nomics	Education	Fine Arts	History	Library & Information Science	Mass Communication	Philosophy	Political Science	Social Work	Sociology
2001	0	0	1	0	1	0	1	1	0	0	2
2002	0	1	1	0	3	0	0	2	1	0	0
2003	0	2	0	0	3	0	0	0	0	0	0
2004	0	0	4	0	0	0	0	0	1	0	1
2005	0	0	0	0	3	0	2	0	2	0	0
2006	0	1	1	0	2	0	0	0	0	0	0
2007	0	0	0	0	2	0	0	3	0	0	0
2008	0	1	1	0	0	0	0	2	1	4	1
2009	0	2	2	0	2	0	7	0	2	1	3
2010	0	2	6	0	3	0	0	0	1	1	0
2011	0	7	4	0	1	0	4	2	3	1	0
2012	0	6	9	2	4	0	6	1	3	1	3
2013	0	8	17	1	5	0	2	0	2	5	11
2014	0	7	6	7	12	0	24	2	3	5	5
2015	0	6	19	4	5	0	7	4	8	1	7
2016	0	10	8	2	12	7	10	4	12	4	5
2017	0	4	11	0	5	1	6	3	4	2	1
2018	2	5	11	0	3	2	7	0	4	3	2
2019	3	7	4	5	4	0	6	1	11	2	5
2020	0	3	3	5	4	0	3	3	10	3	0
2021	3	5	5	3	8	1	13	0	9	10	4
2022	0	2	3	4	8	2	6	4	6	3	6
2023	0	1	2	3	1	4	1	0	1	3	2
Total	**8**	**80**	**118**	**36**	**91**	**17**	**105**	**32**	**84**	**49**	**56**

Table 5: Theses Uploaded by Assam University under Language and Literature

Year	Assamese (Diphu Campus)	Assamese (Main Campus)	Arabic	Bengali	English	Foreign Language	Hindi	Indian Comparative Literature	Linguistics	Manipuri	Sanskrit
2001	0	0	0	12	0	0	0	0	0	0	0
2002	0	0	0	4	0	0	0	0	2	0	0
2003	0	0	0	5	0	0	0	0	0	0	0
2004	0	0	0	4	0	0	0	0	0	0	0
2005	0	0	1	18	0	0	0	0	0	0	1
2006	0	0	1	6	0	0	0	0	0	1	3
2007	0	0	1	8	0	0	0	0	0	0	2
2008	0	0	0	9	0	0	0	0	0	1	0
2009	0	0	3	6	0	0	0	0	0	2	0
2010	0	0	2	9	2	0	0	0	0	3	4
2011	0	0	3	4	2	0	0	0	2	3	4
2012	0	0	5	9	0	0	0	0	2	3	3
2013	0	0	6	16	3	0	0	0	4	4	7
2014	0	0	19	43	6	0	0	0	8	11	3
2015	0	0	6	16	7	0	0	0	8	4	7
2016	0	0	3	10	7	1	1	0	7	5	5
2017	2	0	2	12	11	0	0	6	6	1	4
2018	1	0	0	13	9	0	1	5	5	0	2
2019	1	0	1	23	4	0	4	4	4	4	2
2020	1	0	1	10	2	0	1	1	0	2	2
2021	0	0	2	20	4	0	1	3	1	7	6
2022	0	2	0	5	6	0	0	0	1	5	3
2023	0	0	0	2	2	0	0	1	1	0	0
Total	**5**	**2**	**56**	**264**	**65**	**1**	**8**	**20**	**51**	**56**	**58**

Table 6: Theses Uploaded by Assam University Under Business Administration, Commerce and Law

Year	Business Administration	Commerce	Law
2001	0	1	0
2002	0	0	0
2003	1	1	0
2004	0	1	0
2005	0	0	0
2006	0	2	0
2007	0	2	0
2008	2	1	0
2009	0	1	0
2010	3	2	0
2011	2	3	0
2012	1	1	0
2013	2	5	0
2014	4	10	0
2015	1	5	0
2016	5	7	0
2017	5	18	2
2018	1	7	2
2019	4	10	5
2020	7	2	3
2021	8	4	2
2022	2	4	4
2023	1	0	2
Total	**49**	**87**	**20**

6. Data Analysis and Interpretation

A) Year - Wise Distribution of Thesis of Assam University

In the study, we have taken under consideration the research activities of various Departments of Assam University from the year 2001 to 2023.

Table 1 shows the distribution of thesis of Assam University according to the year. From the table, it is found that in the year 2014, highest nos. of thesis have been completed and uploaded in Shodhganga. The nos. of theses are 240 nos. (11.79%) and the year which have lowest nos. of thesis, i,e. 14 nos. (0.69%) is 2002. The year which has second highest nos. of thesis in Assam University is the year 2015 and the nos. of theses are 185 (9.09%). The years 2017, 2013 and 2021 have secured the 3rd, 4th and 5th position in terms of nos.

of thesis completed by Assam University. The nos. of thesis are 175 (8.6%), 173 (8.5%) and 163 (8%) respectively.

B) Thesis Uploaded by Assam University Under the Departments of Natural Science

There are twelve (12) nos. of departments in Assam University under Natural Science which have uploaded their thesis in Shodhganga. From the Table 2, it can be seen that Ecology and Environmental Science Department has highest no of thesis from the year 2001 to 2023. The total nos. of thesis are 180. Department of Life Science has secured the 2nd position, which has uploaded 177 nos. of thesis. Departments securing 3rd, 4th and 5th position are Biotechnology (69), Chemistry (60) and Physics (56) respectively. The department which has lowest nos. of thesis is Statistics under natural Science which has only 5nos. of thesis.

C) Thesis Uploaded by Assam University Under the Departments of Engineering

In the case of engineering departments, there are only two departments which have uploaded their thesis in Shodhganga. The departments are Agricultural Engineering and Computer Science & Engineering. Table 3 shows that the Department of Agricultural Engineering has 6 nos. of thesis and the Department of Computer Science & Engineering has 20 nos. of thesis.

D) Thesis Uploaded by Assam University Under the Departments of Arts and Humanities

Under the department of Arts & Humanities, there are 11 nos. of department which have uploaded their theses in Sodhganga for the year 2001 to 2023. From the Table 4 it can be seen that, Education department has highest nos. of thesis. The nos. of thesis of Education department are 118. Mass communication has secured the 2nd position, which has 105 nos. of thesis. The departments of History, Political Science and Economics have secured 3rd, 4th and 5th position in terms of thesis under the department of Arts & Humanities having 91nos., 84 nos. and 80 nos. of thesis respectively. Department of Anthropology has uploaded least nos. of thesis, i.e., 08 nos. of thesis

E) Thesis Uploaded by Assam University Under the Departments of Language and Literature

In case of Language and Literature, there are 11 nos. of departments uploading thesis in Sodhganga. Table 5 shows that Bengali Department has highest nos. of thesis which is 264 no's following by Department of English which have 65 nos. of thesis. Department of Sanskrit has secured 3rd position having 58 nos. of thesis. There are two departments which have secured the 4rd position. The departments are Arabic (56) and Manipuri (56). In the 5[th]postion includes department of Linguistics which have 51 nos. of thesis.

F) Thesis Uploaded by Assam University under the Departments of Business Administration, Commerce and Law

Under this Category, Table 6 shows that Commerce has hightest nos.of thesis (87), followed by Business Administration (49), and Low (20) respectively.

7. Results and Discussion

Results

1. **Year-Wise Distribution of Theses of Assam University:** The highest number of theses was completed and uploaded in Shodhganga in 2014, with 240 theses (11.79%). The lowest number of theses was in 2002, with 14 theses (0.69%). The year with the second-highest number of theses was 2015, with 185 theses (9.09%).The years 2017, 2013, and 2021 had the 3rd, 4th, and 5th highest number of theses, with 175 (8.6%), 173 (8.5%), and 163 (8%) theses, respectively.
2. **Thesis Uploaded by Assam University under the Departments of Natural Science:** The Ecology and Environmental Science Department had the highest number of theses with 180.The Department of Life Science had the second-highest number of theses with 177.The Departments of Biotechnology, Chemistry, and Physics had the 3rd, 4th, and 5th highest number of theses with 69, 60, and 56, respectively. The Department of Statistics had the lowest number of theses with only 5.
3. **Thesis Uploaded by Assam University under the Departments of Engineering:** The Department of Agricultural Engineering had 6 theses. The Department of Computer Science & Engineering had 20 theses.
4. **Thesis Uploaded by Assam University under the Departments of Arts and Humanities:** The Education Department had the highest number of theses with 118.The Mass Communication Department had the second-highest number of theses with 105.The Departments of History, Political Science, and Economics had the 3rd, 4th, and 5th highest number of theses with 91, 84, and 80, respectively. The Department of Anthropology had the lowest number of theses with 8.
5. **Thesis Uploaded by Assam University under the Departments of Language and Literature:** The Bengali Department had the highest number of theses with 264.The Department of English had the second-highest number of theses with 65.The Department of Sanskrit had

the 3rd highest number of theses with 58.The Departments of Arabic and Manipuri had the 4th highest number of theses with 56 each. The Department of Linguistics had the 5th highest number of theses with 51.

6. **Thesis Uploaded by Assam University under the Departments of Business Administration, Commerce, and Law:** The Commerce Department had the highest number of theses with 87.The Business Administration Department had the second-highest number of theses with 49.The Law Department had the lowest number of theses with 20.

Discussion

The findings from the study provide valuable insights into the research activities and the distribution of theses across various departments at Assam University. The year-wise analysis reveals that the university has experienced fluctuations in the number of theses completed, with the highest being in 2014 and the lowest in 2002.

The analysis of theses uploaded by the Departments of Natural Science, Engineering, Arts and Humanities, Language and Literature, and Business Administration, Commerce, and Law highlights the research strengths and focus areas of the university. The Ecology and Environmental Science, Life Science, and Bengali Departments appear to be the most active in terms of research output, as reflected by the high number of theses uploaded.

8. Conclusion

The bibliometric analysis of Assam University's thesis data from Shodhganga (2001-2023) provides a comprehensive overview of the university's scholarly productivity and research trends. The tables from 1 to 6 highlight the different types of publications in various domains and disciplines. The findings reveal significant year-over-year fluctuations in thesis output, with the natural sciences, particularly Ecology and Environmental Science, Life Science, and Biotechnology, demonstrating the strongest research performance, hence increasing its productivity. While some departments have shown remarkable expansion, others lag, indicating the need for a more balanced and strategic approach to research development within the university. The data can inform strategic planning, resource allocation, and the implementation of targeted initiatives to enhance the overall research ecosystem of Assam University and maintain its position as a leading center of academic excellence in the region.

References

Bopapurkar, P. K. (2003). A Bibliometric Analysis of Citations of Doctoral Dissertations in Economics: A Comparative Study of Universities of Madhya Pradesh. [Doctoral dissertation, Pt. Ravishankar Shukla Unlversity]. Shodhganga Repository. http://hdl.handle.net.10603/28752.

Davarpanah, M. R., &Aslekia, S. (2008). A scientometric analysis of international LIS journals: Productivity and characteristics. Scientometrics, 77(1), 21-39. DOI: 10.1007/s11192-007-1803-z

Dutta, K., & Balaraman, S. (2024). Analysis of research output of Presidency University, Kolkata through shodhganga theses and dissertation (ETD) database. IMPACT: International Journal of Research in Humanities, Arts and Literature (IMPACT: IJRHAL). 12(2), 37–44.

Gunasekaran, S., Batcha, M. Sadik, & Sivaraman, P. (2006). Mapping chemical science research in india: a bibliometric study. Annals of Library and Information Studies, 53(2), 83-95. http://nopr.niscpr.res.in/handle/123456789/4680

Hazarika, R. (2022). Doctoral research in Assamese language and literature. ILIS Journal of Librarianship and Informatics, 5(2), 7–15. https://www.researchgate.net/publication/375837471

Hazarika, T., & Goswami, R. (2016). Mapping research contributions of Assam University: A scientometric study, 286-293. https://ir.inflibnet.ac.in/bitstream/1944/2038/1/34.pdf

Jeyapragash, B., Rajkumar, T., & Muthuraj, A. (2016). Research output analysis of electronic thesis and dissertations with special reference to Shodhganga. Journal of Current Trends in Library and Information Science, 3(1&2), 16-20. http://www.jctl.org/index.php/jctl/article/view/30

Kumar, Nagesh (2015). Electronic theses & dissertations in Indian university libraries and national level repository: Positions and challenges. International Journal of Research in Library Science 1(2), 18-26. https://www.ijrls.in/journal_category/90-volume-1-issue-2/162-electronic-theses-dissertations-in-indian-university-libraries-and-national-level-repository-positions-and-challenges

Kumar, S., & Naqvi, H. (2010). Research output in the field of natural sciences: A bibliometric case study of Jamia Millia Islamia University, New Delhi. IFLA Journal, 36(4), 317-324. http://dx.doi.org/10.1177/0340035210388242

Mondal, H., & Khan, A. (2021). A bibliometric mapping of LIS research through the shodhganga reservoir of theses: Special reference to West Bengal. Journal of the Maharaja Sayajirao University of Baroda, 55(1), 670–673. https://www.researchgate.net/publication/353210639

Nongrang, K. (2016). Research output of faculty members of North Eastern Hill University (NEHU): A bibliometric study of periodical literature (2000-2010) [Doctoral dissertation, North-Eastern Hill University]. Shodhganga Repository. http://hdl.handle.net/10603/217103.

Pritchard, A. (1969). Statistical bibliography or bibliometrics. Journal of documentation, 25(4), 348-349.

Research output of Tezpur University, during 1995-2023: A bibliometric analysis using Web of Science (2024). Annals of Library and Information Studies, 71(02), 172–179. https://doi.org/10.56042/alis.v71i2.7104

Sevukan, R., Nagaranjan M., & Sharma, J. (2007) Research output of faculties of plant sciences in central universities of India: A bibliometric study. Annals of Library and Information Studies, 54(3), 129-139. https://nopr.niscpr.res.in/handle/123456789/3229

Siddiqui, J., & Kumar, G. &. (2023). Bibliometric analysis of the publications on cloud computing in India from 2011-2017. Journal for Basic Sciences, 23(2), 473-396. https://www.researchgate.net/publication/368756203_Bibliometric_Analysis_of_the_Publications_on_Cloud

Singh, P., & Nandy, A. (2022). A bibliometric analysis of research output from Indian Institutes of Management. DESIDOC Journal of Library & Information Technology, 42(6), 364-376. DOI: 10.14429/djlit.42.6.18316

Sivakumaren, K.S. (2015). Electronic Thesis and Dissertations (ETDs) by Indian Universities in Shodhganga Project: A study. Journal of Advances in Library and Information Science4(1), 62-66. http://jalis.in/pdf/4-1/Sivamit.pdf.

Suryawanshi, S.S. (2024). A biblometrics study on electronic PhD theses and dissertations of the faculty of law in INFLIBNET Shodhganga repository. Jilin Daxue Xuebao (Gongxueban)/Journal of Jilin University (Engineering and Technology Edition). 43, 87–97. https://doi.org/10.5281/ZENODO.10686054

UGC (2009). The University Grants Commission (Minimum Standards and Procedure for Award of M.Phil./PhD Degree) Regulations, 2009. New Delhi: University Grants Commission.

Vijayakumar, J. K., & Vijayakumar, M. (2007). Importance of electronic theses and dissertations (ETD) with special reference to India. In Proceedings of the 14th National Convention for Students, Doctoral Scholars and Young Librarians on eTheses and Digital Libraries in the Internet Age, 3(2), 67-75.

35

The Evolution of Climate Change Research: A Scientometric Perspective

Muthuraj Anbalagan

Assistant Librarian, Amrita School of Engineering, Amrita Vishwa Vidyapeetham Amaravati Campus, Guntur - 522 503, Andhra Pradesh, India
E-mail: a_muthuraj@av.amrita.edu

Abstract

This paper investigates the productivity of climate change research from 2017 to 2022. The data were downloaded from the Scopus database. Every year, the amount of climate change research grows exponentially. The growth of publications has been accelerating since 1970. The six years of study from 2017 to 2022 were used to analyze the increase in research on climate change. The researcher selected "Article type," and the search string used to search the database was "Climate Change." The search string was used in the "Subject" field, and the period 2017–2022 was used to download climate change records. It is found that 36,573 research publications are published, and 26,685 articles received 372,616 citations. The present study aims to analyze the year-wise publications, top ten authors, source-wise publications, keyword and country-wise publications, language- and funding-wise publications, different types of documents, and top ten highly cited research papers and their sharing on social media. It is found that in 2021, 7,587 publications were published in the field of climate change; Zielinski, C., published 76 articles, and the Sustainability Switzerland journal published the majority of 806 climate change articles. It is further found that the National Natural Science Foundation of China is funding 2,336 articles, and the article "Biodiversity redistribution under climate change: Impacts on ecosystems and human well-being" published in the Science journal received 1,410 citations from the Scopus database.

Keywords: *Climate change, Research output, Research impact, Highly cited papers, Scientometrics.*

1. Introduction

Climate change includes higher temperatures, changes in rainfall patterns, alterations in the frequency and distribution of weather events such as droughts, storms, floods, and heatwaves, sea-level rise, and the consequent impacts on human and natural systems (Riedy, 2018). Climate is the average weather in a given area over a longer time. A description of climate includes information on, for example, "the average temperature in different seasons, rainfall, and sunshine. Also, a description of the chance of experiencing extremes is often included. Climate change is any systematic change in the long-term statistics of climate variables such as temperature, precipitation, pressure, or wind sustained over several decades or longer. Climate change can be due to natural external forcings (changes in solar emission or changes in the Earth's orbit, natural internal processes of the climate system), or it can be human-induced" (Buonocore, 2022).

Scientometrics is concerned with the quantitative features and characteristics of science and scientific research (Kiado). Scientometrics is the science of measuring and analyzing science. In practice, scientometrics is often done using bibliometrics, which is a measurement of the impact of publications. Modern scientometrics is mostly based on the work of Derek J. de Solla Price and Eugene Garfield (Definitions). Scientometrics has been defined as the "quantitative study of science, communication in science, and science policy" (Hess, 1997) (OECD).

2. Review of Literature

Sangam and Savitha analyzed "climate change and global warming literature from 2001–2016 from the Web of Science database. The study analyzed various parameters such as growth trend, authorship pattern, collaborative index, collaborative coefficient, degree of collaboration, division of forms, most preferred authors, country-wise distribution, and others". Caka Fjolle and Nebi Caka explored "the evolution of the scientific output publication trend, current research focus, and promising ideas for future related research from a meta-perspective. Since specific climate change research in Kosovo is scarce, this study extended the scope of research output by encompassing related studies on sustainable development, environment, agriculture, forestry, spatial planning, transportation, buildings, energy, waste, air pollution, etc". Kavitha attempted to find the year-wise, language-wise, document-type-wise, and keyword-wise distribution of articles in climate change research and to study the authorship pattern and collaboration in climate change research. Using the search term "climate change," the data were retrieved from the Web of Science from 1994 to 2018. Fu and Waltman described a large-scale bibliometric analysis based on approximately 120,000 climate change publications between 2001 and 2018 to examine how climate change is studied in scientific research. Our analysis provided an overview of scientific knowledge, shifts in research hotspots, the global geographical distribution of research, and the focus of individual countries. Shabahat and Mushtaq analyzed 17,266 publication records on different aspects of climate change published during 2009–2013. The study also assessed the citation patterns, including the country's citations, citations of affiliated institutes, and average, self, and unique citations received in select years. Saravanan *et al.* attempted "to highlight quantitatively the growth and development of climate change literature in terms of publication output as per Web of Science (1991–2012, September). The focus of this analysis is to study the literature on climate change published in five developing countries, namely Argentina, Brazil, China, India, and Mexico. This paper is a comparative study on year-wise, document-type-wise, most productive authors, subject-wise, journal-wise, institution-wise, and language-wise distributions. 7065 records have been retrieved for climate change in the studied countries". Venkatesan described the research productions on climate change were analyzed using the online database of the 'Web of Science'. There were 94,756 records contributed worldwide from 1999 to 2012. The distribution of publications based on the year of production, country, language, and document type of the publications was studied. The relative growth rate of the publications and doubling time were calculated.

3. Research Methodology

The information was gathered from the Scopus database (Scopus). The researcher selected the "article title, abstracts, and keywords" type and used the search term "climate change" to find results in 392,493 documents for the period 1967–2022. Previous studies covered the years 1990–2015; after that, no studies identified the period. The researcher focused only on the 2017–2022 period, selected "Article type," and used the search term "Climate Change" to download the data for the present study. It was found that 36,573 research publications were published, and 26,685 articles received 372,616 citations. The present study aims to analyze the year-wise publications, source-wise publications, keyword and country-wise publications, language- and funding-wise publications, different types of documents, year-wise publications *vs* single author and multiple authors, single author *vs* multiple authors, year-wise publications *vs* cited and uncited papers, top ten Indian and International authors, and top ten highly cited research papers and their sharing on social media. The study utilized SPSS and Excel for data analysis and calculations.

4. Objectives of the Study

The following are the major objectives of the study:

- To determine the year-wise and source-wise publications
- To analyze keyword- and country-wise publications
- To assess the number of publications based on their sources
- To identify language- and funding-wise publications
- To categorize various types of documents
- To identify highly cited research papers and their social media shares
- To evaluate single vs. multiple authors
- To analyze year-wise publications vs. single authors and multiple authors
- To measure year-wise publications vs. cited and uncited papers
- To identify cited and uncited publications
- To observe types of documents vs. cited and uncited publications
- To identify the top 10 leading Indian and international scientists in the field of climate change
- To assess highly cited research papers
- To conduct altmetric analysis of highly cited research papers

5. Results and Discussion

5.1. Year-Wise Publications

Table1: Year wise publications

S. No	Year	Nos.	Percent
1	2017	4885	13.4
2	2018	5446	14.9
3	2019	5746	15.7
4	2020	6462	17.7
5	2021	7587	20.7
6	2022	6447	17.6
	Total	**36573**	**100**

The table 1 presents the distribution of publications on climate change across different years from 2017 to 2022. In 2017, there were 4885 publications, accounting for 13.4% of the total. This number increased to 5446 (14.9%) in 2018 and further to 5746 (15.7%) in 2019. The year 2020 saw a notable rise with 6462 publications, constituting 17.7% of the total. Subsequently, 2021 recorded the highest number of publications at 7587 (20.7%), followed closely by 2022 with 6447 publications (17.6%). Overall, the total number of publications analyzed in this study was 36,573, representing various aspects of climate change research during the specified period.

5.2. Source-Wise Publications

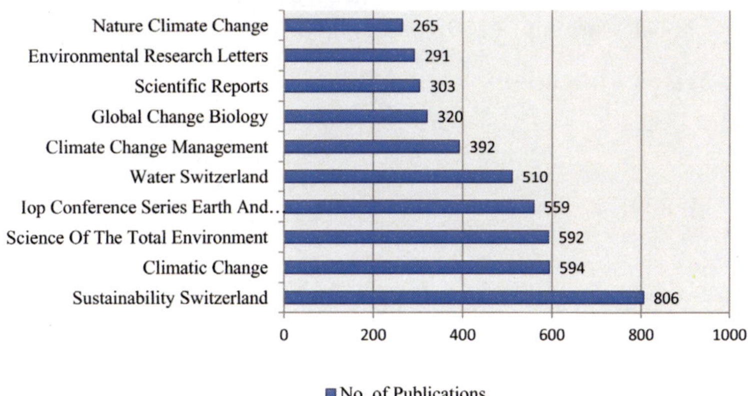

Fig. 1: Source-wise publications

Figure 1 presents details about the top ten source-wise publications. The Sustainability Switzerland journal published the highest number of climate change articles, with 806 publications, followed by the Climate Change journal with 594 articles. Furthermore, the Science of the Total Environment journal published 592 articles, while the IOP Conference Series: Earth Science journal contributed 559 articles. The Water Switzerland journal published 510 articles, followed by the Climate Change Mxanagement journal with 392 articles. Additionally, the Global Change Biology journal produced 320 articles, Scientific Reports published 303 articles, Environmental Research Letters contributed 291 articles, and Nature Climate Change released 265 articles.

5.3. Keyword-Wise Publications

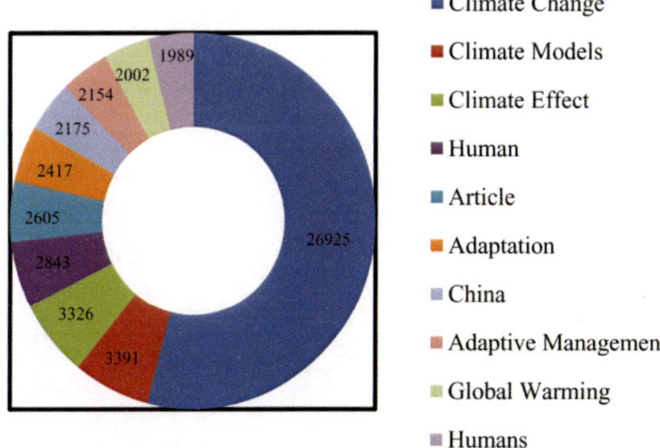

Fig. 2: Keyword-wise publications

Figure 2 illustrates the top ten keywords. The most frequently used keyword is "Climate Change" (26925 times), followed by "Climate Models" (3391 times), "Climate Effect" (3326 times), "Human" (2843 times), "Article" (2843 times), "Adaptation" (2417 times), "China" (2175 times), "Adaptive Management" (2154 times), "Global Warming" (2002 times), and "Humans" (1989 times).

5.4. Country-Wise Publications

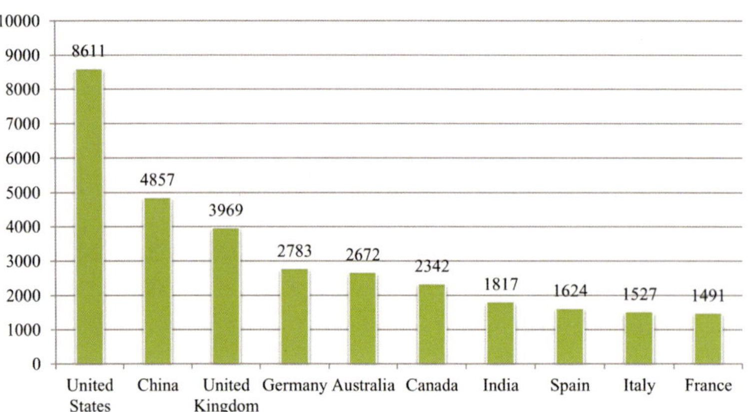

Fig. 3: Country-wise publications

Figure 3 illustrates the top ten countries. The United States published the most climate change research papers (8611), followed by China (4857), the United Kingdom (3969), Germany (2783), Australia (2672), Canada (2342), India (1817), Spain (1624), Italy (1527), and the fewest contributions by France (1491).

5.5. Various Types of Documents

Table 2: Various types of Documents

S. No	Document Type	Nos.	Percent
1	Article	25755	70.4
2	Book	310	0.8
3	Book Chapter	3407	9.3
4	Conference Paper	2152	5.9
5	Conference Review	45	0.1
6	Data Paper	25	0.1
7	Editorial	704	1.9
8	Erratum	422	1.2
9	Letter	279	0.8
10	Note	767	2.1

11	Retracted	8	0
12	Review	2595	7.1
13	Short Survey	98	0.3
14	Undefined	6	0
	Total	**36573**	**100**

Table 2 explains the various types of documents. The highest published document type is articles, with 25755 (70.4%), followed by Book chapter with 3407 (9.3%), review with 2595 (7.1%), Conference paper with 2152 (5.9%), Note with 767 (2.1%), Editorial with 704 (1.9%), Erratum with 422 (1.2%), Book with 310 (0.8%), letter with 279 (0.8%), and Short Survey with 119 (0.3%). The least published types of documents are Conference Review with 45 (0.1%), Data Paper with 25 (0.1%), and Retracted with 8 (0%), and Undefined with 6 (0%).

5.6. Language-Wise Publications

Table 3: Language-wise publications

S. No	Language	Nos.
1	English	35107
2	Chinese	569
3	Spanish	300
4	German	270
5	French	135
6	Portuguese	96
7	Russian	85
8	Italian	41
9	Persian	14
10	Japanese	13

Table 3 describes the top ten languages. There are 35107 articles published in English, followed by 569 articles published in Chinese, 300 articles published in Spanish, 270 articles published in German, 135 articles published in French, 96 articles published in Portuguese, 85 articles published in Russian, 41 articles published in Italian, 14 articles published in Persian, and 13 articles published in Japanese.

5.7. Single *vs* Multiple Authors

The contributions of single and multiple authors are summarized in Table 4. The highest number of publications, 30148 (82.4%), were provided by multiple authors, while the least number of publications, 6425 (17.6%), were provided by a single author.

Table 4: Single *vs* Multiple Authors

S. No	Description	Nos.	Percent
1	Single Author	6425	17.6
2	Multiple Authors	30148	82.4
	Total	**36573**	**100.0**

5.8. Year-Wise Publications *vs* Single Author and Multiple Authors

Table 5: Year-wise publications vs Single Author and Multiple Authors

S. No	Year	Single –Multiple Authors		Total
		Single Author	Multiple Authors	
1	2017	1010	3875	4885
2	2018	1087	4359	5446
3	2019	1083	4663	5746
4	2020	1141	5321	6462
5	2021	1257	6330	7587
6	2022	847	5600	6447
Total		**6425**	**30148**	**36573**

Table 5 details the contributions of single authors and multiple authors' publications year-over-year. In 2021, the highest number of publications was 7587 articles, with 1257 being single-authored and 6330 being multiple-authored. This was followed by 6462 articles in 2020, with 1141 being single-authored and 5321 being multiple-authored. In 2022, there were 6447 articles, with 847 being single-authored and 5600 being multiple-authored. In 2019, there were 5746 articles, with 1083 being single-authored and 4663 being multiple-authored. In 2018, there were 5446 articles, with 1087 being single-authored and 4359 being multiple-authored. In 2017, there were 4885 articles, with 1010 being single-authored and 3875 being multiple-authored.

5.9. Year-Wise Publications *vs* Single Author and Multiple Authors

Table 6: Year-wise publications vs Single Author and Multiple Authors

S. No	Year	Auth1-6						Total
		1	2	3	4	5	6	
1	2017	1010	989	929	645	445	867	4885
2	2018	1087	1071	969	727	550	1042	5446
3	2019	1083	1069	1024	839	578	1153	5746
4	2020	1141	1169	1141	944	662	1405	6462
5	2021	1257	1326	1321	1126	859	1698	7587
6	2022	847	1050	1143	955	720	1732	6447
Total		**6425**	**6674**	**6527**	**5236**	**3814**	**7897**	**36573**

Table 6 details the contributions of Single Author and Multiple Authors' publications Yearly wise. In 2021, the highest number of publications was 7587 articles, which included 1257 articles by the 1st author, 1326 articles by the 2nd author, 1321 articles by the 3rd author, 1126 articles by the 4th author, 859 articles by the 5th author, and 1698 articles by the 6th author. This was followed by 6462 articles in 2020, including 1141 articles by the 1st author, 1169 articles by the 2nd author, 1141 articles by the 3rd author, 944 articles by the 4th author, 662 articles by the 5th author, and 1405 articles by the 6th author. In 2022, there were 6447 articles, including 847 articles by the 1st author, 1050 articles by the 2nd author, 1143 articles by the 3rd author, 955 articles by the 4th author, 720 articles by the 5th author, and 1732 articles by the 6th author. In 2019, there were 5746 articles, including 1083 articles by the 1st author, 1069 articles by the 2nd author, 1024 articles by the 3rd author, 839 articles by the 4th author, 578 articles by the 5th author, and 1153 articles by the 6th author. In 2018, there were 5446 articles, including 1087 articles by the 1st author, 1071 articles by the 2nd author, 969 articles by the 3rd author, 727 articles by the 4th author, 550 articles by the 5th author, and 1042 articles by the 6th author. In 2017, there were 4885 articles, including 1010 articles by the 1st author, 989 articles by the 2nd author, 929 articles by the 3rd author, 645 articles by the 4th author, 445 articles by the 5th author, and 867 articles by the 6th author.

5.10. Year-Wise Publications *vs* Cited and Uncited Papers

Table 7: Year-wise publications vs Cited and Uncited papers

S. No	Year	CUC		Total
		Uncited	Cited	
1	2017	694	4191	4885
2	2018	770	4676	5446
3	2019	910	4836	5746
4	2020	1188	5274	6462
5	2021	2252	5335	7587
6	2022	4075	2372	6447
Total		**9889**	**26684**	**36573**

Table 7 compares year-wise publications and cited and uncited research papers. The total is 36,573, which includes 9,889 uncited and 26,684 cited articles in the research productivity of climate change. The greatest number of publications, 7587, was published in 2021, including 2252 uncited papers and 5335 cited papers, followed by 6462 publications published in 2020, including 1188 uncited papers and 5274 cited papers. The 5746 publications produced in 2019 include 910 uncited papers and 4836 cited papers, while the 5446 publications produced in 2018 include 770 uncited papers and 4676 cited papers. Additionally, the 4885 publications produced in 2017 include 694 uncited papers and 4191 cited papers. The 6447 research publications published in 2022 include 4075 uncited papers and 2372 cited papers.

5.11. Cited and Uncited Publications

Table 8: Cited and Uncited publications

S. No	Description	Nos.	Percent
1	Uncited	9889	27.0
2	Cited	26684	73.0
	Total	**36573**	**100.0**

Table 8 describes the cited and uncited publications on climate change from 2017 to 2022. The most cited publications are 26,684 (73.0%), while the fewest are uncited publications, 9,889 (27.0%).

5.12. Types of documents *vs* Cited and Uncited Publications

Table 9: Types of documents vs Cited and Uncited publications

S. No	Document Type	CUC		Total
		Uncited	**Cited**	
1	Article	5201	20554	25755
2	Book	112	198	310
3	Book Chapter	1713	1694	3407
4	Conference Paper	1184	968	2152
5	Conference Review	45	0	45
6	Data Paper	8	17	25
7	Editorial	360	344	704
8	Erratum	325	97	422
9	Letter	91	188	279
10	Note	361	406	767
11	Retracted	2	6	8
12	Review	444	2151	2595
13	Short Survey	43	55	98
14	Undefined	0	6	6
Total		**9889**	**26684**	**36573**

Table 9 explains the cited and uncited publications of various types of documents related to climate change during 2017–2022. The highest published document type is an article with a total of 25,755, which includes 5,201 uncited and 20,554 cited publications, followed by Book Chapter with 3,407, which includes 1,713 uncited and 1,694 cited publications. Additionally, Data Paper with 25 publications, including 8 uncited and 17 cited papers; Retracted with 8 publications, including 2 uncited and 6 cited papers; Undefined with 6 publications, all cited and no uncited publications; and Conference Review with 45 publications, all uncited papers, are among the least publishable types of documents.

5.13. Top 10 Leading International Scientists in the Field of Climate Change

Table 10: Top 10 leading International Scientists in the field of Climate change

S. No	Name	Latest Affiliation	Latest Affiliation Country	Number of matching documents	Total citations	Total documents	h-index	First publication year
1	Ciais, Philippe	Laboratoire des Sciences du Climat et de l'Environnement	France	161	60940	892	153	1992
2	Penuelas, Josep	Consejo Superior de Investigaciones CientÃficas	Spain	159	45307	1059	125	1983
3	Piao, Shilonog	Peking University	China	136	30805	436	112	2001
4	Camarero, JesÃºs Julio	Instituto Pirenaico de EcologÃa	Spain	127	8577	438	65	1996
5	Ebi, Kristie L.	Uganda Martyrs University	Uganda	125	16200	306	68	1996
6	Shahid, Shamsuddin	Universiti Teknologi Malaysia	Malaysia	118	4899	334	54	1999
7	Smith, Pete	University of Aberdeen	United Kingdom	111	33978	599	115	1994
8	Laybourn-Langton, Laurie	University of Exeter	United Kingdom	109	139	189	4	2016
9	Singh, Vijay Pal	Texas A&M University	United States	106	26911	1410	93	1975
10	Chen, Deliang	GÃ¶teborgs Universitet	Sweden	103	11868	420	65	1995

Table 10 indicates the Top 10 leading International Scientists in the field of Climate Change. Philippe Ciais, affiliated with Laboratoire des Sciences du Climat et de l' Environnement from France, has 161 matching documents out of 892 total documents, with a total of 60,940 citations and the highest h-index of 153. Shilong Piao, affiliated with Peking University from China, has 136 matching documents out of 436 total documents, with a high citation count of 30,805. Vijay Pal Singh, affiliated with Texas A&M University from the United States, has 106 matching documents out of 1,410 total documents, with the highest citation count of 26,911 and an h-index of 93. Deliang Chen, affiliated with Göteborgs Universitet from Sweden, has 103 matching documents out of 420 total documents, with the highest citation count of 11,868 and an h-index of 65.

5.14. Top 10 Leading Indian Scientists in the Field of Climate Change

Table 11: Top 10 leading Indian scientists in the field of Climate Change

S. No	Name	Latest Affiliation	Number of matching documents	Total cita-tions	Total docu-ments	h-index	First publica-tion year
1	Sejian, Veerasamy	ICAR - National Institute of Animal Nutrition and Physiology, Bengaluru	40	1785	176	30	2005
2	Dimri, Ashok Priyadarshan	Indian Institute of Geomagnetism	37	2109	172	34	1999
3	Jat, Mangi Lal	International Maize and Wheat Improvement Centre (CIMMYT)	34	3429	188	41	2000
4	Pal, Subodh Chandra	The University of Burdwan	32	878	133	27	2016
5	Goyal, Manish Kumar	Indian Institute of Technology Indore	31	1835	104	29	2007
6	Mall, R. K.	Banaras Hindu University	29	1412	88	19	1991
7	Pathak, Himanshu	Indian Council of Agricultural Research	29	7901	215	55	1996
8	Behera, Mukunda Dev	Indian Institute of Technology Kharagpur	29	1400	112	23	2000
9	Singh, Chandni	Indian Institute for Human Settlements	28	669	39	16	2016
10	Kulkarni, Anil V.	Indian Institute of Science	28	2752	116	30	1991

Table 11 indicates the top ten Indian authors based on matching documents since 2017. Veerasamy Sejian, affiliated with ICAR - National Institute of Animal Nutrition and Physiology, Bengaluru, has 40 matching documents out of 176 total documents, with the highest citation count of 1785 and an h-index of 30. Ashok Priyadarshan Dimri, affiliated with the Indian Institute of Geomagnetism, has 37 matching documents out of 172 total documents, with the highest citation count of 2109 and an h-index of 34. Mangi Lal Jat, affiliated with the International Maize and Wheat Improvement Centre (CIMMYT), has 34 matching documents out of 188 total documents, with the highest citation count of 3429 and an h-index of 41. Subodh Chandra Pal, affiliated with The University of Burdwan, has 31 matching documents out of 104 total documents, with the highest citation count of 1835 and an h-index of 27. Manish Kumar Goyal, affiliated with the Indian Institute of Technology Indore, has 31 matching documents out of 104 total documents, with the highest citation count of 1835 and an h-index of 29. R. K. Mall, affiliated with Banaras Hindu University, has 29 matching documents out of 88 total documents, with the highest citation count of 1412 and an h-index of 19. Himanshu Pathak, affiliated with the Indian Council of Agricultural Research, has 29 matching documents out of 215 total documents, with the highest citation count of 7901 and an h-index of 55. Mukunda Dev Behera, affiliated with the Indian Institute of Technology Kharagpur, has 29 matching documents out of 112 total documents, with the highest citation count of 7901 and an h-index of 23. Chandni Singh, affiliated with the Indian Institute for Human Settlements, has 28 matching documents out of 669 total documents, with the highest citation count of 669 and an h-index of 16. Anil V. Kulkarni, affiliated with the Indian Institute of Science, has 28 matching documents out of 116 total documents, with the highest citation count of 2752 and an h-index of 30.

5.15. Funding-Wise Publications

Table 12: Funding wise publications

S. No	Funding sponsor	Nos.
1	National Natural Science Foundation of China	2336
2	National Science Foundation	1358
3	National Key Research and Development Program of China	675
4	Horizon 2020 Framework Programme	636
5	Chinese Academy of Sciences	587
6	European Commission	569
7	Natural Environment Research Council	438
8	Natural Sciences and Engineering Research Council of Canada	410
9	Conselho Nacional de Desenvolvimento CientÃfico e TecnolÃ³gico	403
10	CoordenaÃ§Ã£o de AperfeiÃ§oamento de Pessoal de NÃvel Superior	309

Table 12 shows the top ten funding publications. The National Natural Science Foundation of China is highly funded, supporting 2336 articles, followed by the National Science Foundation, which funded 1358 articles. Additionally, the Coordenação de Aperfeiçoamento de Pessoal de Nível Superior has funded the least number of articles, with 309 articles.

5.16. Highly Cited Research Papers

Table 13: Highly Cited Research Papers

S. No	Title	Source title	Year	Citations
1	Biodiversity redistribution under climate change: Impacts on ecosystems and human well-being	Science	2017	1410
2	Forest disturbances under climate change	Nature Climate Change	2017	992
3	The Global Syndemic of Obesity, Under nutrition, and Climate Change: The Lancet Commission report	The Lancet	2019	954
4	The 2019 report of The Lancet Countdown on health and climate change: ensuring that the health of a child born today is not defined by a changing climate	The Lancet	2019	589
5	Scientistsâ€™ warning to humanity: microorganisms and climate change	Nature Reviews Microbiology	2019	573
6	The Lancet Countdown on health and climate change: from 25 years of inaction to a global transformation for public health	The Lancet	2018	558
7	The 2020 report of The Lancet Countdown on health and climate change: responding to converging crises	The Lancet	2021	522
8	Climate-changeâ€“driven accelerated sea-level rise detected in the altimeter era	Proceedings of the National Academy of Sciences of the United States of America	2018	500
9	The role of CO2 capture and utilization in mitigating climate change	Nature Climate Change	2017	491
10	Climate change and interconnected risks to sustainable development in the Mediterranean	Nature Climate Change	2018	487

Table 13 indicates the top ten Highly Cited Research Papers. The article titled "Biodiversity redistribution under climate change: Impacts on ecosystems and human well-being," published in the Science journal, received the highest number of citations (1410) from the Scopus database. It is followed by the article titled "Forest disturbances under climate change," published in the Nature Climate Change journal, which received 992 citations. Furthermore, the article titled "Climate change and interconnected risks to sustainable development in the Mediterranean," published in the Nature Climate Change journal, has received the least number of citations, with 487.

5.17. Altmetric Analysis of Highly Cited Research Papers

Table 14: Altmetric analysis of highly cited research papers

S. No	Title	Citations	Usage	Captures	Mentions	Social Media
1	Biodiversity redistribution under climate change: Impacts on ecosystems and human well-being	1,467	16,172	3,782	99	5,750
2	Forest disturbances under climate change	1033	77	1758	15	159
3	The Global Syndemic of Obesity, Undernutrition, and Climate Change: The Lancet Commission report	994	970	2206	164	5447
4	The 2019 report of The Lancet Countdown on health and climate change: ensuring that the health of a child born today is not defined by a changing climate	641	556	1299	294	3033
5	Scientistsâ€™ warning to humanity: microorganisms and climate change	584	70	2372	76	819
6	The Lancet Countdown on health and climate change: from 25 years of inaction to a global transformation for public health	621	370	1440	95	1672
7	The 2020 report of The Lancet Countdown on health & climate change: responding to converging crises	557		1017	132	3758
8	Climate-changeâ€"driven accelerated sea-level rise detected in the altimeter era	533	774	1126	97	472
9	The role of CO2 capture and utilization in mitigating climate change	508	18	803	14	250
10	Climate change and interconnected risks to sustainable development in the Mediterranean	506	21	832	17	

Table 14 shows the Altmetric analysis of highly cited research papers. "Biodiversity redistribution under climate change: Impacts on ecosystems and human well-being" has received the highest number of citations, 1410, and people have shared this article on social media 16,172 times, with 3,782 captures, 99 mentions, and 5,750 social media shares. Following this, "Scientists' warning to humanity: microorganisms and climate change" has received 584 citations and has been shared 70 times on social media, with 2,372 captures, 76 mentions, and 819 social media shares. Additionally, "Climate change and interconnected risks to sustainable development in the Mediterranean" has received the least number of citations (506) and the lowest share of social media usage (21 usages, 832 captures, and 17 mentions).

6. Findings and Conclusion

The study reveals several key findings:

- The productivity of climate change publications has shown a consistent increase year after year since 1970. The analysis includes 36,573 research publications, with 26,685 articles receiving a total of 372,616 citations.
- Zielinski, C., emerges as a prolific author, with 76 published articles on climate change, indicating significant contributions to the field.
- Sustainability Switzerland journal stands out as the leading publisher in climate change research, having published 806 articles, reflecting its commitment to advancing knowledge in this area.
- The United States leads in climate change research output, with 8,611 published papers, underscoring the country's significant contribution to the global understanding of climate change.
- The study highlights the significant support provided by the National Natural Science Foundation of China, sponsoring 2,336 climate change articles, indicating the nation's investment in this critical area of research.

In conclusion, the findings underscore the growing importance of climate change research globally. With increasing awareness of the impacts of climate change on societies worldwide, the need for robust scientific inquiry and publication is more critical than ever. The study's insights into publication trends, prolific authors, leading publishers, and funding sources provide valuable information for researchers, policymakers, and stakeholders working to address the challenges posed by climate change. Moving forward, continued efforts to expand research initiatives and disseminate findings through high-quality publications will be essential in advancing our understanding and response to this pressing global issue.

7. Declaration of Conflicting Interests

The author declared no potential conflicts of interest with respect to the research, authorship, and/or publication of this article.

8. Funding

The author received no financial support for the research, authorship, and/or publication of this article.

References

Buonocore M. (2022). What is climate? What is climate change? climateurope. Accessed December 17, 2022. https://www.climateurope.eu/what-is-climate-and-climate-change/

Caka Fjolle, & Nebi Caka (2022). A scientometric analysis of climate change research in Kosovo. The International Journal of Climate Change: Impacts and Responses, 15(1), 73-83. doi:10.18848/1835-7156/CGP/v15i01/73-83.

Fu H. Z., & Waltman L. (2021). A large-scale bibliometric analysis of global Climate Change research between 2001 and 2018. Research Square. doi:10.21203/rs.3.rs-759304/v1

Kavitha, E.S. (2022). Research trends on climate change publications in BRICS countries: A scientometrics analysis. International Journal of Information Science & Management, 20 (4), 33-48.

Kiado, A. (2022). Scientometrics. Springer. Accessed December 17, 2022. https://www.springer.com/journal/11192

Riedy, C. (2008). Climate change. In Ritzer, G. (ed). The Blackwell Encyclopedia of Sociology. Wiley-Blackwell.https://www.researchgate.net/publication/311301385_Climate_Change Sangam, S. L., Savitha, K. S. (2023). Climate change and global warming: A scientometric study. COLLNET Journal of Scientometrics and Information Management, 13(1), 199-212. DOI: 10.1080/09737766.2019.1598001

Saravanan, G., Rajan, V.R., Prasad S., & Muthusankar, G. (2014). Climate change research (1991–2012): comparative scientometric study of Argentina, Brazil, China, India and Mexico. Library Philosophy and Practice (e-jurnal). Paper 1134. (hal-01353603)

Scopus.com. Accessed December 17, 2022. http://scopus.com

Scientometrics- OECD. Oecd.org. Accessed December 17, 2022. https://www.oecd.org/sti/inno/scientometrics.htm

Shabahat, H., Mushtaq, M. (2015). Scientometric assessment of climate change data. Library Herald, 53(4), 379-388. DOI: 10.5958/0976-2469.2015.00034.2

Shabahat, H., Mushtaq, M. (2015). Research assessment of climate change data: A scientometric construct. Qualitative and Quantitative Methods in Libraries (QQML) Special Issue Bibliometrics and Scientometrics, 183-194. Accessed December 16, 2022. http://qqml.net/papers/Special_Issue_January_2015_Bibliometrics/4S18_Journal_2015_SpecialIssueBibliometrics_HusainandMushtaq_183-194.pdf

Venkatesan, M., Gopalakrishnan, S., & Gnanasekaran, D. (2013). Growth of literature on climate change research: A scientometric study. Journal of Advances in Library and Information Science, 2(4), 236-242. Accessed December 16, 2022. https://www.jalis.in/pdf/Pdf2-4/Venkatesan.pdf

What does SCIENTOMETRICS mean? Definitions.net. Accessed December 17, 2022. https://www.definitions.net/definition/SCIENTOMETRICS

36

RFID Technology: An Emerging Library Management Tool

***Subrata Chakraborty*[1], *Sutapa Chakraborty*[2] and *Subhankar Ray Chowdhury*[3]**

[1]*Library Assistant, NSHM Knowledge Campus, Kolkata - 700 041, West Bengal, India*
E-mail: subrata311289@gmail.com

[2]*MLISc., Rabindra Bharati University, Kolkata - 700 050, West Bengal, India*
E-mail: sutapapaul333@gmail.com

[3]*Senior Library Assistant, NSHM Knowledge Campus, Kolkata - 700 053*
West Bengal, India - 700041, E-mail: src2509@gmail.com

Abstract

With the rapid advancement of information technology, libraries are continually seeking innovative solutions to streamline their operations and enhance user experience. Radio Frequency Identification (RFID) technology has emerged as a promising tool in the field of library management, revolutionizing traditional library practices. This article provides a comprehensive review of the utilization of RFID technology as an emerging library management tool.The article begins by introducing the fundamental concepts of RFID technology, including its components and working principles. It explores the various applications of RFID technology in libraries, focusing on key areas such as circulation management, inventory control, security systems, and user services. The benefits of RFID implementation are discussed, highlighting improved operational efficiency, enhanced patron experience, and increased library security.

Overall, this article provides a comprehensive overview of the use of RFID technology as an emerging library management tool. It serves as a valuable resource for librarians, administrators, and researchers seeking to understand the benefits, challenges, and future directions of RFID implementation in libraries.

Keywords: *RFID, Radio Frequency Identification, RFID libraries.*

1. Introduction

Organizing and disseminating the amount of information being generated in the age of information explosion is the challenge for library staff. According to Ranganathan's fourth law of library science it is very important factor today, that how quickly library provides its information to the users. So, in this modern era to do this work beautifully, it needs the help of advance technology. Not only that, but also requires staff training.Even the library continuously changes its concepts to keeping the modernity, like library known as 'Knowledge Resource Centre', the books and non-book materials known as information and so on.Now a day the library has made its services better and better by using several technologies. From this point of view, the use of RFID technology is very important. RFID consists of four components *vis* Tags, Antenna, RFID Reader and Software. Though its use, we make the work of circulation easier, we can identify any missing documents very easily. After all, we can say that no document can be taken out of the library without a proper way. So there is acceptance of stock verification work.

2. What is RFID?

RFID is Radio Frequency Identification another way it can say that it is a wireless communication. Its use is mainly seen in the beginning in large industries.Currently it is also used in multiplexes, Tolltax and libraries. Every material is well defined by the RFID's four components. Tag is attached on a material, into the tag there has a microchip and the microchip surrounds by antenna. The chip is loaded minimum information of the particular materials. RFID reader read the particular information of the material and sends it to the database through the network.

3. Objectives of the Study

i. To identify benefits of RFID Technology
ii. To know how much it is compatible then previous system
iii. To know how to maintain the document through RFID
iv. How to save the document from stealing by using of RFID
v. To find out how much it is user friendly and we able to know how it is saving the time of the users as well as the library staff.

4. Features of RFID

The RFID technology is to provide a reliable and efficient method for identifying, tracking, and managing objects or individuals through the use of radio frequency communication. RFID (Radio Frequency Identification) technology offers several features that make it widely used in various industries and applications. Here are some key features of RFID technology:

Contactless Communication: RFID operates through radio waves, allowing for contactless communication between the RFID reader and the RFID tag. This means that the reader does not need a direct line of sight or physical contact with the tag to read or write data.

Automatic Identification: RFID technology aims to enable automatic identification and data capture without the need for direct line-of-sight or physical contact between the RFID reader and the RFID tag. This allows for seamless and non-intrusive identification of objects or individuals, improving operational efficiency.

Unique Identification: Each RFID tag contains a unique identification number or code, known as the Electronic Product Code (EPC). This unique identifier enables individual items to be tracked and distinguished from one another, providing item-level visibility and traceability.

Durability: RFID tags are designed to withstand various environmental conditions, such as moisture, heat, cold, and physical stress. They are available in different form factors, including labels, cards, and encapsulated tags, suitable for different types of objects and environments.

Read/Write Capability: Unlike traditional barcodes, RFID tags can be both read from and written to by the RFID reader. This allows for dynamic data updates, such as modifying information or adding additional data to the tag, making it adaptable for various applications.

Real-Time Tracking and Visibility: RFID enables real-time tracking and visibility of assets, products, or individuals throughout their lifecycle. By attaching RFID tags to items, their locations can be monitored and tracked, providing accurate and up-to-date information about their whereabouts.

Security Features: RFID technology offers security features to protect data and prevent unauthorized access. Encryption and authentication mechanisms can be implemented to ensure data integrity and confidentiality, making it suitable for applications where data security is crucial.

5. How does RFID Technology Works

This technology works in this following way:

Process Efficiency and Automation: RFID technology aims to streamline and automate various processes, such as inventory management, supply chain logistics, and asset tracking. It reduces manual labour, improves accuracy, and enables faster data capture and analysis, leading to increased operational efficiency.

Improved Data Accuracy: RFID technology helps to minimize human error associated with manual data entry or barcode scanning. RFID tags can store and transmit data more accurately, reducing the risk of data inconsistencies and improving overall data accuracy.

Enhanced Security and Authentication: RFID technology can provide enhanced security features by embedding unique identification codes or encryption mechanisms within RFID tags. This helps prevent counterfeiting, unauthorized access, and tampering, ensuring better authentication and data integrity.

Integration with Existing Systems: RFID technology aims to seamlessly integrate with existing systems, such as enterprise resource planning (ERP) software, supply chain management systems, and customer relationship management (CRM) systems. This integration facilitates the exchange of data and information between different systems, enabling efficient data analysis and decision-making.

Cost Reduction and Improved ROI: RFID technology has the potential to reduce costs associated with manual labour, inventory inaccuracies, lost or stolen assets, and inefficient processes. By improving operational efficiency and providing accurate data, RFID technology can lead to a better return on investment (ROI) for businesses.

Enhanced Customer Experience: RFID technology can enhance the customer experience by enabling faster and more accurate checkout processes, personalized interactions, and improved inventory availability. This can lead to increased customer satisfaction and loyalty.

Overall, the objective of RFID technology is to provide an advanced and efficient method for identifying, tracking, and managing objects or individuals, leading to improved operational efficiency, enhanced data accuracy, increased security, and better overall business performance.

6. Usage of RFID in Library

To bring the traditional library to automated, it's necessary to use some new technology. One of the uses of new technology is RFID. Though this, the location of library materials can be determined very easily and it is also focus to the safety of the materials. By using this technology, it saves the time of the library staff as well as the users on the other hand very easily fulfil their needs and received them at the right time.

This technology is widely used in libraries for several reasons. Here are some key reasons why RFID is employed in library settings:

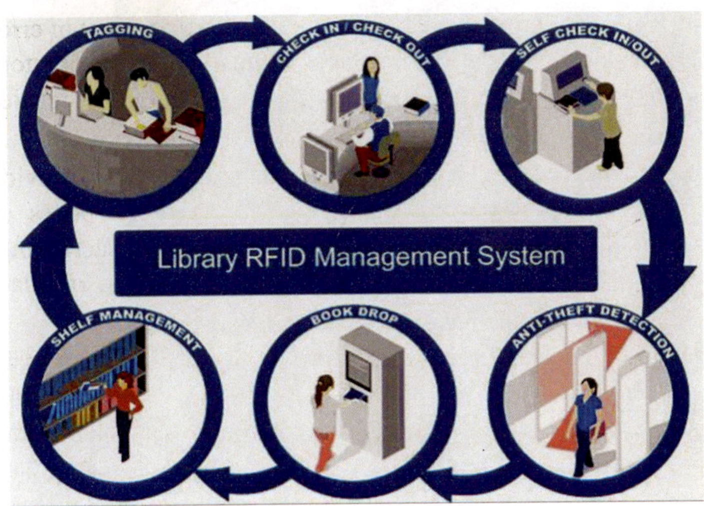

Check-in and Check-out: RFID technology allows for faster and more efficient check-in and check-out processes. Library items equipped with RFID tags can be quickly scanned using RFID readers, enabling simultaneous scanning of multiple items. This reduces the time required for each transaction and improves the overall circulation process.

Self-Service Options: RFID enables self-service options for library patrons. Self-checkout kiosks equipped with RFID readers allow patrons to independently check out items by scanning their library cards and the RFID-tagged items. This promotes self-sufficiency and enhances the user experience by reducing reliance on library staff.

Inventory Management: RFID technology simplifies inventory management in libraries. RFID handheld scanners or automated inventory systems can be used to perform rapid and accurate inventory checks. Librarians can quickly scan shelves or storage areas, and the RFID system identifies which items are present or missing. This helps maintain accurate inventory records and improves the efficiency of collection management.

Security and Anti-Theft Measures: RFID plays a crucial role in library security. RFID gates installed at library exits can detect items that have not been properly checked out. If a patron attempts to take an item with an active RFID tag beyond the gates, an alarm is triggered, alerting library staff to a potential theft. This helps prevent unauthorized removal of library materials and enhances the overall security of the library's collection.

Shelf Management and Automation: RFID technology can automate shelf management tasks. RFID readers installed on bookshelves can identify the precise location of RFID-tagged items. This assists librarians in quickly and accurately shelving items and in locating misplaced or misfiled materials.

Patron Tracking: RFID tags can be embedded in library cards to enable patron tracking within the library premises. This feature allows libraries to monitor patron movements, collect data on library usage patterns, and gain insights for space planning and resource allocation.

Accessibility and Searchability: RFID technology facilitates easy accessibility and searchability of library materials. RFID readers can quickly locate specific items within the library's collection, making it easier to retrieve requested items and assist patrons with finding resources. This improves overall efficiency and enhances the user experience.

Special Collections and Archives: RFID can be used to manage special collections and archival materials. RFID tags can be attached to rare books, manuscripts, or other valuable items to enable efficient tracking, security, and retrieval of these unique materials.

Data Collection and Analysis: RFID systems provide valuable circulation data, enabling libraries to gather insights into borrowing patterns, resource utilization, and collection development decisions.

7. Conclusion

In conclusion, RFID (Radio Frequency Identification) technology has become an essential tool in library settings, revolutionizing various aspects of library operations. RFID offers numerous benefits and advantages, making it a popular choice for libraries around the world. Overall, RFID technology has transformed library operations, improving efficiency, security, and user satisfaction. By automating processes, enhancing inventory management, and providing better access to resources, RFID has become an indispensable tool for modern libraries, supporting their mission of providing excellent services and facilitating knowledge dissemination.

References

Anuragi, M. (2014). RFID technology for libraries: An Indian scenario. International Journal for Research in Applied Science and Engineering Technology (IJRAS ET), 2(2), 60–65.

Aydın, Kenan, &Yildirim, Seda (2012). Case study about RFID system in library services. International Journal of Synergy and Research, 1(2), 91-102.

Berek, L. (2016). Implementing RFID technology for libraries in the field of library security. Óbuda University E-Bulletin, 6(1), 17–20.

Gupta, P., & Madhusudhan, M. (2017). RFID technology in libraries: A review of literature of Indian perspective. DESIDOC Journal of Library &Information Technology, 37(1), 58–63.

Jaska, P., Johnson, D. B. A., Nalla, J., Reddy, N. V. K., & Tadisina, R. (2010). Improved customer service using RFID technology. Review of Business Information Systems, 14(3), 11–18.

Khanna, S. (2014). Impact of RFID Technology on Library Services: A case study of A.C. Joshi Library, Panjab University, Chandigarh. International Journal of Digital Library Services, 4(2), 117–126.

Kotecha, D.H. (2008). Automation in library using RFID technology. 6th Convention PLANNER, 06-07 November, 76-83.

Nagalakshmi, L., & Radha Alias, (2011). Deployment of RFID (Radio Frequency Identification) at Indian academic libraries: Issues and best practice.

Pal, A. K., Tripathi, A., & Saigal, A. (2017). RFID technology: An overview. International Journal of Research, 5(12), 176–182.

Pal, N., & Sharma, A.K. (2017). Implementation of RFID technology in library. International Journal of Digital Library Services, 7(2), 70–79.

Pandey, Prabhat. (2010). Application of RFID technology in libraries and role of librarian. 12th MANLIBNET Convention.

Rahaman, Wasim (2016). Enhancing library services using barcode, QR code and RFID technology: A case study in Central Library National Institute of Technology, Rourkela. International Journal of Digital Library Services, 6(3), 39-50.

Shahid, S. Md. (2005). Use of RFID technology in libraries: A new approach to circulation, tracking, inventorying, and security of library materials. Library Philosophy and Practice (e-journal), 8(1), 1–9.

Singh, J., Brar, N., & Fong, C. (2006). The state of RFID applications in libraries. Information Technology and Library, 24–32.

Singh, N. K., Dhawan, S. M., & Gupta, R. (2016). RFID technology and libraries: A bibliometric assessment of global literature during 2002-14. SRELS Journal of Information Management, 53(2), 89–98.

Singh, Neerajkumar, & Mahajan, Preeti (2014). Application of RFID technology in libraries. International Journal of Library and Information Studies, 4(2), 1–9.

Sumi, Suman, & Kumar, Jatinder (2007). Application of RFID technology in libraries. International CALIBER, 08-10 February, 459-467.

Younis, M.I. (2012). SLMS: a smart library management system based on an RFID technology. International Journal of Reasoning-Based Intelligent Systems, 4(4), 186–191.

Index

R

S

T